Hilbert Space, Boundary Value Problems and Orthogonal Polynomials

Allan M. Krall

Springer Basel AG

Author:

Allan M. Krall
Department of Mathematics
The Pennsylvania State University
University Park, PA 16802
USA
e-mail: krall@math.psu.edu

2000 Mathematics Subject Classification 47E05, 33C45, 33C47, 34B05, 37J99

A CIP catalogue record for this book is available from the
Library of Congress, Washington D.C., USA

Deutsche Bibliothek Cataloging-in-Publication Data

Krall, Allan M.:
Hilbert Space, boundary value problems and orthogonal polynomials /
Allan M. Krall. - Basel ; Boston ; Berlin : Birkhäuser, 2002
  (Operator theory ; Vol. 133)
  ISBN 978-3-0348-9459-3   ISBN 978-3-0348-8155-5 (eBook)
  DOI 10.1007/978-3-0348-8155-5

9 8 7 6 5 4 3 2 1   www.birkhauser-science.com

*To Professor and Mrs. F. V. Atkinson,
whose constant encouragement throughout the years
has been a vital source of inspiration.*

Contents

Preface xiii

Part 1

I Hilbert Spaces

  1 Linear Spaces 1

  2 Hermitian Forms 3

  3 Hilbert Spaces 6

  4 Projections 9

  5 Continuous Linear Functionals 12

  6 Orthonormal Sets 13

  7 Isometric Hilbert Spaces 15

II Bounded Linear Operators on a Hilbert Space

  1 Bounded Linear Operators 17

  2 The Adjoint Operator 19

  3 Projections 21

  4 Some Spectral Theorems 23

  5 Operator Convergence 28

  6 The Spectral Resolution of a Bounded
Self-Adjoint Operator 33

  7 The Spectral Resolution of Bounded Normal
and Unitary Operators 37

     7.1 Normal Operators 37

     7.1 Unitary Operators 39

III Unbounded Linear Operators on a Hilbert Space

  1 Unbounded Linear Operators 41

  2 The Graph of an Operator 42

  3 Symmetric and Self-Adjoint Operators 44

  4 The Spectral Resolution of an Unbounded
Self-Adjoint Operator 46

Part 2

IV Regular Linear Hamiltonian Systems

1	The Representation of Scalar Problems	51
2	Dirac Systems	54
3	S-Hermitian Systems	56
4	Regular Linear Hamiltonian Systems	57
5	The Spectral Resolution of a Regular Linear Hamiltonian Operator	64
6	Examples	70

V Atkinson's Theory for Singular Hamiltonian Systems of Even Dimension

1	Singular Hamiltonian Systems	74
2	Existence of Solutions in $L_A^2(a,b)$	75
3	Boundary Conditions	79
4	A Preliminary Greens Formula	81

VI The Niessen Approach to Singular Hamiltonian Systems

1	Boundary Values of Hermitian Forms	88
2	The Eigenvalues of $\mathcal{A}(x)$	91
3	Generalization of the Second Weyl Theorem	92
4	Singular Boundary Value Problems	94
5	The Green's Function	95
6	Self-Adjointness	97
7	Modification of the Boundary Conditions	100
8	Other Boundary Conditions	102
9	The Limit Point Case	102
10	The Limit m Case	103
11	The Limit Circle Case	104
12	Comments Concerning the Spectral Resolution	106

VII Hinton and Shaw's Extension of Weyl's $M(\lambda)$ Theory to Systems

1	Notations and Definitions	107
2	The $M(\lambda)$ Matrix	109
3	M Circles	111
4	Square Integrable Solutions	115
5	Singular Boundary Conditions	117
6	The Differential Operator L	118
7	Extension of the Boundary Conditions	122
8	The Extended Green's Formula with One Singular Point	125

9   Self-Adjoint Boundary Value Problems with
Mixed Boundary Conditions 131

10 Examples 132

VIII Hinton and Shaw's Extension with Two Singular Points

1   $M(\lambda)$ Functions, Limit Circles, L^2 Solutions 138

2   The Differential Operator 141

3   The Resolvent, The Green's Function 142

4   Parameter Independence of the Domain 144

5   The Extended Green's Formula with Two Singular Points 145

6   Examples 148

    6.1   The Jacobi Boundary Value Problem 149

    6.2   The Legendre Boundary Value Problem 150

    6.3   The Tchebycheff Problem of the First Kind 150

    6.4   The Tchebycheff Problem of the Second Kind 151

    6.5   The Generalized Laguerre Boundary Value Problem 152

    6.6   The Ordinary Laguerre Boundary Value Problem 152

    6.7   The Hermite Boundary Value Problem 152

    6.8   Bessel Functions 153

    6.9   The Legendre Squared Problem 154

    6.10 The Laguerre-Type Problem 155

IX The $M(\lambda)$ Surface

1   The Connection Between the Hinton-Shaw and
Niessen Approaches 159

2   A Direct Approach to the $M(\lambda)$ Surface 162

3   Examples 164

**X The Spectral Resolution for Linear Hamiltonian Systems
with One Singular Point**

1   The Specific Problem 167

2   The Spectral Expansion 168

3   The Converse Problem 176

4   The Relation Between $M(\lambda)$ and $P(\lambda)$ 181

5   The Spectral Resolution 182

6   An Example 184

7   Subspace Expansions 185

8   Remarks 187

XI The Spectral Resolution for Linear Hamiltonian Systems with Two Singular Points

1 The Specific Problem 189

2 The Spectral Expansion 190

3 The Converse Problem 199

4 The Relation Between M_a, M_b, and $P(\lambda)$ 203

5 The Spectral Resolution 205

XII Distributions

1 Test Functions with Compact Support, D; Distributions Without Constraint, D' 207

2 Limits of Distributions 211

3 Test Functions of Rapid Decay, S; Distributions of Slow Growth, S' 212

4 Test Functions of Slow Growth, P; Distributions of Rapid Decay, P' 213

5 Test Functions Without Constants, E; Distributions of Compact Support, E' 214

6 Distributional Differential Equations 215

Part 3

XIII Orthogonal Polynomials

1 Basic Properties of Orthogonal Polynomials 223

2 Orthogonal Polynomials, Differential Equations, Symmetry Factors and Moments 226

XIV Orthogonal Polynomials Satisfying Second Order Differential Equations

1 The General Theory 237

2 The Jacobi Polynomials 239

3 The Legendre Polynomials 243

4 The Generalized Laguerre Polynomials 245

5 The Hermite Polynomials 249

6 The Generalized Hermite Polynomials 252

  6.1 The Generalized Hermite Polynomials of Even Degree 253

  6.2 The Generalized Hermite Polynomials of Odd Degree 255

7 The Bessel Polynomials 257

XV Orthogonal Polynomials Satisfying Fourth Order Differential Equations

1 The General Theory 261
2 The Jacobi Polynomials 262
3 The Generalized Laguerre Polynomials 264
4 The Hermite Polynomials 265
5 The Legendre-Type Polynomials 265
6 The Laguerre-Type Polynomials 270
7 The Jacobi-Type Polynomials 274

XVI Orthogonal Polynomials Satisfying Sixth Order Differential Equations

1 The H.L. Krall Polynomials 281
2 The Littlejohn Polynomials 287
3 The Second Littlejohn Polynomials 289
4 Koekoek's Generalized Jacobi Type Polynomials 290

XVII Orthogonal Polynomials Satisfying Higher Order Differential Equations

1 The Generalized Jacobi-Type Polynomials 291
2 The Generalized Laguerre-Type Polynomials $\{L_n^{\alpha M}(x)\}_{n=0}^\infty$ 295
3 The Generalized Laguerre-Type Polynomials $\{L_n^{2(1/R)}(x)\}_{n=0}^\infty$ 296

XVIII Differential Operators in Sobolev Spaces

1 Regular Second Order Sobolev Boundary Value Problems 302
2 Regular Sobolev Boundary Value Problems
  for Linear Hamiltonian Systems 307
3 Singular Second Order Sobolev Boundary Value Problems 312

XIX Examples of Sobolev Differential Operators

1 Regular Second Order Operators 327
2 Regular Hamiltonian Systems 328
3 Singular Second Order Sobolev Boundary Value Problems 330
  3.1 The Laplacian Operators 330
  3.2 The Bessel Operators 331
  3.3 The Jacobi Operator 333
  3.4 The Generalized Laguerre Operator 334
  3.5 The Hermite Operator 336
  3.6 The Generalized Even Hermite Operator 336
  3.7 The Generalized Odd Hermite Operator 336

XX The Legendre-Type Polynomials and the Laguerre-Type Polynomials in a Sobolev Space

1   The Legendre-Type Polynomials 339
2   The Laguerre-Type Polynomials 340
3   Remarks 341

Closing Remarks 343

Index 345

Preface

The following tract is divided into three parts: Hilbert spaces and their (bounded and unbounded) self-adjoint operators, linear Hamiltonian systemsand their scalar counterparts and their application to orthogonal polynomials. In a sense, this is an updating of E. C. Titchmarsh's classic Eigenfunction Expansions.

My interest in these areas began in 1960–61, when, as a graduate student, I was introduced by my advisors E. J. McShane and Marvin Rosenblum to the ideas of Hilbert space. The next year I was given a problem by Marvin Rosenblum that involved a differential operator with an "integral" boundary condition. That same year I attended a class given by the Physics Department in which the lecturer discussed the theory of Schwarz distributions and Titchmarsh's theory of singular Sturm–Liouville boundary value problems. I think a Professor Smith was the instructor, but memory fails. Nonetheless, I am deeply indebted to him, because, as we shall see, these topics are fundamental to what follows.

I am also deeply indebted to others. First F. V. Atkinson stands as a giant in the field. W. N. Everitt does likewise. These two were very encouraging to me during my younger (and later) years. They did things "right." It was a revelation to read the book and papers by Professor Atkinson and the many fine fundamental papers by Professor Everitt. They are held in highest esteem, and are given profound thanks.

Charles Fulton was responsible for reintroducing the author to the Titchmarsh theory in the late 1970's. His continued support has been invaluable. Angelo Mingarelli, likewise, was also a spur at the right time. Don Hinton and Ken Shaw provided me with the linear Hamiltonian theory, already developed, and it has proved to be invaluable. At the same time, and ever since, Lance Littlejohn and Dick Brown have been marvelous sounding boards, colleagues and friends.

During the mid 1980's, Hans Kaper provided several opportunities at the Argonne National Laboratory. Indeed, he invited me to collaborate with him, Tony Zettl, Derek Atkinson and a number of visitors during the 1986–1987 special year for differential equations at Argonne. It was a tremendous experience. In 1984 I met with Heinz-Dieter Niessen, whose work on differential systems, while slightly different from that of Hinton–Shaw, provided new insight into the nature of boundary value problems.

I am also indebted to my late father H. L. Krall (1907–1994), who instructed me in the theory of orthogonal polynomials, and, in the late 1970's, mentioned that he had worked out incompletely three sets of orthogonal polynomials and their accompanying 4-th order differential equations. When the author reactivated them, they caused a sensation, especially in Europe. Likewise he mentioned that he had investigated polynomials satisfying a 6-th order differential equation, but had forgotten the details and had never written anything about them. This led to the Lance Littlejohn dissertation. The Europeans took note, and the subject exploded.

Finally the two "little" girls of the late 1980's, Mojdeh Hazmirzaahmad and Alouf Jirari, stimulated me at the right time; they, together with W. N. Everitt and Lance Littlejohn assisted in the development of the "left definite" or Sobolev problems.

I would also like to thank Kathy Wyland for doing a beautiful job typing the manuscript. The mistakes are mine, not hers. The remaining typos, and there must be many, are "deliberately" left so the readers may find them.

Allan M. Krall
University Park, PA
August 2001

Chapter I

Hilbert Spaces

Most mathematics books have an introductory chapter which mentions those subjects the author deems necessary for understanding the material which is the subject of the book. All too frequently this chapter is not read, largely because it tends to be disconnected and incomplete, and, therefore, not a coherent platform on which to base the remainder of the book. Nonetheless such a chapter is essential because it sets the tone for both author and reader. For a beginner in the subject it can prove to be useful.

In the opening chapter for this book, we list the various facts we shall need concerning Hilbert spaces, including several standard examples, as well as the settings used here. Since our coverage is by no means complete, we include at the end of the chapter a list of excellent references for those who need additional exposure. Perhaps that is one of the most important tools that an author can provide for a reader.

I.1   Linear Spaces

I.1.1. Definition. A complex linear space, which we will denote by X, is a collection of elements X together with the operations of addition, and scalar multiplication, i.e. if x and y are in X and α is a complex number, then $x + y$ and αx are defined and in X. In addition, the following hold.

(1) Addition is associative and commutative. If x, y, z are in X, then

$$(x + y) + z = x + (y + z),$$
$$x + y = y + x.$$

(2) There exists a unique element O such that $x + O = x$ for all $x \in X$.

(3) For each $x \in X$ there is a unique element $-x \in X$ such that $x + (-x) = O$.

(4) If α and β are complex numbers, and x and y are in X, then

$$\alpha(x + y) = \alpha x + \alpha y,$$
$$\alpha(\beta x) = (\alpha\beta)x,$$
$$1x = x,$$
$$0x = O,$$
$$(\alpha + \beta)x = \alpha x + \beta x.$$

Subtraction is now defined by setting $x - y = x + (-1)y$. Further, that $-x = (-1)x$, $\alpha O = O$, $x+y = x+z$ implies $y = z$, are all well-known computations.

Be sure to distinguish between 0, the complex number, and O, the element in X. $0x = O$ uses both.

I.1.2. Definition. A subset, M, of a linear space X is called a linear manifold if when x and y are in M and α is a complex number, then $x + y$ and αx are in M.

Most people also use the term subspace synonymously with linear manifold. In the back of the author's mind, however, 'subspace' implies closure, while 'linear manifold' does not. Without the concept of closure, therefore, there is no difference, but with it, we shall use these words in the manner described in this paragraph. Subspaces are automatically closed. Linear manifolds are not necessary (see [1]).

I.1.3. Examples. There are many natural examples. We list some of the more important in what follows.

(1) The complex numbers, K, form a linear space. More generally, K^n, the collection of n-dimensional vectors with complex components, forms a linear space. Hyperplanes in K^n, those vectors whose components satisfy a homogeneous linear equation, form a linear manifold.

(2) The previous example can be extended. Infinite-dimensional vectors with complex components form a linear space K^∞. Those vectors with only finitely many nonzero components form a linear manifold.

(3) More interesting examples can be found by considering functions. The continuous functions form a linear space. Differentiable functions form a linear manifold. k times differentiable functions, $k = 1, 2, \ldots$, form a whole collection of linear manifolds, each containing the former.

Still other examples may be found by considering Riemann integrable functions, or square integrable functions. In fact, those functions f, defined on $[a, b]$, satisfy

$$R \int_a^b |f(x)|^p \, dx < \infty \quad , \quad 1 \le p < \infty,$$

for a linear space.

If the Lebesgue integral is used instead, then those complex functions which are measurable and satisfy

$$L \int_a^b |f(x)|^p \, dx < \infty \quad , \quad 1 \le p < \infty,$$

form a linear space. Those functions using the Riemann integral form a linear manifold within.

(4) The ideas of vectors and integrals can be combined. Let X consist of those n-dimensional vectors x whose components are complex valued measurable functions, satisfying

$$L \int_a^b x^* A x \, dx < \infty,$$

where A is a positive, measurable $n \times n$ matrix valued function. It is this linear space we shall use, in one form or another, throughout much of the book.

Within the space of A-square integrable vector functions, those vector functions that are differentiable form a linear manifold.

I.2 Hermitian Forms

The concept of distance between elements of a linear space X is important. Measurement of distance is done through the use of a "matrix," or more strongly a "norm." We give the definition, then show how many norms may be introduced through the concept of a Hermitian form.

I.2.1. Definition. Let x and y be elements in a linear space X, and let α be a complex number. A norm on X, denoted by $\| \cdot \|$ is a real valued function, defined on X, satisfying

(1) $\|x + y\| \le \|x\| + \|y\|$,

(2) $\|\alpha x\| = |\alpha| \, \|x\|$,

(3) $\|x\| \ge 0$ and $\|x\| = 0$ if and only if $x = 0$.

The distance between x and y is given by $\|x - y\|$.

I.2.2. Examples. Norms may be put on each of our earlier examples.

(1) The absolute value $|\cdot|$ is a norm on K. If $y = (y_1, \ldots, y_n)^T$ is in K^n, then

$$\|y\| = \left[\sum_{k=1}^n |y_k|^p \right]^{1/p} \quad , \quad 1 \le p < \infty,$$

is a norm on K^n.

$$\|y\| = \max_k |y_k|$$

is also a norm. It is the limit of the p-norms as $p \to \infty$. (See [3]).

(2) The expressions

$$\|y\| = \left[\sum_{k=1}^{\infty} |y_k|^p \right]^{\frac{1}{p}} \quad , \quad 1 \leq p < \infty,$$

or

$$\|y\| = \sup_k |y_k|$$

serve as norms on K^∞, the space of infinite-dimensional vectors with complex entries, vectors of the form $y = (y_1, \ldots, y_k, \ldots)^T$. Of course, the vectors must now be restricted so that the sum above converges.

(3) The classic norm on continuous functions, defined on $[a, b]$, is

$$\|F\| = \sup_{x \in [a,b]} |f(x)|.$$

On the spaces of integrable functions, defined on (a, b), $-\infty < a < b < \infty$,

$$\|f\| = \left[\int_a^b |f(x)|^p dx \right]^{1/p} \quad , \quad 1 \leq x < p,$$

or

$$\|f\| = \operatorname*{essentialsup}_x |f(x)|,$$

serve as norms. Care must now be given however, to identify (to think of as a single element) those functions whose difference has norm zero. This is a nontrivial requirement.

For our purposes those norms introduced by Hermitian forms are most interesting, since they generate Hilbert spaces.

I.2.3. Definition. A Hermitian form on a linear space is a function, f, defined on $x \otimes X$, with values in the complex numbers K, satisfying

(1) $f(x + y, z) = f(x, z) + f(y, z), \quad f(\alpha x, y) = \alpha f(x, y).$

(2) $f(x, y) = \overline{f(y, x)}$, where the bar denotes complex conjugate,

   for all $x, y, z \in X$ and $\alpha \in K$.

The assumptions (1) and (2) above imply that

$$f(x, y + z) = f(x, y) + f(x, z),$$
$$f(x, \alpha y) = \overline{\alpha} f(x, y)$$

as well. So f is linear in the first variable and conjugate linear in the second.

I.2.4. Examples. A bit of care is now required in order to illustrate the idea of a Hermitian form.

(1) On the complex numbers K

$$f(x,y) = x\,\overline{y}$$

is a Hermitian form. On K^n

$$f(x,y) = \sum_{k=1}^{n} x_k\,\overline{y}_k = y^*x,$$

or, more generally,

$$f(x,y) = \sum_{j=1}^{n}\sum_{k=1}^{n} a_{jk}x_k\overline{y}_j = y^*Ax,$$

where $a_{jk} = \overline{a}_{kj}$, serve as Hermitian forms. The dimension n may be replaced by ∞ provided convergence requirements are met.

(2) For functions defined on $[a,b]$, $-\infty < a < b < \infty$,

$$f(x,y) = \int_a^b x(t)\,\overline{y(t)}\,w(t)dt,$$

where w is real, generates a Hermitian form. Here again x, y and w must be restricted so that the integral is always defined.

(3) For vector valued functions $x = (x_1,\ldots,x_n)^T$, $y = (y_1,\ldots,y_n)^T$, (T denotes transpose),

$$f(x,y) = \int_a^b y^*(t)\,A(t)\,x(t)dt,$$

where A is an appropriate $n \times n$ matrix ($A = A^*$ among other things), generates a Hermitian form. Measures $d\mu$ other than Lebesgue measure dt may be used as well. In particular a matrix valued measure dP will be encountered later.

$$f(x,y) = \int_a^b y^*(t)\,dP(T)\,x(t)$$

is the form in this instance. Note that since x, y and P are all matrices, the order is important.

I.2.5 Theorem. *(Polarization). If f is a Hermitian form on $X \otimes X$, then*

(1) *When restricted to real entries (X is a real linear space with scalar α's real numbers)*

$$4f(x,y) = f(x+y,x+y) - f(x-y,x-y).$$

(2) *In general,*

$$4f(x,y) = f(x+y, x+y) - f(x-y, x-y) + if(x+iy, x+iy) - if(x-iy, x-iy).$$

The proof consists of expanding the right sides and cancelling terms.

I.2.6. Definition. A Hermitian form f on $X \otimes X$ is positive if $f(x,x) \geqq 0$ for all $x \in X \otimes X$. If $f(x,x) > 0$ for all $x \neq 0$, then f is positive definite.

I.2.7. Theorem. *(Schwarz's Inequality). If f is a positive Hermitian form on X, then*

$$|f(x,y)|^2 \leqq f(x,x)f(y,y).$$

The proof follows by expanding $f(x + \lambda y, x + \lambda y) \geqq 0$, then letting $\lambda = -f(x,y)/f(y,y)$ when $f(y,y) \neq 0$. If $f(y,y) = 0$, but $f(x,x) \neq 0$, interchange x and y. If both are 0, let $\lambda = -f(x,y)$.

I.2.8. Theorem. *(Minkowski's Inequality). If f is a positive Hermitian form on X, then*

$$f(x+y, x+y)^{\frac{1}{2}} \leqq f(x,x)^{\frac{1}{2}} + f(y,y)^{\frac{1}{2}}.$$

The proof follows upon expanding $f(x+y, x+y)$ and using Schwarz's inequality to overestimate the mixed x, y terms.

I.2.9. Theorem. *(Parallelogram law). If f is a positive Hermitian form on X, then*

$$2f(x,x) + 2f(y,y) = f(x+y, x+y) + f(x-y, x-y).$$

This follows immediately upon expanding the right side.

I.2.10. Definition. An inner product space H is a linear space X together with a positive definite Hermitian form f. For each x, y in H we call the expression $f(x,y)$ the inner product between x and y. We use the notation $\langle x, y \rangle$ to denote $f(x,y)$. For each x in H, the norm of x is

$$\|x\| = \langle x, x \rangle^{\frac{1}{2}} = f(x,x)^{\frac{1}{2}}.$$

I.3 Hilbert Spaces

I.3.1 Definition. A Hilbert space H is an inner product space which is complete under its norm $\|x\| = \langle x, x \rangle^{\frac{1}{2}}$.

This means that every Cauchy sequence $\{x_n\}_{n=1}^{\infty}$ (sequences for which arbitrary $\epsilon > 0$ there exists an $N > 0$ such that if $m, n > N$, then $\|x_n - x_m\| < \epsilon$) has a limit x_0 in H.

Inherited from two- and three-dimensional spaces is the concept of orthogonality.

I.3.2. Definition. If x, y are in H and $\langle x, y \rangle = 0$, then x and y are orthogonal. A collection A in H is orthogonal if every pair x, y in A is an orthogonal pair. An element x is normal if $\|x\| = 1$. A collection A that is orthogonal and normal is called an orthonormal set.

The expressions previously derived now have a new notational form.

Polarization Identity. *When H is a real inner product space with real scalar numbers α,*

$$\langle x, y \rangle = (1/4)(\|x + y\|^2 - \|x - y\|^2).$$

In general,

$$\langle x, y \rangle = (1/4)(\|x + y\|^2 - \|x - y\|^2 + i\|x + iy\|^2 - i\|x - iy\|^2).$$

Schwarz's Inequality.

$$|\langle x, y \rangle| \leqq \|x\| \, \|y\|.$$

Minkowski's Inequality.

$$\|x + y\| \leqq \|x\| + \|y\|.$$

Parallelogram Law.

$$2\|x\|^2 + 2\|y\|^2 = \|x + y\|^2 + \|x - y\|^2.$$

Further, inherited from two-dimensional space,

Pythagorean Theorem. *If x and y are orthogonal, then*

$$\|x + y\|^2 = \|x\|^2 + \|y\|^2.$$

I.3.3. Examples. All the examples of Hermitian forms generate inner product and Hilbert spaces provided they are positive definite.

(1) On the complex numbers K,

$$\langle x, y \rangle = x\,\overline{y} \quad , \quad \|x\| = |x|$$

makes K an inner product and Hilbert space. On K^n,

$$\langle x, y \rangle = \sum_{k=1}^{n} x_k \overline{y}_k \quad , \quad \|x\| = \left[\sum_{k=1}^{n} |x_k|^2 \right]^{\frac{1}{2}}$$

does also. A more general expression

$$\langle x, y \rangle = \sum_{j=1}^{n} \sum_{k=1}^{n} a_{jk} x_k y_j = y^x A x$$

does also, provided A is positive definite. If A is singular and merely positive, then elements with norm 0 must be "factored out." That is, if $\|x - y\| = 0$, then x and y are thought of as being the same, even though they may be different vectors. The collection of equivalent vector sets is a Hilbert space.

The index n may be ∞, provided convergence requirements are met. The space of elements $x = (x_1, \ldots, x_n, \ldots)^T$ with inner product and norm

$$\langle x, y \rangle = \sum_{k=1}^{\infty} x_k \, \overline{y}_k,$$

$$\|x\| = \left[\sum_{k=1}^{\infty} |x_k|^2 \right]^{\frac{1}{2}}$$

is denoted by ℓ^2.

(2) For functions the situation is more complex. On $[a, b]$ with the Riemann integral in use, the continuous and piecewise continuous functions form an inner product space $R^2(a, b)$ under the inner product and norm

$$\langle x, y \rangle = R \int_a^b x(t) \, \overline{y(t)} \, w(t) dt,$$

$$\|x\| = R \int_a^b |x(t)|^2 \, w(t) dt,$$

when w is piecewise continuous and > 0. If w is merely ≥ 0, then functions f and g for which $\|f - g\| = 0$ must be identified. But even with this convention, the space R^2 is not complete. In order to have completeness, the Lebesgue integral is required, and the completed space is called $L^2(a, b)$. With other positive Lebesgue–Stieltjes measures μ, under which

$$\langle x, y \rangle = L \int_a^b x(t) \, y(t) \, d\mu(t),$$

$$\|x\| = \left[L \int_a^b |x(t)|^2 d\mu(t) \right]^{1/2},$$

the resulting complete space is denoted by $L^2_\mu(a, b)$.

(3) For vector valued functions and a positive matrix A or a positive matrix measure P, for which

$$\langle x, y \rangle = L \int_a^b y^*(t) \, A(t) \, x(t) dt,$$

$$\|x\| = \left[L \int_a^b x^*(t) \, A(t) \, x(t) dt \right]^{1/2},$$

or

$$\langle x, y \rangle = L \int_a^b y^*(t) \, dP(t) \, x(t),$$

$$\|x\| = \left[L \int_a^b x^*(t) \, dP(t) \, x(t) \right]^{1/2} ;$$

the resulting complete space of equivalence classes of vectors is denoted by $L_A^2(a, b)$ or $L_p^2(a, b)$. Again the Lebesgue or Lebesgue–Stieltjes integral is required.

We state the following without proof. For those interested in its verification we suggest "Applied Analysis" as a guide, but recommend the reader write his own.

I.3.4. Theorem. *Let (X, f) be an inner product space. Then there is a Hilbert space H in which (X, f) is densely embedded.*

The idea is to first identify elements x, y in X for which $\|x - y\| = f(x - y, x - y)^{\frac{1}{2}} = 0$. Then from the resulting space of equivalence classes, define a new space of Cauchy sequences $H = \{x : x = \{x_n\}_{n=1}^\infty$ a Cauchy sequence$\}$. A fair amount of labor is involved, but the result is a Hilbert space. Since mathematicians are frequently lazy, they will often drop the language of "Cauchy sequences of equivalence classes of" and speak merely of functions. As long as everyone understands this shorthand, there is no difficulty, but it should not be forgotten.

I.4 Projections

The concept of orthogonality can be extended still further, leading to the definition of a projection operator.

I.4.1. Definition. Let M be a closed linear manifold in an inner product space H (in M every Cauchy sequence has a limit in M). Then M is called a subspace of H.

The following is a consequence of the parallelogram law.

I.4.2 Theorem. *Let M be a subspace in an inner product space H. Let x be in H and let $\delta = \inf_{y \in M} \|x - y\|$. Then there exists a unique y_0 in M such that $\|x - y_0\| = \delta$.*

Proof. Let $\{y_n\}_{n=1}^\infty$ be a sequence in M and satisfy $\lim_{n \to \infty} \|y_n - x\| = \delta$. In the parallelogram law,

$$2\|x\|^2 + 2\|y\|^2 = \|x + y\|^2 + \|x - y\|^2,$$

replace x by $x - y_n$ and y by $x - y_m$ to find

$$\|y_n - y_m\|^2 + 4\left\|x - \frac{1}{2}(y_n + y_m)\right\|^2 = 2\|x - y_n\|^2 + 2\|x - y_m\|^2,$$

or

$$\|y_n - y_m\|^2 = 2\|x - y_n\|^2 + 2\|x - y_m\|^2 - 4\left\|x - \frac{1}{2}(y_n + y_m)\right\|^2.$$

Since M is a linear manifold, $\frac{1}{2}(y_n + y_m)$ is in M, and $\left\|x - \frac{1}{2}(y_n + y_m)\right\| \geqq \delta$. Thus

$$\|y_n - y_m\|^2 \leqq 2\|x - y_n\|^2 + 2\|x - y_m\|^2 - 4\delta^2.$$

As n, m approach ∞, the right side approaches $2\delta^2 + 2\delta^2 - 4\delta^2 = 0$. Thus $\{y_n\}_{n=1}^\infty$ is a Cauchy sequence in M. Since M is closed, $\{y_n\}_{n=1}^\infty$ approaches a limit y_0 in M. Since

$$\|x - y_0\| \leqq \|x - y_n\| + \|y_n - y_0\|,$$

and $\|y_n - y_0\|$ approaches 0, $\|x - y_0\| \leqq \delta$. But y_0 is in M, so $\|x - y_0\| \geqq \delta$. Thus $\|x - y_0\| = \delta$.

If two elements, y_0 and y_0', exist in M and satisfy $\|x - y_0\| = \delta$, $\|x - y_0'\| = \delta$, then

$$\|y_0 - y_0'\|^2 = 2\|x - y_0\|^2 + 2\|x - y_0'\|^2 - 4\left\|x - \frac{1}{2}(y_0 + y_0')\right\|^2,$$

$$\leqq 2\delta^2 + 2\delta^2 - 4\delta^2,$$

$$= 0.$$

So $y_0 = y_0'$. $\hfill\square$

The theorem is not necessarily true in other spaces.

An extension is immediate.

I.4.3. Theorem. *Let M and N be subspaces in an inner product space H, with $M \subset N$, $M \neq N$. Then there exists an element $z \in N$, $z \neq 0$ such that z is orthogonal to M.*

Proof. If x is in $N - M$, and $\delta = \inf_{y \in M} \|y - x\|$, then there exists a unique y_0 in M such that $\|y_0 - x\| = \delta$. Let $z = y_0 - x$. If y is in M, then

$$\|z + \lambda y\| = \|y_0 - x + \lambda y\| \geqq \delta = \|z\|.$$

So

$$0 \leqq \|z + \lambda y\|^2 - \|z\|^2$$

for all y in M and any λ. Expanding,

$$0 \leqq \|z\|^2 + \lambda\langle y, z\rangle + \overline{\lambda}\langle z, y\rangle + |\lambda|^2\|y\|^2 - \|z\|^2.$$

If $\langle y, z\rangle \neq 0$, let $\lambda = -r|\langle y, z\rangle|/\langle y, z\rangle$, where $r > 0$, then

$$0 \leqq -r|\langle y, z\rangle\| - r|\langle y, z\rangle| + r^2\|y\|^2,$$

and

$$2|\langle y, z\rangle| \leqq r\|y\|^2.$$

As r approaches 0, we arrive at a contradiction. Thus $\langle y, z \rangle = 0$ for y in M, and z is orthogonal to M.

So far we have considered the possibility of elements in H being orthogonal to subspaces. $\qquad\square$

I.4.4. Definition. Let M be a subset of an inner product space H. Then $M^\perp$ is the set of all elements x in H such that x is orthogonal to M.

I.4.5. Theorem. *If H is a Hilbert space, then*

(1) $M^\perp$ *is a subspace of H.*

(2) $M \cap M^\perp = \{0\}$ *when 0 is in M.*

(3) $M \subset M^{\perp\perp}$.

(4) *If $M \subset N$, then $N^\perp \subset M^\perp$.*

(5) $M^\perp = M^{\perp\perp\perp}$.

(6) *If M is a subspace of H, then $M = M^{\perp\perp}$.*

I.4.6. Definition. If M and N are subspaces of an inner product space H and $M \perp N$, then
$$M + N = \{x + y,\ x \in M,\ y \in N\}.$$

I.4.7. Theorem. *If M and N are orthogonal subspaces in an inner product space H, then $M + N$ is a subspace of H.*

If H is a Hilbert space, then
$$M + M^\perp = H.$$

Proof. $M+N$ is clearly a linear manifold. Let $\{z_n = x_n + y_n : x_n \in M,\ y_n \in N\}_{n=1}^{\infty}$ be a Cauchy sequence in $M + N$. Then, since
$$\|z_n - z_m\|^2 = \|x_n - x_m\|^2 + \|y_n - y_m\|^2,$$
$$\|x_n - x_m\|^2 \leqq \|z_n - z_m\|^2,$$
$$\|y_n - y_m\|^2 \leqq \|z_n - z_m\|^2.$$

Thus $\{x_n\}_1^\infty$ and $\{y_n\}_{n=1}^\infty$ are Cauchy sequences in M and N. Since M and N are closed, $\lim_{n \to \infty} x_n = x_0$ in M, $\lim_{n \to \infty} y_n = y_0$ in N, and $\lim_{n \to \infty} z_n = x_0 + y_0$ is in $M + N$. $M + N$ is closed. $\qquad\square$

The implications of the last statement are far reaching. If $M + M^\perp = H$. Then each x in H has the representation
$$x = y + z \quad,\quad z \in M,\ z \in M^\perp.$$

Further y and z are unique. These and other results may be formalized in the following.

I.4.8. Theorem. *Let M be a subspace of a Hilbert space H. Let x be in H and have the representation $x = P_M x + (I - P_M)x$, where $P_M x$ is in M, $(I - P_M)x$ is in $M^\perp$. Then P_M is linear, continuous and satisfies*

$$\|P_M x\| \leqq \|x\|$$

for all $x \in H$. The range of P_M is M, $P_M^2 = P_M$, and $\langle P_M x, y \rangle = \langle x, P_M y \rangle$ for all $x, y \in H$.

I.4.9. Definition. In the context of Theorem I.4.8, P_M is called the projection of H onto M. $I - P_M = P_{M^\perp}$.

The last equation is more theorem than definition. The equations $P_M^2 = P_M$ and $\langle P_M x, y \rangle = \langle x, P_M y \rangle$ completely characterize the idea of a projection operator, for, if they hold, for any operator P, then P projects H onto PH. The subspaces $M = PH$ and $M^\perp = (I - P)H$ are mutually orthogonal and $M + M^\perp = H$.

I.5   Continuous Linear Functionals

This section is included more for esthetic reasons than utility. The concept of a continuous linear functional is an old one. In a Hilbert space setting it is completely characterized by the remarkable theorem of Riesz and Frechet.

I.5.1. Definition. A continuous linear functional, L, on a Hilbert space H, is a continuous, linear function with domain H and range K.

Consequently if x, y are in H and x is complex, then L satisfies

$$L(x + y) + Lx + Ly,$$
$$L(\alpha x) = \alpha(Lx).$$

As a consequence of continuity, the expression $|L(x)|/\|x\|$ is bounded, for if not, there would be a sequence $\{x_n\}_{n=1}^\infty$ such that $|L(x_n)|/\|x_n\| > n$. If we let $y_n = x_n/(\sqrt{n}\,\|x_n\|)$, then $y_n \to 0$, but $\|L(y_n)\|$ approaches ∞, a contradiction. Therefore, the norm of L is defined by

$$\|L\| = \sup_{x \neq 0} |L(x)|/\|x\|.$$

I.5.2. Theorem. *(Riesz–Frechet). Let L be a continuous linear functional on a Hilbert space H. Then there exists a unique $y_L \in H$ such that*

$$L(x) = \langle x, y_L \rangle .$$

Further

$$\|L\| = \sup_{x \neq 0} |L(x)|/\|x\| = \|y_L\|.$$

The proof is surprisingly easy. Let $M = \{x : L(x) = 0\}$. If $M = H$, let $y_L = 0$. If not, choose y_0 in $M^\perp - \{0\}$. Note that $x - \{L(x)y_0/L(y_0)\}$ is in M and orthogonal to y_0. Solve

$$\langle x - \{L(x)y_0/L(y_0)\}, y_0\rangle = 0$$

for $L(x)$. Uniqueness follows from the inner product representation, letting $x = y_{L_1} - y_{L_2}$, if two such y_L's exist.

The reader should give his own examples in each of the various Hilbert spaces listed as examples.

I.6 Orthonormal Sets

We have already given the definition of an orthonormal set. We show here how they may be used to represent arbitrary elements in a Hilbert space.

I.6.1. Theorem. *(Gram–Schmidt). Let $\{x_j\}_{j=1}^\infty$ be an infinite collection in an inner product space H with the property that every finite subset is linearly independent. From $\{x_j\}_{j=1}^\infty$ we can extract an orthonormal set $\{y_j\}_{j=1}^\infty$ such that for each n, $n = 1,\ldots,\infty$, $\{x_j\}_{j=1}^n$ and $\{y_j\}_{j=1}^n$ span the same linear manifold.*

The proof depends on the recursive formula

$$y_n = \left[x_n - \sum_{j=1}^{n-1}(x_n, y_j)y_j\right] \Big/ \left\|x_n - \sum_{j=1}^{n-1}(x_n, y_j)y_j\right\|.$$

Each y_n is orthogonal to $y_1,\ldots,y_{n-1}$, and each y_n is a linear combination of $x_1,\ldots,x_n$.

Given an orthonormal set, we can write what is called the generalized Fourier series of an arbitrary element in a Hilbert space H. Let $\{y_j\}_{j=1}^\infty$ be the orthonormal set and let x be an arbitrary element in H. We attempt to minimize the expression

$$\left\|x - \sum_{j=1}^n \lambda_j y_j\right\|$$

by choosing appropriate coefficients λ_j, $j = 1,\ldots,n$. Note that

$$\left\|x - \sum_{j=1}^n \lambda_j y_j\right\|^2 = \|x\|^2 - \sum_{j=1}^n |\langle x, y_j\rangle|^2 + \sum_{j=1}^n |\lambda_j - \langle x, y_j\rangle|^2.$$

This clearly shows the minimum occurs when $\lambda = \langle x, y_j\rangle$. Since the last term is zero, we have

I.6.2. Theorem. *(Bessel's Inequality). Let $\{y_j\}_{j=0}^{\infty}$ be an orthonormal set in an inner space H. Let x be in H, then*

$$\left\| x - \sum_{j=1}^{\infty} \langle x, y_j \rangle y_j \right\|^2 = \|x\|^2 - \sum_{j=1}^{\infty} |\langle x, y_j \rangle|^2,$$

and

$$\|x\|^2 \geq \sum_{j=1}^{\infty} |\langle x, y_j \rangle|^2.$$

Since the infinite series converges, the individual terms (x, y_j) approach 0 as j approaches ∞. This is the Riemann–Lebesgue lemma.

Projections take on an elegant appearance if orthonormal sets are used.

I.6.3. Theorem. *Let M be a linear manifold in a Hilbert space H. Let $\{y_j\}_{j=1}^{\infty}$ span M. (Every element $x \in M$ can be written $x = \sum_{j=1}^{\infty} \langle x, y_j \rangle y_j$). Then for each $x \in H$,*

$$P_M x = \sum_{j=1}^{\infty} \langle x, y_j \rangle y_j$$

is the projection on to M of x.

The proof is elementary. If

$$x = \sum_{j=1}^{\infty} \langle x, y_j \rangle y_j + \left(x - \sum_{j=1}^{\infty} \langle x, y_j \rangle y_j \right),$$

the first term is in M, the second orthogonal to it. Note that $P_M^2 = P_M$ and

$$\langle P_M x, y \rangle = \langle x, P_m y \rangle.$$

Remarkably the converse is true as well.

I.6.4. Theorem. *(Riesz–Fischer) . If $\{y_j\}_{j=1}^{\infty}$ is an orthonormal set in a Hilbert space H, let $\{\lambda_j\}_{j=1}^{\infty}$ be a collection of complex numbers satisfying $\sum_{j=1}^{\infty} |\lambda_j|^2 < \infty$. Then $\sum_{j=1}^{\infty} \lambda_j y_j$ converges to an element in H. In particular $\sum_{j=1}^{\infty} \langle x, y_j \rangle y_j$ converges for each x in H.*

There is still an open question as to whether the series $\sum_{j=1}^{\infty} \langle x, y_j \rangle y_j$ equals x.

I.6.5. Definition. An orthonormal set $\{y_j\}_{j=1}^{\infty}$ is complete in a Hilbert space H if whenever $\langle x, y_j \rangle = 0$, $j = 1, \ldots, \infty$, for an element in H, then $x = 0$.

I.6.6. Theorem. *$\{y_j\}_{j=1}^{\infty}$ is a complete orthonormal set in a Hilbert space H if and only if for any $x \in H$,*

$$x = \sum_{j=1}^{\infty} \langle x, y_j \rangle y_j.$$

Proof. If $\{y_j\}_{j=1}^{\infty}$ is complete, let $z = x - \sum_{j=1}^{\infty} \langle x, y_j \rangle y_j$. Then $\langle z, y_j \rangle = 0$ for all j and $x = 0$, $x = \sum_{j=1}^{\infty} \langle x, y_j \rangle y_j$.

Conversely if $x = \sum_{j=1}^{\infty} \langle x, y_j \rangle y_j$ for all x, then when $\langle x, y_j \rangle = 0$ for all y_j, $x = 0$ and $\{y_j\}_{j=1}^{\infty}$ is complete. $\square$

I.6.7 Theorem. *(Parseval's Equality).* $\{y_j\}_{j=1}^{\infty}$ *is a complete orthonormal set if and only if*

$$\|x\|^2 = \sum_{j=1}^{\infty} |\langle x, y_j \rangle|^2.$$

Proof. Bessel's inequality shows

$$0 \leqq \left\| x - \sum_{j=1}^{\infty} \langle x, y_j \rangle y_j \right\|^2 = \|x\|^2 - \sum_{j=1}^{\infty} |\langle x, y_j \rangle|^2.$$

If $\{y_j\}_{j=1}^{\infty}$ is complete, the middle term is 0. Hence so is the last.

If Parseval's equality holds, then the middle term is again 0, and $\{y_j\}_{j=1}^{\infty}$ is complete. $\square$

I.7 Isometric Hilbert Spaces

One final idea concerning Hilbert spaces themselves is needed for further reference.

I.7.1. Definition. Let H_1 and H_2 be Hilbert spaces with the scalar field of complex numbers. H_1 and H_2 are isomorphic and isometric if there exists a linear transformation U sending H_1 onto H_2 satisfying

$$\|Ux\|_{H_2} = \|x\|_{H_1}$$

for all x in H_1.

I.7.2. Theorem. *If Hilbert spaces H_1 and H_2 are isomorphic and isometric through U, then*

$$\langle Ux_1, Ux_2 \rangle_{H_2} = \langle x_1, x_2 \rangle_{H_1}$$

for all x_1, x_2 in H_1. Further U^{-1} exists and

$$\langle U^{-1}y_1, U^{-1}y_2 \rangle_{H_1} = \langle (y_1, y_2)_{H_2}$$

for all y_1, y_2 in H_2.

The proof is through the polarization identity.

I.7.3. Theorem. *Let H be a Hilbert space.*

(1) *If H has finite dimension (has a finite complete orthonormal set $\{y_j\}_{j=1}^n$) and a complex scalar field, then H is isomorphic and isometric to K^n.*

(2) *If H is countably infinite in dimension (has a complete orthonormal set $\{y_j\}_{j=1}^\infty$) and a complex scalar field, then H is isomorphic and isometric to ℓ^2.*

If x is in H, then $x = \sum \langle x, y_j \rangle y_j$. The transformation U defined by

$$Ux = (\langle x, y_1 \rangle, \langle (x, y_2) \rangle, \dots)^T$$

has all the required properties.

References

[1] N. I. Akhiezer and I. M. Glazman, **Theory of Linear Operators in Hilbert Space, vol. I and II**, Frederick Ungar, New York, 1961.

[2] A. M. Krall, **Applied Analysis**, D. Reidel, Dordrecht, 1986.

[3] F. Riesz and B. Sz.-Nagy, **Functional Analysis**, Frederick Ungar, New York, 1955.

[4] A. E. Taylor, **Introduction to Functional Analysis**, John Wiley and Sons, New York, 1958.

Chapter II

Bounded Linear Operators On a Hilbert Space

Everyone is familiar with linear operators. Multiplication by a constant is a linear operator. Multiplication of vectors by matrices generates an operator. Integration usually generates another, depending upon the setting.

In virtually every setting those mentioned above are **bounded** operators. Further, when the underlying space is a Hilbert space they are **self-adjoint**. For such operators there is a beautiful theory, an integral representation, which reduces them to a kind of multiplication. This chapter develops that theory.

We hope the sketch following is not too brief. The references at the end of the chapter provide further details.

Our goal, actually, is to extend results for bounded operators so that they apply for unbounded operators, but that is the subject of the next chapter. We ask the reader to be patient.

II.1 Bounded Linear Operators

II.1.1. Definition. Let A be a linear operator on a Hilbert space H with domain D and range R. We say that A is bounded if for all $x \in D$, there is a number M such that
$$\|Ax\| \leq M\|x\|.$$

It is easy to show that D can be closed by taking limits. Since for $x, y \in D$,

$$\|Ax - Ay\| \leq M\|x - y\|,$$

Cauchy sequences in D lead to Cauchy sequences in R. In H such sequences have limits, and so, using these limits, D can be closed.

Once D is closed, if it is not all of H, then we can write

$$H = D + D^{\perp}.$$

Extending A to $D^\perp$ by setting $Ax = 0$ if x is in $D^\perp$, we have now an operator whose domain is all of H. Further, letting $x = y + z$, where x is arbitrary in H, $y \in D$, $z \in D^\perp$ we see that

$$\begin{aligned}
\|Ax\| &= \|Ay + Az\| \\
&= \|Ay\| \\
&\leq M\|y\| \\
&\leqq M\|x\|.
\end{aligned}$$

So the extended A is also bounded by M.

II.1.2. Definition. 1. We define the operator norm of A, $\|A\|$, by setting

$$\|A\| = \sup_{\substack{x \in H \\ x \neq 0}} \|Ax\|/\|x\|.$$

We leave as an exercise the verification that

$$\|A\| = \inf\{M : \|Ax\| \leq M\|x\|, \text{ all } x \in H\}.$$

2. We denote the collection of bounded linear operators on H by $\mathcal{L}$.

Associated with any linear operator A are numbers called the spectrum of A. These numbers are good or bad, depending upon what is being studied or attempted.

II.1.3. Definition. Let H be a Hilbert space. Let A be in $\mathcal{L}$.

(1) The spectrum of A, $\sigma(A)$, is the set of all complex numbers λ such that $(\lambda I - A)^{-1}$ does not exist as a bounded operator on H. (I denotes the identity operator.)

(2) The resolvent of A, $\rho(A)$, is the complement of $\sigma(A)$ in the complex plane.

(3) If $\lambda \in \rho(A)$, then $(\lambda I - A)^{-1}$ is the resolvent operator.

II.1.4. Theorem. *Let H be a Hilbert space, let A be in $\mathcal{L}$. Then*

(1) *$\sigma(A)$ is a compact set of complex numbers.*

(2) *If $\lambda \in \sigma(A)$, then $|\lambda| \leq \|A\|$.*

(3) *$\rho(A)$ is an open set of complex numbers.*

Proof. We show 2, then 3. To prove 2, we note that when $|\lambda| > \|A\|$, the Neumann series

$$(\lambda I - A)^{-1} = \sum_{n=0}^{\infty} A^n/\lambda^{n+1}$$

converges, and so λ is in $\rho(A)$.

To show 3, we note that if λ is in $\rho(A)$, then $(\lambda I - A)^{-1}$ exists as a bounded operator. If $|\epsilon|$ is less than this bound, then

$$((\lambda + \epsilon)I - A)^{-1} = -\sum_{n=0}^{\infty} \epsilon^n [(A - \lambda I)^{-1})]^{n+1}$$

also converges. Hence λ is surrounded by an ϵ neighborhood of numbers, also in $\rho(A)$. $\qquad\square$

II.2 The Adjoint Operator

For a bounded linear operator $A \in \mathcal{L}$, the definition of an adjoint operator A^* poses no problem.

II.2.1. Definition. Let H be a Hilbert space. Let A be in $\mathcal{L}$, and let x, y be in H. Then we define the adjoint of A, A^*, by setting

$$\langle Ax, y \rangle = \langle x, A^*y \rangle.$$

The existence of A^*y is guaranteed once it is noticed that

$$f_y(x) = \langle Ax, y \rangle$$

is a continuous linear functional on H. A^*y is the element z for which

$$f_y(x) = \langle x, z \rangle.$$

The linearity of A^* is left as an easy exercise.

II.2.2. Theorem. *Let H be a Hilbert space. Let A and B be in $\mathcal{L}$. Then*

(1) $(A^*)^* = A$.

(2) $(A + B) = A^* + B^*$.

(3) $(AB)^* = B^* A^*$.

(4) $I^* = I$, $0^* = 0$ *(The identity and zero operators)*.

(5) $\|A\| = \|A^*\|$.

(6) $\|A^* A\| = \|AA^*\| = \|A\|^2$.

(7) $A^* A = 0$ *if and only if* $A = 0$.

(8) *If A^{-1} exists as a bounded operator, then $(A^*)^{-1}$ exists as well, and $(A^*)^{-1} = (A^{-1})^*$.*

II.2.3. Definition. Let H be a Hilbert space. Let $A \in \mathcal{L}$.

(1) If $A = A^*$, then A is self-adjoint.

(2) If $A^* A = AA^*$, then A is normal.

(3) If $A^* = A^{-1}$, then A is unitary.

Roughly speaking, self-adjoint operators are analagous to real numbers, normal operators to complex numbers, and unitary operators to complex numbers with modulus 1. This will become more apparent when the spectral integrals are exhibited.

II.2.4. Examples. 1. Let H be the space K^n of n-dimensional complex vectors $x = (x_1, \ldots, x_n)^T$. Let A be represented by the matrix (a_{ij}) through

$$Ax = \left(\sum_{j=1}^{n} a_{1j}x_j, \ldots, \sum_{j=1}^{n} a_{nj}x_j \right)^T,$$

or

$$Ax = \begin{pmatrix} a_{11} & \cdots & a_{1n} \\ \vdots & & \vdots \\ a_{n1} & \cdots & a_{nn} \end{pmatrix} \begin{pmatrix} x_1 \\ \vdots \\ x_n \end{pmatrix}.$$

A is self-adjoint if $a_{ij} = \overline{a_{ji}}$.

To tell when A is normal is tricky, even when $n = 2$.

A is unitary when the rows and columns form an orthonormal set.

2. Let $H = L^2(a, b; \sigma)$, the Hilbert space generated by the inner product

$$\langle x, y \rangle = \int_a^b x(t)\overline{y}(t)d\sigma(t).$$

If A is defined by

$$Ax(t) = \int_a^b K(t, s)x(s)d\sigma(s)$$

where

$$\int_a^b \int_a^b |K(t, s)|^2 \, d\sigma(s)d\sigma(t) < \infty,$$

then

$$A^*y(s) = \int_a^b \overline{K(t, s)}y(t)d\sigma(t).$$

A is self-adjoint when $K(t, s) = \overline{K(s, t)}$. If

$$\int_a^b \overline{K(u, t)}K(u, s)d\sigma(u) = \int_a^b K(t, u)\overline{K(s, u)}d\sigma(u),$$

then A is normal. If both the above integrals equal the Dirac delta function $\delta(t-s)$, then A is unitary. The delta function is discussed in Chapter XII.

Self-adjoint operators are our special interest. For such operators there are three ways to compute the operator norm.

II.2.5. Theorem. *Let H be a Hilbert space. Let A be self-adjoint in $\mathcal{L}$. Then $\|A\|$ can be computed from any one of the following.*

(1) $\|A\| = \sup\limits_{x \neq 0} \|Ax\|/\|x\|$, *(definition)*.

(2) $\|A\| = \sup\limits_{\|x\|=1} \|Ax\|$.

(3) $\|A\| = \sup\limits_{\|x\|=1} \|\langle Ax, x\rangle\|$.

Proof. Only the third requires any proof. Let $\lambda = \sup\limits_{\|x\|=1} |\langle Ax, x\rangle|$. Then when $\|x\| = 1$,

$$|\langle Ax, x\rangle| \leq \|A\|\,\|x\|^2 = \|A\|.$$

So $\lambda \leq \|A\|$.

To see the reverse inequality, let $\|y\| = 1$, $\|z\| = 1$. Since (Ax, x) is real, we find

$$\langle A[y+z], [y+z]\rangle \leq \lambda\|y+z\|^2,$$
$$\langle A[y-z], [y-z]\rangle \geq \lambda\|y-z\|^2.$$

Subtracting these inequalities yields

$$4\mathrm{Re}\,\langle Ay, z\rangle \leqq 2\lambda[\|y\|^2 + \|z\|^2 = 4\lambda,$$
$$\mathrm{Re}\,\langle Ay, z\rangle \geq \lambda.$$

Now let $z = Ay/\|Ay\|$. When $Ay \neq 0$, this yields $\|Ay\| \leq \lambda$. If we take the supremum over y, we have $\|A\| \leq \lambda$. $\qquad\square$

II.3 Projections

The measures used to represent self-adjoint, normal and unitary operators as integrals are generated by projection operators.

II.3.1. Definition. Let H be a Hilbert space. A projection operator P is a bound linear operator on H satisfying

(1) $P = P^*$.

(2) $P^2 = P$.

II.3.2. Theorem. *Let P be a projection on a Hilbert space H. Let the range of P be M. Then M is a closed linear manifold and $P = P_M$, the projection of H onto M.*

Proof. If y, y_1, y_2 are in the range of P and α is a complex (real) number, then there exists (at least one) x, x_1, x_2 such that $y = Px$, $y_1 = Px_1, y_2 = Px_2$. Thus

$$y_1 + y_2 = Px_1 + Px_2 = P(x_1 + x_2).$$
$$\alpha y = \alpha Px = P(\alpha x).$$

So $y_1 + y_2$ and αy are also in M, the range of P. M is a linear manifold.

Next we note that each x in H has the representation

$$x = Px + (I - P)x.$$

Since $\langle Px, [I - P]x \rangle = \langle x, P[I - P]x \rangle = 0$, by the Pythagorean theorem,

$$\|x\|^2 = \|Px\|^2 + \|[I - P]x\|^2,$$

and we conclude $\|Px\| \leq \|x\|$, $\|P\| \leq 1$. Now let $\{y_n\}_{n=1}^{\infty}$ be a Cauchy sequence in M. Since $\{y_n\}_{n=1}^{\infty}$ is in H, $\lim_{n \to \infty} y_n = y_0$ exists in H. Since for each y_n, there exists an x_n such that $Px_n = y_n$, we find $Py_n = P^2 x_n = Px_n = y_n$. Thus we have

$$y_0 = y_n + (y_0 - y_n),$$
$$Py_0 = y_n + P(y_0 - y_n),$$

and

$$y_0 - Py_0 = (I - P)(y_0 - y_n).$$

Taking norms, we find

$$\begin{aligned}
\|y_0 - Py_0\| &= \|(I - P)(y_0 - y_n)\| \\
&\leq \|I - P\| \, \|y_0 - y_n\| \\
&\leq [\|I\| + \|P\|] \|y_0 - y_n\| \\
&\leq 2\|y_0 - y_n\|.
\end{aligned}$$

Since the left side is fixed, and the last expression on the right approaches 0, $y_0 = Py_0$, and y_0 is also in M. Thus M is closed. That $P = P_M$ is now obvious. $\qquad\square$

II.3.3. Theorem. *Let P_1 and P_2 be projection operators on a Hilbert space H. Then*

(1) *$P_1 + P_2$ is a projection if and only if $P_1 P_2 = P_2 P_1 = 0$.*

(2) *$P_1 P_2$ is a projection if and only if $P_1 P_2 = P_2 P_1$.*

(3) *$P_1 P_2 = P_2$ if and only if $\langle P_1 x, x \rangle \geq \langle P_2 x, x \rangle$ for all $x \in H$.*

The inequality

$$\langle P_1 x, x \rangle \geq \langle P_2 x, x \rangle$$

is more briefly written $P_1 \geq P_2$.

Proof. 1. If $P_1 P_2 = P_2 P_1 = 0$, then $(P_1 + P_2)^2 = P_1 + P_2$. $(P_1 + P_2)^* = P_1^* + P_2^* = P_1 + P_2$, and $P_1 + P_2$ is a projection.

Conversely, $(P_1 + P_2)^2 = P_1 + P_2$ implies that $P_1 P_2 + P_2 P_1 = 0$. Now if x is in H,

$$\begin{aligned}
\langle P_1 P_2 x, x \rangle &= \langle P_1 P_2^2 x, x \rangle = -\langle P_2 P_1 P_2 x, x \rangle \\
&= -\langle P_1 P_2 x, P_2 x \rangle = -\langle P_1^2 P_2 x, P_2 x \rangle \\
&= -\|P_1 P_2 x\|^2 \leq 0.
\end{aligned}$$

But

$$\langle P_1 P_2 x, x \rangle = -\langle P_2 P_1 x, x \rangle = -\langle P_2 P_1^2 x, x \rangle$$
$$= \langle P_1 P_2 P_1 x, x \rangle = \langle P_2 P_1 x, P_1 x \rangle$$
$$= \langle P_2^2 P_1 x, P_1 x \rangle = \| P_2 P_1 x \|^2$$
$$\geqq 0.$$

Thus $P_1 P_2 = P_2 P_1 = 0$.

2. Clearly $P_1 P_2 = P_2 P_1$ implies that $P_1 P_2$ is a projection. Conversely if $P_1 P_2$ is a projection, then

$$(P_1 P_2) = (P_1 P_2)^* = P_2^* P_1^* = P_2 P_1.$$

3. If $P_1 P_2 = P_2$, then $P_2 P_1 = (P_1 P_2)^* = (P_2)^* = P_2$. So $P_2 P_1 = P_2$ also. Then

$$(P_1 - P_2)^2 = P_1^2 - P_1 P_2 - P_2 P_1 + P_2^2$$
$$= P_1 - P_2,$$

and $P_1 - P_2$ is also a projection. Since for all x in H,

$$\langle [P_1 - P_2] x, x \rangle = \langle [P_1 - P_2]^2 x, x \rangle$$
$$= \| [P_1 - P_2] x \|^2$$
$$\geqq 0,$$

we have $\langle P_1 x, x \rangle \geqq \langle P_2 x, x \rangle$ for all x in H.

Conversely if $\langle P_1 x, x \rangle \geqq \langle P_2 x, x \rangle$ for all x in H, then

$$\langle [I - P_2] x, x \rangle \geqq \langle [I - P_1] x, x \rangle.$$

Thus

$$\| (I - P_1) P_2 x \|^2 = \langle [I - P_1]^2 P_2 x, P_2 x \rangle$$
$$= \langle [I - P_1] P_2 x, P_2 x \rangle$$
$$\leqq \langle [I - P_2] P_2 x, P_2 x \rangle$$
$$= 0. \qquad \square$$

II.4 Some Spectral Theorems

No discussion of operators would be complete without at least a rudimentary discussion of the divisions of the spectrum of an operator. We begin with what is commonly called the spectral mapping theorem.

II.4.1. Theorem. *Let H be a Hilbert space. Let A be in $\mathcal{L}$. If $p(\lambda) = \sum_{j=0}^{n} a_j \lambda^j$ is a polynomial, we can define $p(A)$ by setting $p(A) = \sum_{j=0}^{n} a_j A^j$, a polynomial operator. Then*

(1) $\sigma(p(A)) = p(\sigma(A)) = \{p(\lambda) : \lambda \in \sigma(A)\}$.

(2) *If A^{-1} exists, then*

$$\sigma(A^{-1}) = [\sigma(A)]^{-1} = \{1/\lambda : \lambda \in \sigma(A)\}.$$

(3) $\sigma(A^*) = \overline{\sigma(A)}$.

Proof. 1. For any fixed number λ_0,

$$p(\lambda) - p(\lambda_0) = (\lambda - \lambda_0)q(\lambda),$$

where $q(\lambda)$ is also a polynomial. If λ_0 is in $\sigma(A)$, then $A - \lambda_0$ does not have an inverse. Since

$$p(A) - p(\lambda_0)I = (A - \lambda_0 I)q(A),$$

then $p(A) - p(\lambda_0)I$ does not have an inverse either, for if it did, then $[p(A) - p(\lambda_0)I]^{-1}q(A)$ would be the inverse of $A - \lambda_0 I$, which is impossible. Thus λ_0 in $\sigma(A)$ implies that $p(\lambda_0)$ is in $\sigma(p(A))$, or

$$p(\sigma(A)) \subset \sigma(p(A)).$$

Conversely, let λ_0 be in $\sigma(p(A))$. If

$$p(\lambda) - \lambda_0 = \alpha(\lambda - \lambda_1)\ldots(\lambda - \lambda_n),$$

then

$$p(A) - \lambda_0 I = \alpha(A - \lambda_1 I)\ldots(A - \lambda_n I).$$

Since $p(A) - \lambda_0 I$ does not have an inverse, one of the terms $A - \lambda_j I$ does not, $j = 1,\ldots,n$. That value λ_j is in $\sigma(A)$. Letting $\lambda = \lambda_j$, we find $p(\lambda_j) = \lambda_0$, and

$$\sigma(p(A)) \subset p(\sigma(A)).$$

Thus

$$\sigma(p(A)) = p(\sigma(A)).$$

2. If A has an inverse, then 0 is not in $\sigma(A)$. So $[\sigma(A)]^{-1}$ is well defined. If λ is not in $\sigma(A)$ and $\lambda \neq 0$, then the equation

$$A^{-1} - (1/\lambda)I = [\lambda I - A](1/\lambda)A^{-1}$$

shows that $1/\lambda$ is not in $\sigma(A^{-1})$. If $1/\lambda$ is in $\sigma(A^{-1})$, then it implies λ is in $\sigma(A)$, or $\lambda = 0$. In other words,

$$\sigma(A^{-1}) \subset [\sigma(A)]^{-1}.$$

Conversely, if we apply this result to A^{-1} we find

$$[\sigma(A)]^{-1} \subset \sigma(A^{-1}).$$

Thus

$$[\sigma(A)]^{-1} = \sigma(A^{-1}).$$

3. If λ is not in $\sigma(A)$, then $A - \lambda I$ has an inverse. Thus $A^* - \overline{\lambda} I$ does and $\overline{\lambda}$ is not in $\sigma(A^*)$. The contrapositive of this statement shows

$$\sigma(A^*) \subset \overline{\sigma(A)}.$$

Applying this to A^*,

$$\overline{\sigma(A)} \subset \sigma(A^*).$$

Thus

$$\sigma(A^*) = \overline{\sigma(A)}. \qquad \square$$

There are several ways to divide the spectrum of an operator into subsets. Perhaps the definition following is preferred.

II.4.2. Definition. Let H be a Hilbert space. let $A \in \mathcal{L}$.

(1) The set of points $\lambda \in \sigma(A)$ for which $(\lambda I - A)$ is not $1 - 1$ is called the point spectrum , $\sigma_p(A)$. $\lambda \in \sigma_p(A)$ if and only if $Ax = \lambda x$ for some nonzero $x \in H$.

(2) The set of all points $\lambda \in \sigma(A)$ for which $(\lambda I - A)$ is $1 - 1$ and $(\lambda I - A)H$ is dense in H, but $(\lambda I - A)H \neq H$, is called the continuous spectrum, $\sigma_c(A)$.

(3) The set of all points $\lambda \in \sigma(A)$ for which $(\lambda I - A)$ is $1 - 1$, but for which $(\lambda I - A)H$ is not dense in H, is called the residual spectrum $\sigma_r(A)$.

It is easy to show that these are all mutually disjoint and their union is $\sigma(A)$.

II.4.3. Examples. 1. On $L^2(0,1)$ let Ax be defined by $Ax(t) = m(t)x(t)$, where

$$\begin{aligned} m(t) &= 2t \quad , \quad 0 \le t \le 1/2, \\ &= 1 \quad , \quad 1 \le t \le 1. \end{aligned}$$

Here $\sigma(A) = [0,1]$; $\sigma_p(A) = \{1\}$; $\sigma_c(A) = [0,1)$; $\sigma_r(A) = \phi$, the empty set.

2. Let $\{e_n\}_{n=1}^\infty$ be a complete orthonormal set in an arbitrary Hilbert space H. For any $a \in H$, $x = \sum_{n=1}^\infty (x, e_n)e_n$. Now define A by setting

$$Ax = \sum_{n=1}^\infty (x, e_n)e_{n+1}.$$

Since e_1 is missing from the expansion of Ax, the range of A is $e_1^\perp$. Therefore $Ax = e_1$ has no solution and 0 is in $\sigma(A)$. But 0 is not in the point or continuous spectrum. So 0 is in the residual spectrum.

A^* is defined by

$$A^* y = \sum_{n=2}^\infty \langle y, e_n \rangle e_{n-1}.$$

We find that $A^*e_1 = 0\,e_1$, and so 0 is in the point spectrum of A^*. This is true in general: If λ is in $\sigma_r(A)$, then $\overline{\lambda}$ is in $\sigma_p(A^*)$. This follows from noting that under these circumstances $(A - \lambda I)H$ is not dense in H, and so there is an element orthogonal to $(A - \lambda I)H$. That is, if x is in H, then there is an element y such that

$$\langle (A - \lambda I)x, y \rangle = 0.$$

This implies

$$\langle x, (A^* - \overline{\lambda} I)y \rangle = 0.$$

Since x is arbitrary, let $x = (A^* - \overline{\lambda})y$ to see that $A^*y = \overline{\lambda}y$.

II.4.4. Theorem. *Let H be a Hilbert space. Let A be self-adjoint in $\mathcal{L}$. Then $\sigma(A)$ is a compact set of real numbers.*

Proof. We already know $\sigma(A)$ is bounded and closed. We must only show it is real.

Suppose λ is complex and is in $\sigma(A)$. Let $\lambda = \mu + i\nu$, $\nu \neq 0$. Then

$$\|(A - \lambda I)x\|^2 = \|(A - \mu I)x\|^2 + \nu^2 \|x\|^2,$$

and

$$\|(A - \lambda I)x\| \geq |\mu|\,\|x\|.$$

This implies that $A - \lambda I$ is $1-1$, so $(A - \lambda I)^{-1}$ exists. If $(A - \lambda I)x = f$, $x = (A - \lambda I)^{-1}f$, then substitution above yields

$$\|(A - \lambda I)^{-1}f\| \leqq \frac{1}{|\nu|}\|f\|,$$

and so $(A - \lambda I)^{-1}$ is a bounded operator.

If $(A - \lambda I)H$ is not dense in H, then there is a z orthogonal to it. So for all x in H, $\langle (A - \lambda I)x, z \rangle = 0$. This implies

$$\langle x, (A - \overline{\lambda} I)z \rangle = 0.$$

Letting $x = (A - \overline{\lambda} I)z$, we see $(A - \lambda I)z = 0$. Applying the first inequality above shows $z = 0$.

If $(A - \lambda I)H$ is dense in H, let f be any element of H. Then there exists a sequence x_n such that $\lim(A - \lambda I)x_n = f$. Since $(A - \lambda I)^{-1}$ is bounded, $x_n = (A - \lambda I)^{-1}f_n$ is a Cauchy sequence, and $\lim\limits_{n \to \infty} x_n = x$. Therefore we find

$$(A - \lambda I)x = (A - \lambda I)\lim_{n \to \infty} x_n$$
$$= \lim_{n \to \infty}(A - \lambda I)x_n$$
$$= \lim_{n \to \infty} f_n$$
$$= f.$$

Thus $(A - \lambda I)^{-1}$ is defined on all of H and λ is not in $\sigma(A)$, but is instead in $\rho(A)$.

$$\square$$

II.4.5. Corollary. *Let H be a Hilbert space. Let A be self-adjoint in $\mathcal{L}$. Then $\sigma_r(A)$ is empty.*

Proof. If λ is in $\sigma_r(A)$, then there is an element z orthogonal to $(A - \lambda I)H$. Then for all $x \in H$,

$$\langle (A - \lambda I)x, z \rangle = 0,$$

and

$$\langle x, (A - \lambda I)z \rangle = 0.$$

Let $x = (A - \lambda I)z$ to see that λ is in $\sigma_p(A)$. This is impossible. $\qquad\square$

II.4.6. Theorem. *Let U be a unitary operator on a Hilbert space H. If $\lambda \in \sigma(U)$, then $|\lambda| = 1$.*

Proof. If $|\lambda| > 1$, let $(\lambda I - U)x = f$. This has a solution

$$x = \sum_{n=0}^{\infty} \left(\frac{1}{\lambda} \right)^{n+1} U^n f,$$

and λ is in $\rho(U)$.

If $|\lambda| < 1$, then

$$x = - \sum_{n=0}^{\infty} \lambda^n (U^*)^{n+1} f,$$

and λ is in $\rho(U)$. Thus $\sigma(U)$ is on the unit circle. $\qquad\square$

We conclude with yet another characterization of the operator norm.

II.4.7. Theorem. *Let H be a Hilbert space. Let A be self-adjoint in $\mathcal{L}$. Then*

$$\|A\| = \sup\{|\lambda| : \lambda \in \sigma(A)\}.$$

Thus $\sigma(A)$ is not empty.

Proof. If $|\lambda| > \|A\|$, we know that λ is in $\rho(A)$, so if λ is in $\sigma(A)$, $|\lambda| \leq \|A\|$. Hence

$$\sup\{|\lambda| : \lambda \in \sigma(A)\} \leqq \|A\|.$$

To see the converse, we note there exists a sequence $\{x_n\}_{n=1}^{\infty}$ such that $\|x_n\| = 1$ and $\lim_{n \to \infty} \|Ax_n\| = \|A\|$. Then with $\lambda = \|A\|$,

$$\|A^2 x_n - \lambda^2 x_n\|^2 = \|A^2 x_n\|^2 - 2\lambda^2 \|Ax_n\|^2 + \lambda^4 \|x_n\|^2$$
$$\leqq \|A\|^2 \|Ax_n\|^2 - 2\lambda^2 \|Ax_n\|^2 + \lambda^4 \|x_n\|^2.$$

As n approaches ∞, the last expression approaches 0. This implies λ^2 is in the spectrum of A^2, $\sigma(A^2)$, which implies that either λ or $-\lambda$ is in the spectrum of A. Thus

$$\|A\| \leqq \sup\{|\lambda| : \lambda \text{ in } \sigma(A)\}. \qquad\square$$

II.4.8. Corollary. *Let H be a Hilbert space. Let A be self-adjoint in $\mathcal{L}$. Let $p(\lambda)$ be a polynomial with real coefficients. Then*

$$\|p(A)\| = \sup\{|p(\lambda)| : \lambda \in \sigma(A)\}.$$

II.4.9. Examples. 1. Let H be n-dimensional Euclidean space. Let the operator A be represented by an $n \times n$ matrix M, so that $Ax = Mx$. Then $\|A\|$ is the maximum of the absolute values of the eigenvalues of M.

2. Let $H = L^2(0,1)$. Let A be defined by setting $Ax(t) = m(t)x(t)$, where m is real and continuous on $[0,1]$. Then $\|A\| = \sup_{t \in [0,1]} |m(t)|$ and $\sigma(A) = \text{range } m(t)$.

II.5 Operator Convergence

In order to develop an integral representation for self-adjoint operators in which the operator is reduced to a kind of multiplication, it is necessary to employ several kinds of operator comparisons. Each has its uses. In most instances we do the best we can, using the "strongest" kind of comparison possible.

II.5.1. Definition. Let H be a Hilbert space and let $\{A_n\}_{n=1}^{\infty}$ be a sequence of operators in $\mathcal{L}$. Then

(1) The sequence $\{A_n\}_{n=1}^{\infty}$ converges uniformly to an operator $A \in \mathcal{L}$ when

$$\lim_{n \to \infty} \|A_n - A\| = 0.$$

(2) The sequence $\{A_n\}_{n=1}^{\infty}$ converges strongly to an operator $A \in \mathcal{L}$ when

$$\lim_{n \to \infty} \|A_n x - Ax\| = 0$$

for all $x \in H$.

(3) The sequence $\{A_n\}_{n=1}^{\infty}$ converges weakly to an operator $A \in \mathcal{L}$ when

$$\lim_{n \to \infty} \langle A_n x - Ax, y \rangle = 0$$

for all $x, y \in H$.

II.5.2. Theorem. *Uniform convergence implies strong convergence. Strong convergence implies weak convergence.*

We leave the proof as an exercise. It depends upon the definition of operator norm and Schwarz's inequality.

The concept of an order relation among operators is also useful.

II.5.3. Definition. Let H be a Hilbert space, and let A and B be self-adjoint and in $\mathcal{L}$. Then

(1) If $\langle Ax, x \rangle > \langle Bx, x \rangle$ for all $x \in H$, we say $A > B$.

(2) If there exist constants m and M such that $mI \leq A \leq MI$, then m and M are lower and upper bounds for A.

(3) If $A \geq 0$, the 0 operator, then A is positive. If $A > 0$, then A is positive definite.

II.5.4. Theorem. *Let H be a Hilbert space. Then every bounded, monotonic sequence of self-adjoint operators $\{A_n\}_{n=1}^{\infty} \subset \mathcal{L}$ converges strongly to a self-adjoint operator $A \in \mathcal{L}$.*

The proof is surprisingly complicated. Since it is important for what follows, we present it here.

Proof. We assume without loss of generality that

$$0 \leq A_1 \leq A_2 \cdots \leq A_n \leq I.$$

We let $A_{ij} = A_j - A_i$ when $j > i$.

If $A_0 \geq 0$ is a self-adjoint operator in $\mathcal{L}$, then $\langle A_0 x, x \rangle \geq 0$ for all $x \in H$. If $x = y + \lambda \langle A_0 y, z \rangle z$, y and $z \in H$, then for real λ,

$$0 \leqq \langle A_0 y, y \rangle + 2\lambda |\langle A_0 y, z \rangle|^2 + \lambda^2 |\langle A_0 y, z \rangle|^2 \langle A_0 z, z \rangle.$$

The only way for this to be true for all real λ is for

$$|\langle A_0 y, z \rangle|^2 \leq \langle A_0 y, y \rangle \langle A_0 z, z \rangle,$$

for all y and z in H.

Now replace A_0 by A_{ij}, y by x and z by $A_{ij} x$. Then

$$\begin{aligned}
\|A_{ij} x\| &= \langle A_{ij} x, A_{ij} x \rangle^2 \\
&\leq \langle A_{ij} x, x \rangle \langle A_{ij}^2 x, A_{ij} x \rangle \\
&\leq \langle A_{ij} x, x \rangle \|A_{ij}\|^3 \|x\|^2.
\end{aligned}$$

Since $\|A_{ij}\| \leq 1$, this implies

$$\|A_j x - A_i x\|^4 \leq |\langle A_j x, x \rangle - \langle A_i x, x \rangle| \|x\|^2.$$

Since $\{(A_n x, x)\}_{n=1}^{\infty}$ is a bounded monotonic sequence, it is convergent. This implies that $\{A_n x\}_{n=1}^{\infty}$ converges strongly to a limit Ax. The self-adjointness of A follows from the self-adjointness of each A_n. $\qquad\square$

We use this to show that each positive operator has a positive square root.

II.5.5. Theorem. *Let H be a Hilbert space. Let $A \geq 0$ be in $\mathcal{L}$. Then there exists a unique $A^{\frac{1}{2}} \geq 0$.*

Proof. We assume $\|A\| \leq 1$, and let $A = I - B$. Then the equation $(I - Y)^2 = A$ implies

$$Y = \frac{1}{2}(B + Y^2),$$

which we solve by successive approximation. Note that $0 \leq A \leq I$ and $0 \leq B \leq I$, so $B \geq 0$. Note further that $B^n \geq 0$: If $n = 2m + 1$, then

$$\langle B^n x, x \rangle = \langle B[B - x], [B^m x] \rangle \geq 0.$$

If $n = 2m$, then

$$\langle B^n x, x \rangle = \langle B^m x, B^m x \rangle \geq 0.$$

This implies that all polynomials in B with positive coefficients are positive.

Now let $Y_0 = 0$, $Y_1 = B/2$,

$$Y_{j+1} = (B + Y_j^2)/2.$$

By induction we find that each Y_j is a polynomial in B with positive coefficients, so $Y_j \geq 0$ for all j;

$$\|Y_{j+1}\| \leq [\|B\| + \|Y_j\|^2]/2 \leqq 1,$$

and

$$Y_{j+1} - Y_j = (Y_j + Y_{j-1})(Y_j - Y_{j-1}) \geq 0.$$

So $Y_{j+1} \geq Y_j$. The collection $\{Y_j\}_{j=1}^{\infty}$ is monotonic increasing and bounded. Hence there is an element C of $\mathcal{L}$ such that

$$\lim_{y \to \infty} Y_j = C.$$

C is positive and self-adjoint. C commutes with B and hence with A. Since

$$C = (B + C^2)/2,$$
$$A = (I - C)^2,$$

and $A^{\frac{1}{2}} = I - C$.

To see that $A^{\frac{1}{2}}$ is unique, let X be an additional positive square root. Since $XA = X^3 = AX$, X commutes with A, B, and $A^{\frac{1}{2}}$. Let Z and W be square roots of $A^{\frac{1}{2}}$ and X. Then for any $x \in H$, and $y = (A^{\frac{1}{2}} - X)x$,

$$\begin{aligned}
\|Zy\|^2 + \|Wy\|^2 &= \langle Z^2 y, y \rangle + \langle W^2 y, y \rangle \\
&= \langle [A^{\frac{1}{2}} + X]y, y \rangle \\
&= \langle [A^{\frac{1}{2}} + X][A^{\frac{1}{2}} - X]x, y \rangle \\
&= 0.
\end{aligned}$$

Therefore both Zy and Wy are 0, $A^{\frac{1}{2}}y = Z(Zy) = 0$, $Xy = W(Wy) = 0$, and

$$\|[A^{\frac{1}{2}} - X]x\|^2 = \langle [A^{\frac{1}{2}} - X]^2 x, x \rangle$$
$$= \langle [A^{\frac{1}{2}} - X]y, x \rangle$$
$$= 0.$$

Thus $X = A^{\frac{1}{2}}$ strongly. $\qquad\square$

II.5.6. Corollary. *Let H be a Hilbert space. Let A and B be positive commuting operators in $\mathcal{L}$. Then AB is positive and self-adjoint.*

Proof. Let x be in H. Then

$$\langle ABx, x \rangle = \langle AB^{\frac{1}{2}}x, B^{\frac{1}{2}}x \rangle \geq 0. \qquad\square$$

II.5.7. Theorem. *Let H be a Hilbert space. Let A be a bounded, self-adjoint operator in $\mathcal{L}$ satisfying $m\,I \leq A \leq M\,I$. Let $p(\lambda) = \sum_{j=0}^{n} c_j \lambda^j$ be a polynomial with real coefficients for which $p(\lambda) \geq 0$ when $m \leq \lambda \leq M$. Then $p(A) = \sum_{j=0}^{n} c_j A^j \geqq 0$.*

Proof. Let $p(\lambda)$ be written in factored form

$$p(\lambda) = c\,\Pi\,(\lambda - \alpha_j)\Pi(\beta_j - \lambda)\Pi([\lambda - \gamma_j]^2 + \delta_j^2),$$

where $c > 0$, $\alpha_j \leq m$, $\beta_j \geq M$ and the γ_j, δ_j terms represent complex conjugate zeros of $p(\lambda)$. Then

$$p(A) = c\,\Pi(A - \alpha_j I)\Pi(\beta_0 I - A)\Pi([A - \gamma_j I]^2 + \delta_j^2 I).$$

Each term is positive. Since they commute with each other, $p(A)$ is positive. $\quad\square$

We continue to extend the idea of operators which are functions of A by taking monotonic limits.

II.5.8. Definition. Let S denote the class of non-negative piecewise continuous functions, defined on the interval $[m, M]$.

II.5.9. Theorem. *For each element in $u \in S$ there exists a monotonic decreasing sequence of polynomials p_n such that for all $\lambda \in [m, M]$, $\lim_{n \to \infty} p_n(\lambda) = u(\lambda)$.*

Proof. Since u has only a finite number of finite discontinuities we may modify u in linear fashion near those discontinuities to find a sequence of continuous functions $\{q_j(\lambda)\}_{j=1}^{\infty}$ which converge to u at each point of $[m, M]$. Then for each j, $q_j + (1/j)$ can be approximated uniformly to within $1/2^j$ by a polynomial $p_j(\lambda)$. As j approaches ∞, the sequence $\{p_1(\lambda)\}_{j=1}^{\infty}$ decreases monotonically to $u(\lambda)$ when j is sufficiently large. $\qquad\square$

II.5.10. Theorem. *Let H be a Hilbert space. Let A be self-adjoint in $\mathcal{L}$ satisfying $m\,I \leq A \leq M\,I$. Let $\{p_j(\lambda)\}_{j=0}^{\infty}$ be an arbitrary sequence of monotonically decreasing polynomials which approach the piecewise continuous function $u(\lambda)$ in S from above on $[m, M]$. Then the sequence of operators $\{p_j(A)\}_{j=1}^{\infty}$ converges strongly to a limit $u(A)$ which is self-adjoint, positive and in $\mathcal{L}$.*

Proof. It is easy to see that each sequence $\{p_j(A)\}_{j=1}^{\infty}$ converges to a limit. To see that that limit is unique, suppose $\{q_j(A)\}_{j=1}^{\infty}$ is another sequence. Then there exist r and s such that

$$p_s(\lambda) \leq q_r(\lambda) + 1/r,$$
$$q_s(\lambda) \leq p_r(\lambda) + 1/r,$$

for all λ in $[m, M]$. Thus

$$p_s(A) \leq q_r(A) + (1/r)I,$$

and

$$q_s(A) \leq p_r(A) + (1/r)I.$$

Letting s approach ∞ with r fixed, then letting r approach ∞, we find

$$\lim_{s\to\infty} p_s(A) \leq \lim_{r\to\infty} q_r(A),$$
$$\lim_{s\to\infty} q_s(A) \leq \lim_{r\to\infty} p_r(A),$$

and the limits are the same.   □

II.5.11. Theorem. *Let H be a Hilbert space and let A be self-adjoint and in $\mathcal{L}$. Let $u_1\ u_2$, u be in S, and let $\alpha \geq 0$. Then*

$$(u_1 + u_2)(A) = u_1(A) + u_2(A),$$
$$(\alpha u)(A) = \alpha u(A).$$

$S(A)$ the class of operators which correspond to functions in S is additive and multiplicative by positive real numbers.

II.5.12. Definition. Let T denote the class of all piecewise continuous functions on $[m, M]$.

II.5.13. Theorem. *Each element $w \in T$ can be written as the difference of two elements u and $v \in S$.*

Proof. Let $u = \sup[w, 0]$, $v = \sup[-w, 0]$. Then $w = u - v$.

This decomposition is not unique.

$$
\begin{aligned}
u &= w + a, \ w > a \\
  &= \quad a \ \ , w \leq a, \\
v &= \quad a \ \ , w > a \\
  &= -w + a, \ w \leq a, \qquad a > 0,
\end{aligned}
$$

also satisfies $w = u - v$, u and v are in S.   □

II.5.14. Theorem. *Let H be a Hilbert space. Let A be self-adjoint in $\mathcal{L}$ bounded by m and M. Then for each $w \in T$, there exists a unique operator $w(A)$ which is the difference of operators $u((A)$ and $v(A)$ in $S(A)$. $w(A)$ is the strong limit of polynomial operators $\{p_j(A)\}_{j=1}^{\infty}$, where $\{p_j(\lambda)\}_{j=1}^{\infty}$ approaches $w(\lambda)$ for all $\lambda \in [m, M]$.*

We need only to verify that $w(A)$ is *uniquely* defined. Let $w(\lambda) = u_1(\lambda) - v_1(\lambda) = u_2(\lambda) - v_2(\lambda)$. Then $u_1(\lambda) + v_2(\lambda) = u_2(\lambda) + v_1(\lambda)$ are in S, and $u_1(A) + v_2(A) = u_2(A) + v_1(A)$. Hence

$$w(A) = u_1(A) - v_2(A) = u_2(A) - v_2(A).$$

II.5.15. Theorem. *Let H be a Hilbert space. Let A be self-adjoint in $\mathcal{L}$. Then $T(A)$, the space of all linear operators corresponding to piecewise continuous functions $T(\lambda)$ is a real, normed linear subspace of $\mathcal{L}$ under the operator norm.*

II.6 The Spectral Resolution of a Bounded Self-adjoint Operator

We have reached the stage where we can show that any bounded self-adjoint operator A can be written as an integral, where integration takes place over $\sigma(A)$ on the real axis. The integral is the strong limit of Riemann-type sums. It exhibits the operator in a way that the action of the operator is generated by multiplication.

The measure used in the integral is projection valued. We develop it first.

II.6.1. Theorem. *Let H be a Hilbert space. Let A be a self-adjoint operator in $\mathcal{L}$, bounded by m and M. Let $e(\lambda) \in S$ satisfy $e(\lambda) = 0$ or 1 when $\lambda \in [m, M]$. The the operator $e(A)$ corresponding to $e(\lambda)$ is a projection.*

Proof. Since $e(\lambda)$ is real valued, $e(A)$ is self-adjoint. Since $e(\lambda)^2 = e(\lambda)$, $e(A)^2 = e(A)$. $\qquad\square$

II.6.2. Theorem. *Let H be a Hilbert space. Let A be a self-adjoint operator in $\mathcal{L}$, bounded by m and M. Let*

$$e_\mu(\lambda) = 1, \lambda \le \mu$$
$$= 0, \lambda > \mu,$$

and let $E(\mu)$ be its corresponding projection. Then

(1) $E(\mu)E(\nu) = E(\mu)$ when $\mu \le \nu$, i.e., $E(\mu) \le E(\nu)$ when $\mu \le \nu$.

(2) $E(\mu) = 0$ when $\mu < m$.

(3) $E(\mu) = I$ when $\mu \ge M$.

(4) $E(\mu)$ is continuous from above.

Proof. The first three follow by construction. To see the fourth, let $p_j(\lambda)$ approach $e(\lambda)$ from above and at the same time satisfy

$$p_j(\lambda) \geq e_{\mu+1/j}(\lambda) \geq e_\mu(\lambda).$$

Then

$$p_j(A) \geq E(\mu + 1/j) \geq E(\mu).$$

As j approaches ∞, $p_j(A)$ approaches $E(\mu)$, and

$$\lim_{j \to \infty} E(\mu + 1/j) = E(\mu). \qquad \square$$

II.6.3. Theorem. *Let H be a Hilbert space. Let A be a self-adjoint operator in $\mathcal{L}$, bounded by m and M. Then there exists a collection of projection operators $\{E(\mu) : \mu \in [m, M]\}$, strong limits of polynomials in A, satisfying*

(1) $E(\mu) \leq E(\nu)$ *when* $\mu \leq \nu$,

(2) $E(\mu) = 0$ *when* $\mu < m$,

(3) $E(\mu) = I$ *when* $\mu \geq M$,

(4) $E(\mu)$ *is continuous from above,*

(5)
$$A = \lim_{\sup |\mu_k - \mu_{k-1}| \to 0} \sum_k \lambda_k [E(\mu_k) - E(\mu_{k-1})],$$

$$= \int_{m-}^{M} \lambda \, d\, E(\lambda), \qquad \textit{(definition)},$$

where λ_k is $m[\mu_{k-1}, \mu_k]$ and $\{\mu_k\}_k$ is a partition of $[m, M]$. The integral exists in the Riemann-Stieltjes sense. The equality between A and its integral representation is in the uniform sense.

Proof. If $\mu \leq \nu$, then

$$\mu[e_\nu(\lambda) - e_\mu(\lambda)] \leqq \lambda[e_\nu(\lambda) - e_\mu(\lambda)] \leq \nu[e_\nu(\lambda) - e_\mu(\lambda)],$$

and

$$\mu[E(\nu) - E(\mu)] \leq A[E(\nu) - E(\mu)] \leq \nu[E(\nu) - E(\mu)].$$

If we let

$$m = \mu_0 \leqq \mu_1 \leqq \cdots \leqq \mu_{n-1} \leqq \mu_n = M,$$

then with $\mu = \mu_{k-1}$, $\nu = \mu_k$, $k = 1, \ldots, n$, and summing, we find

$$\sum_k \mu_{k-1}[E(\mu_k) - E(\mu_{k-1})] \leqq A \sum_k [E(\mu_k) - E(\mu_{k-1})] \leq \sum_k \mu_k[E(\mu_k) - E(\mu_{k-1})].$$

The middle term is $A[E(\mu_n) - E(\mu_0)] = A$. So

$$\sum_k \mu_{k-1}[E(\mu_k) - E(\mu_{k-1})] \leq A \leq \sum_k \mu_k[E(\mu_k) - E(\mu_{k-1})].$$

Now let λ_k be an arbitrary point in $[\mu_{k-1}, \mu_k]$, $k = 1, \ldots, n$. Subtracting

$$\sum_k \lambda_k [E(\mu_k) - E(\mu_{k-1})]$$

from each term,

$$\sum_k (\mu_{k-1} - \lambda_k)[E(\mu_k) - E(\mu_{k-1})] \leqq A - \sum_k \lambda_k [E(\mu_k) - E(\mu_{k-1})],$$
$$\leqq \sum_k (\mu_k - \lambda_k)[E(\mu_k) - E(\mu_{k-1})].$$

If $\epsilon > 0$ is arbitrary and $\sup_k [\mu_k - \mu_{k-1}] < \epsilon$, then $-\epsilon < \mu_{k-1} - \lambda_k$ and $\mu_k - \lambda_k < \epsilon$. So

$$-\epsilon I \leqq A - \sum_k \lambda_k [E(\mu_k) - E(\mu_{k-1})] \leq \epsilon I.$$

As ϵ approaches 0,

$$A = \lim_{\sup[\mu_k - \mu_{k-1}] \to 0} \sum_k \lambda_k [E(\mu_k) - E(\mu_{k-1})]$$
$$= \int_{m-}^{M} \lambda \, dE(\lambda).$$

This can be extended to functions of A:

$$I = \int_{m-}^{M} dE(\lambda) \, , \text{ uniformly.}$$

$$p(A) = \int_{m-}^{M} p(\lambda)dE(\lambda) \, , \text{ uniformly for polynomials } p.$$

$$w(A) = \int_{m-}^{M} w(\lambda)dE(\lambda) \, , \text{ strongly for functions } w \in T.$$

The integral is linear,

$$\alpha \int_{m-}^{M} w(\lambda)dE(\lambda) = \int_{m-}^{M} \alpha w(\lambda)dE(\lambda),$$
$$\int_{m-}^{M} [w_1(\lambda) + w_2(\lambda)]dE(\lambda) = \int_{m-}^{M} w_1(\lambda)dE(\lambda) + \int_{m-}^{M} w_2(\lambda)dE(\lambda),$$

and, what is surprising, multiplicative,

$$\int_{m-}^{M} w_1(\lambda)dE(\lambda) \int_{m-}^{M} w_2(\lambda)dE(\lambda) = \int_{m-}^{M} w_1(\lambda)w_2(\lambda)dE(\lambda).$$

Please note that the correspondence $A \longleftrightarrow \lambda$, $p(A) \longleftrightarrow p(\lambda)$, $w(A) \longleftrightarrow w(\lambda)$ represents a reduction of operation by $w(A)$ to multiplication by $w(\lambda)$. $\quad \square$

II.6.4. Examples. 1. Let H be two-dimensional Euclidean space, and let A be represented by the matrix $\begin{pmatrix} 3 & -2 \\ -2 & 3 \end{pmatrix}$. Then $\sigma(A)$ is discrete, consisting of two eigenvalues $\lambda_1 = 1$, $\lambda_2 = 5$.

Associated with $\lambda_1 = 1$ is the eigenvector

$$v_1 = \begin{pmatrix} 1/\sqrt{2} \\ 1/\sqrt{2} \end{pmatrix} \quad \text{and projection} \quad P_1 = v_1 v_1^* = \begin{pmatrix} 1/2 & 1/2 \\ 1/2 & 1/2 \end{pmatrix}.$$

Associated with $\lambda_2 = 5$ is the eigenvector

$$v_2 = \begin{pmatrix} 1/\sqrt{2} \\ -1/\sqrt{2} \end{pmatrix} \quad \text{and projection} \quad P_2 = v_2 v_2^* = \begin{pmatrix} 1/2 & -1/2 \\ -1/2 & 1/2 \end{pmatrix}.$$

The projection measure $E(\lambda)$ is defined by

$$\begin{aligned} E(\lambda) &= 0 \quad, \quad \lambda < 1, \\ &= P_1 \quad, \quad 1 \le \lambda < 5, \\ &= P_1 + P_2 = I, \quad 5 \le \lambda, \end{aligned}$$

so $dE(\lambda) = 0$ if $\lambda \ne 1, 5$, $dE(1) = P_1$, $dE(5) = P_2$. We then find

$$I = \int_{1-}^{5} dE(\lambda) = P_1 + P_2,$$

$$A = \int_{1-}^{5} \lambda dE(\lambda) = 1\, P_1 + 5\, P_2,$$

$$p(A) = \int_{1-}^{5} p(\lambda) d\, E(\lambda) = p(1)P_1 + p(5)P_2,$$

when p is a polynomial, or for that matter, any function, piecewise continuous from above.

2. On $L^2(-1, 1)$, define A by setting

$$Ax(t) = t\, x(t).$$

For all w in T and x, y in $L^2(-1, 1)$,

$$\begin{aligned} \langle w(A)x, y \rangle &= \int_{-1}^{1} w(t)\, x(t)\overline{y(t)} dt \\ &= \int_{-1}^{1} w(t)\, d\left[\int_{-1}^{1} K_{[-1,t)}(s)x(s)\overline{y(s)} ds \right], \end{aligned}$$

where

$$\begin{aligned} K_{[-1,t)}(s) &= 1 \quad, \quad -1 \le s < t \le 1, \\ &= 0 \quad, \quad -1 \le t \le s \le 1. \end{aligned}$$

Since

$$\langle w(A)x, y\rangle = \int_{-1}^{1} w(t)d[(E(t)x, y)],$$

we see

$$d(E(t)x, y) = d\int_{-1}^{t} x(s)\overline{y(s)}ds,$$
$$= x(t)\overline{y(t)}dt.$$

For any set $\Delta \in [-1, 1]$, the projection $E(\Delta)$ is defined by

$$E(\Delta)\, x(t) = K_\Delta(t)\, x(t),$$

where

$$K_\Delta(t) = 1 \quad , \quad t \in \Delta,$$
$$= 0 \quad , \quad t \notin \Delta.$$

II.7 The Spectral Resolutions of Bounded Normal and Unitary Operators

Every operator A in $\mathcal{L}$, no matter what, possesses what can be called its real and imaginary parts, X and Y, just as complex numbers do.

$$X = [A + A^*]/2,$$
$$Y = [A - A^*]/2i.$$

Then

$$A = X + iY,$$
$$A^* = X - iY.$$

The problem with the analogy to complex numbers is that complex numbers commute while A and A^*, X and Y, in general do not. Such commutativity is essential to extending the concept of a spectral resolution.

II.7.1 Normal Operators

We note that if N is normal, then not only does $NN^* = N^*N$, but $XY = YX$. This implies more. $XY = YX$ guarantees that X commutes with polynomials in Y, as well as functions $w(Y)$, where w is piecewise continuous. Thus X commutes with projections $E_Y(\mu)$ as well. Further the roles of X and Y can be reversed so Y commutes with polynomials in X, functions $w(X)$, where w is piecewise continuous, and also projections $E_X(\lambda)$. Finally we note that all such functions of X commute with all such functions of Y. Let the identity, X and Y have the following representations:

$$I = \int_{m-}^{M} d\,E_X(\lambda),$$

$$X = \int_{m-}^{M} \lambda\, d\,E_X(\lambda),$$

$$I = \int_{n-}^{N} d\,E_Y(\mu),$$

$$Y = \int_{n-}^{N} \mu\, d\,E_Y(\mu).$$

Then

$$N = X + iY$$
$$= XI + iIY$$
$$= \int_{m-}^{M} \lambda\, d\,E_X(\lambda) \int_{n-}^{N} d\,E_Y(\mu) + \int_{m-}^{M} d\,E_X(\lambda) \int_{n-}^{N} i\,\mu\, d\,E_Y(\mu)$$
$$= \int_{m-}^{M} \int_{n-}^{N} (\lambda + i\,\mu) d\,E_X(\lambda) d\,E_Y(\mu).$$

We let $z = \lambda + i\,\mu$, $d\,E(z) = d\,E_X(\lambda) d\,E_Y(\mu)$, and note that $E(z)$ is a family of projections due to the commutativity of X and Y. Then

$$N = \iint_R z\, d\,E(z),$$

and in general, for functions $f(\lambda, \mu)$ which are approximable by polynomials,

$$f(N, N^*) = \iint_R f(z, \bar{z}) d\,E(z).$$

It is possible to use two-dimensional Riemann sums to derive these integrals directly. The family $\{E(z)\}$ has the appropriate two-dimensional order relation properties.

II.7.1.1. An Example. Good nontrivial examples of normal operators are not easy to find. The following one is simple, yet illustrative.

Let H be two-dimensional Euclidean space under the standard inner product. Let N be represented by a 2×2 matrix

$$N = \begin{pmatrix} 1 & 2 \\ 2i & -i \end{pmatrix}.$$

Then

$$X = \begin{pmatrix} 1 & 1-i \\ 1+i & 0 \end{pmatrix} , \; Y = \begin{pmatrix} 0 & 1-i \\ 1+i & -1 \end{pmatrix} .$$

X has eigenpairs $2, \begin{pmatrix} [1-i]/\sqrt{3} \\ 1/\sqrt{3} \end{pmatrix}$ and $-1, \begin{pmatrix} -1/\sqrt{3} \\ [1-i]/\sqrt{3} \end{pmatrix}$.

Consequently the projections associated with 2 and -1 are, respectively,

$$P_2 = \begin{pmatrix} 2/3 & [1-i]/3 \\ [1+i]/3 & 1/3 \end{pmatrix} , \; P_1 = \begin{pmatrix} 1/3 & [-1+i]/3 \\ [-1-i]/3 & 2/3 \end{pmatrix} ,$$

and

$$I = P_{-1} + P_2 \quad , \quad X = -1\,P_{-1} + 2\,P_2.$$

Y has eigenpairs $1, \begin{pmatrix} [1-i]/\sqrt{3} \\ 1/\sqrt{3} \end{pmatrix}$ and $-2, \begin{pmatrix} -1/\sqrt{3} \\ [1+i]/\sqrt{3} \end{pmatrix}$. The projections associated with 1 and -2 are, respectively

$$Q_1 = \begin{pmatrix} 2/3 & [1-i]/3 \\ [1+i]/3 & 1/3 \end{pmatrix} , \; Q_{-2} = \begin{pmatrix} 1/3 & [-1+i]/3 \\ [-1-i]/3 & 2/3 \end{pmatrix} ,$$

and

$$I = Q_{-2} + Q_1 , \; Y = -2Q_{-2} + 1Q_1.$$

Note that $P_{-1} = Q_{-2}$, $P_2 = Q_1$ and $P_{-1}P_2 = Q_{-2}Q_1 = 0$.

So we have

$$\begin{aligned}
N &= X + iY \\
&= XI + iIY \\
&= [-1P_{-1} + 2P_2][Q_{-2} + Q_1] + i[P_{-1} + P_2][-2Q_{-2} + 1Q_1] \\
N &= (-1 - 2i)P_{-1} + (2 + i)P_2.
\end{aligned}$$

The two-dimensional spectral measure is 0 unless $z = -1 - 2i$ or $2 + i$ is contained in $d\,E(z)$.

Then $d\,E(z)$ is P_{-1} or P_2. The integral is a two term sum.

Note $-1 - 2i$ and $2 + i$ are eigenvalues for N.

II.7.2   Unitary Operators

Since every unitary operator is also normal, we merely need to reduce the previous spectral resolution

$$N = \iint\limits_R z\, d\,E(z).$$

Since when $|z| \neq 1$, z is in $p(N)$, those portions of the measure inside and outside the unit circle generate nothing. If the unit circle is reparameterized by setting $z = e^{i\theta}$, then for a unitary operator U,

$$U = \int_{0-}^{1} e^{i\theta} d\,E(\theta).$$

The adjoint and inverse U^* is given by

$$U^* = \int_{0-}^{1} e^{-i\theta} d\,E(\theta).$$

In each of these the spectral measure E_θ is a one parameter family of projections, just as was true for a self-adjoint operator.

It is possible to derive these formulas directly using trigonometric polynomials. Since the technique is similar to that used for self-adjoint operators, we refer the reader to the references at the end of the chapter.

II.7.2.1. An Example. Unitary operators in Euclidean space satisfy $UU^* = U^*U = I$. As a result the rows and columns of U form an orthonormal set. If the previous example is modified by dividing by $\sqrt{5}$, the result is unitary:

$$U = \begin{pmatrix} 1/\sqrt{5} & 2/\sqrt{5} \\ 2i/\sqrt{5} & -i/\sqrt{5} \end{pmatrix}.$$

The eigenvalues change. They are divided by $\sqrt{5}$. Since $U = N/\sqrt{5}$, little else changes.

References

[1]   N. I. Akhiezer and I. M. Glazman, **Theory of Linear Operators in Hilbert space, vol. I and II**, Frederick Ungar, New York, 1961.

[2]   A. M. Krall, **Applied Analysis**, D. Reidel, Dordrecht, 1986.

[3]   F. Riesz and B. Sz.-Nagy, **Functional Analysis**, Frederick Ungar, New York, 1955.

[4]   A. E. Taylor, **Introduction to Functional Analysis**, John Wiley and Sons, New York, 1958.

Chapter III

Unbounded Linear Operators
On a Hilbert Space

So far we have spent a great deal of time focusing our attention on *bounded* operators, when, indeed, our interest is in dealing with differential operators which are *unbounded*. The reason for this is that in order to properly attack the problems associated with them, a foundation concerning bounded operators must be laid first. The spectral resolution for a bounded self-adjoint operator,

$$A = \int_{m-}^{M} \lambda \, d\,E(\lambda), \quad \text{has its unbounded counterpart} \quad A = \int_{-\infty}^{\infty} \lambda \, d\,E(\lambda),$$

but the proof, which is considerably more complicated, is derived by looking at a certain unitary operator, the Cayley transform. Hence the need for the preceding work. We follow the path ingeniously described by John von Neumann.

III.1 Unbounded Linear Operators

If an operator satisfying $\langle Ax, y \rangle = \langle x, Ay \rangle$ is defined for all x, y in a Hilbert space H, or if A is a *closed* operator defined on all of H, then it is possible to show that A is bounded. Therefore in any discussion of unbounded operators, we must be careful to define its domain D. The concept of adjoint must be redefined.

III.1.1. Definition. Let A be a linear operator on a Hilbert space H with a domain D which is dense in H. Let y be an element in H for which there exists another element y^* in H such that

$$\langle Ax, y \rangle = \langle x, y^* \rangle$$

for all $x \in D$. We define the adjoint operator A^* by setting $A^* y = y^*$.

III.1.2. Definition. Let A be a linear operator on a Hilbert space H with a domain D which is dense in H. A is closed if whenever x_n has limit $x \in D$ and Ax_n has limit $y \in H$, then x is in D and $Ax = y$.

This corresponds to the idea of a closed graph in the plane.

III.1.3. Theorem. *Let H be a Hilbert space. Let A be a linear operator on H with domain D which is dense in H. Then A^* is uniquely defined, closed and linear.*

Proof. If $A^*y = y^*$ and $A^*y = z^*$, then $\langle x, y^* - z^* \rangle = 0$ for all x in D. We choose a sequence $\{x_n\}$ with limit $y^* - z^*$ to see that $\|y^* - z^*\| = 0$.

To see that A^* is closed, let y_n approach $y \in D^*$, the domain of A^*, and let $A^*y_n = z_n$ have limit $z \in H$. Then for all $x \in D$,

$$\langle Ax, y \rangle = \lim_{n \to \infty} \langle Ax, y_n \rangle$$
$$= \lim_{n \to \infty} \langle x, A^*y_n \rangle = \langle x, z \rangle.$$

Thus y is in D^*, and $A^*y = z$.

We leave the linearity of A^* as an exercise.

It is possible for the domain of the adjoint A^* of an (a differential) operator to consist only of the 0 element. As we shall show shortly, however, if A is closed, then the domain of A^* is dense in H. $\qquad\square$

III.2   The Graph of an Operator

In order to proceed further, it is convenient to introduce the concept of the graph of an operator. It is analogous to the formal definition of a function or of the graph of a function in the plane.

III.2.1. Definition. Let $\mathcal{H} = H \otimes H$, the set of ordered pairs $\{x, y\}$ of elements in a Hilbert space H. If we define

$$c\{x, y\} = \{cx, cy\},$$

when c is a complex number,

$$\{x_1, y_1\} + \{x_2, y_2\} = \{x_1 + x_2, y_1 + y_2\},$$
$$\langle \{x_1, y_1\}, \{x_2, y_2\} \rangle = \langle x_1, x_2 \rangle + \langle y_1, y_2 \rangle,$$

then $\mathcal{H}$ is a Hilbert space.

III.2.2. Definition. Let H be a Hilbert space. Let A be a linear operator on H with domain D. The collection of all pairs $\{x, Ax\}$ in $\mathcal{H} = H \otimes H$ is called the graph of A, and is denoted by G_A.

We also need two mixing operators on $\mathcal{H}$, U and V.

III.2.3. Definition. We define operators U and V on $\mathcal{H} = H \otimes H$ by setting

$$U\{x, y\} = \{y, x\},$$
$$V\{x, y\} = \{y, -x\}.$$

III.2.4. Theorem. $UV = -VU$, $U^2 = I$, $V^2 = -I$.

Note that the formula
$$\langle Ax, y \rangle = \langle x, y^* \rangle,$$
used in defining A^*, can be written as
$$\langle V\{x, Ax\}, \{y, y^*\} \rangle = 0.$$

III.2.5. Definition. Let H be a Hilbert space and let X be a subset of H. Then by $[X]$ we mean the closed subspace of H generated by X, that is, the closure of all linear combinations of elements in X.

For example, if $H = \ell^2$, the space of all vectors $x = (x_1, \ldots, x_n, \ldots)$ such that $\sum_1^\infty |x_i|^2 < \infty$, and if X is the subset of vectors with only finitely many non-zero components also having $x_1, \ldots, x_{10} = 0$, then $[X]$ would be the orthogonal complement of the finite-dimensional space consisting of vectors with nonzero entries only in the first 10 positions.

III.2.6. Theorem. *Let G_{A^*} denote the graph of A^* in $\mathcal{H}$. Then $G_{A^*} = \mathcal{H} - V[G_A]$.*

Proof. We note that $V[G_A] = [VG_A]$. Therefore $[VG_A]$ is orthogonal to G_{A^*} from remarks made previously. So $\mathcal{H} = G_{A^*} + [VG_A]$. $\qquad\square$

III.2.7. Theorem. *Let H be a Hilbert space. Let A be a linear operator on H with domain D. The existence of A^{-1}, A^* and $(A^{-1})^*$ implies the existence of $(A^*)^{-1}$ and $(A^*)^{-1} = (A^{-1})^*$.*

Proof. First note that $G_{A^{-1}} = U\,G_A$. Then
$$\begin{aligned}
G_{(A^{-1})^*} &= \mathcal{H} - V[G_{A^{-1}}] = \mathcal{H} - VU[G_A] \\
&= U\,[U\mathcal{H} - V[G_A]] = U(\mathcal{H} - V[G_A]) \\
&= U\,G_{A^*} = G_{(A^*)^{-1}},
\end{aligned}$$
which shows that $(A^*)^{-1}$ exists and equals $(A^{-1})^*$.

A problem with adjoints is that their domains may only consist of the zero element, as was pointed out earlier. The following, therefore, is most important if self-adjointness is to be considered. $\qquad\square$

III.2.8. Theorem. *Let H be a Hilbert space. Let A be a closed linear operator on H with domain D which is dense in H. Then the domain of A^*, D^*, is also dense in H. Thus A^{**} exists and $A^{**} = A$.*

Proof. Suppose D^* is not dense in H. There exists, then, an element $x \in H$ orthogonal to D^*. Then $\{0, x\}$ is orthogonal to all elements of the form $\{A^*y, -y\}$, where $y \in D^*$. Hence $\{0, x\}$ is orthogonal to $V\,G_{A^*}$. Since the orthogonal complement of G_{A^*} is $V[G_A]$, the orthogonal complement of $V\,G_{A^*}$ is $V^*[G_A] = -I[G_A] = [G_A] = G_A$, since A is closed. Thus $\{0, x\}$ is in G_A, and $A\,0 = x$. Thus $x = 0$, a contradiction.

A reapplication to A^* shows A^{**} exists. Since $G_{A^{**}}$ is the orthogonal complement of $V\,G_{A^*}$, we see that $G_{A^{**}} = G_A$ and $A^{**} = A$. $\qquad\square$

III.3 Symmetric and Self-adjoint Operators

Consider the operator A defined on appropriate elements of $L^2(0,1)$ by setting $A\,x(t) = (1/t)x(t)$. A is not defined on all of $L^2(0,1)$ because

$$\int_0^1 (1/t)^2\,|x|^2 dt \not< \infty$$

for all elements x (let $x = 1$). Further, if

$$x_n = \sqrt{n(n+1)} \quad , \quad \frac{1}{n+1} < x < \frac{1}{n},$$
$$= 0 \quad , \quad \text{elsewhere,}$$

then $\|x_n\| = 1$ and $\|Ax_n\| = \sqrt{n(n+1)}$. Hence there exists *no* constant K such that

$$\|Ax_n\| \leqq \,|\, < \|x_n\|.$$

A is unbounded.

To alleviate the difficulties near $t = 0$, we let D_0 consist of those elements x in $L^2(0,1)$ which vanish near $t = 0$. A, restricted to D_0, is denoted by A_0. If x is in D_0 and y is in D_0^*, the adjoint domain for A_0^*,

$$\langle A_0 x, y\rangle = \int_0^1 (1/t)x\overline{y}dt = \langle x, A_0^* y\rangle = \int_0^1 x\overline{A_0^* y}dt\,.$$

Comparing the integrals, we see

$$\int_0^1 x\overline{[(1/t)y - A_0^* y]}dt = 0.$$

For any t_0 in $(0,1]$ we let x be a spike near t_0. This tells us that $A_0^* y = (1/t)y$ a.e., from which we conclude that $A_0^* y = (1/t)y$ in $L^2(0,1)$, and that

$$\int_0^1 (1/t)^2|y(t)|^2 dt < \infty.$$

I.e. y is in D, and $A_0^* = A$.

The introduction of D_0, A_0 was really unnecessary. If x and y are in D,

$$\langle Ax, y\rangle = \int_0^1 (1/t)x\overline{y}dt = \langle x, Ay\rangle.$$

We can associate $1/t$ with either x or y. This says that $D \subset D^*$ and $A \subset A^*$.

An argument similar to that given for A_0^* shows that in fact $A^* y = (1/t)y$. As a result $D^* \subset D$ and $A^* \subset A$.

Thus $D = D^*$ and $A = A^*$.

So $\langle A_0 x, y\rangle = \langle x, A_0 y\rangle$, but D_0 is too restrictive; while in the case of A, $\langle Ax, y\rangle = \langle x, Ay\rangle$, when x and y are in and only in D. We are motivated to generalize and make the following definitions.

III.3.1. Definition. Let H be a Hilbert space. Let A be a linear operator on H with domain D.

(1) A is symmetric if D is dense in H and $A \subset A^*$. That is, $D \subset D^*$ and $Ax = A^*x$ for all x in D.

(2) A is self-adjoint if $A = A^*$. This means that $D = D^*$ and $Ax = A^*x$ on their common domain.

It is easy to show, in general, that $A \subseteq A^*$ implies $A^{**} \subseteq A^*$, so

$$A^{**} \subseteq A^* = (A^*)^{**} = (A^{**})^*.$$

Every symmetric operator has a closed symmetric extension. Our interest, of course, will ultimately be to determine when self-adjointness occurs.

III.3.2. Theorem. *Let H be a Hilbert space. Let A be a linear operator on H with domain D. If A is self-adjoint and A^{-1} exists, then A^{-1} is also self-adjoint.*

Proof. We know

$$(A^{-1})^* = (A^*)^{-1} = A^{-1}.$$

The only problem is that the domain of A^{-1} may not be dense in H. If so, then there exists an element z orthogonal to all y in the domain of A^{-1}, i.e., $\langle y, z \rangle = 0$.

But $y = Ax$ for some x in D. Thus $\langle Ax, z \rangle = 0$ for all x in D. This implies that z is in $D^*(= D)$ and $Az = 0$. We apply A^{-1} to conclude that $z = 0$.

In order to extend the spectral resolution theorem for bounded self-adjoint operators to unbounded self-adjoint operators, we need to introduce two special operators for use therein. $\qquad\square$

III.3.3. Theorem. *Let H be a Hilbert space. Let A be a closed linear operator on H with domain D, dense in H. Then the operators*

$$B = (I + A^*A)^{-1},$$
$$C = A(I + A^*A)^{-1}$$

are defined for all x in H. $\|B\| \leq 1$, $\|C\| \leq 1$. B is symmetric and positive.

Proof. We return to the graph of A. Since G_A and $V G_{A^*}$ are orthogonal complements, any element x can be uniquely decomposed by

$$\{x, 0\} = \{y, Ay\} + \{A^*z, -z\}.$$

This implies

$$x = y + A^*z,$$
$$0 = Ay - z$$

can be uniquely solved for y and z. If $y = Bx$ and $z = Cx$, then

$$I = B + A^*C, \quad 0 = AB - C.$$

These yield

$$I = [I + A^*A]B \quad , \quad C = AB.$$

What is more,

$$\|x\|^2 = \|\{x, 0\}\|^2 = \|\{y, Ay\}\|^2 + \{A^*z, -z\}\|^2$$
$$= \|y\|^2 + \|Ay\|^2 + \|A^*z\|^2 + \|z\|^2.$$

Therefore

$$\|x\|^2 \geq \|y\|^2 + \|z\|^2 = \|Bx\|^2 + \|Cx\|^2,$$

so $\|B\| \leq 1$, $\|C\| \leq 1$.

If x is in the domain of $I + A^*A$, then

$$\langle [I + A^*A]x, x \rangle = \|x\|^2 + \|Ax\|^2 \geq \|x\|^2.$$

Thus if $[I + A^*A]x = 0$, then $x = 0$, and $[I + A^*A]^{-1}$ exists. Since

$$B = I,$$
$$B = [I + A^*A]^{-1}.$$

Finally

$$\langle Bx, y \rangle = \langle Bx, [I + A^*A]By \rangle$$
$$= \langle [I + A^*A]Bx, By \rangle = \langle x, By \rangle.$$

B is symmetric, and

$$\langle Bx, x \rangle = \langle Bx, Bx \rangle + \langle ABx, ABx \rangle \geq 0.$$

So B is positive. $\qquad\qquad\qquad\qquad\qquad\qquad\qquad\qquad\qquad\qquad\quad\square$

III.4 The Spectral Resolution of an Unbounded Self-Adjoint Operator

The pieces are now in place to derive the major aim of this first section, to show that the spectral resolution of an unbounded, self-adjoint operator A has the same look as that for a bounded, self-adjoint operator. The procedure must be different, however, because polynomials in A cannot be used to generate projections.

III.4.1. Theorem. *Let H be a Hilbert space. Let A be a self-adjoint operator on H with domain D. Then*

$$R_{\pm i} = (A \pm i\,I)^{-1}$$

exists, is everywhere defined and is bounded.

Proof. We note formally that

$$C \pm iB = (A \pm iI)(I + A^2)^{-1}$$
$$= (A \mp iI)^{-1}.$$

If x is in D, then

$$\|(A \mp iI)x\|^2 = \|Ax\|^2 \mp i\langle x, Ax \rangle \pm i\langle Ax, x \rangle + \|x\|^2$$
$$= \|Ax\|^2 + \|x\|^2.$$

So

$$\|(A \mp iI)x\|^2 \geq \|x\|^2,$$

and $(A \mp iI)^{-1}$ exists. If $y = (A \mp iI)x$, then

$$\|y\|^2 \geq \|(A \mp iI)^{-1}y\|^2.$$

So $R_{\mp i}$ is bounded by 1. We recall from Theorem III.3.3 that both B and C are defined on all of H. So therefore is $R_{\pm i}$. $\square$

III.4.2. Definition. Let H be a Hilbert space. Let A be a self-adjoint operator on H with domain D. The operator

$$T = (A - i\,I)(A + i\,I)^{-1}$$

is called the Cayley transform of A.

III.4.3. Theorem. *The Cayley transform of a self-adjoint operator A on a Hilbert space H is a unitary operator.*

Proof. Note from the proof of Theorem III.4.1 that

$$\|(A + iI)x\| = \|(A - iI)x\|$$

for all $x \in D$. If $y = (A + i\,I)x$, then

$$\|y\| = \|(A - iI)(A + iI)^{-1}y\| = \|T\,y\|.$$

T is isometric. It is defined for all elements of the form $y = (A + iI)x$ by $Ty = (A - iI)x$, where $x \in D$. According to Theorem III.4.1, as x varies through D, y and Ty cover all of H. The range of T is all of H.

Now note that for all y,

$$\langle y, y \rangle = \langle Ty, Ty \rangle = \langle T^*Ty, y \rangle.$$

This implies using the polarization identities that

$$\langle y, z \rangle = \langle T^*Ty, z \rangle$$

for all $y, z \in H$. Let $z = T^*Tg - y$. Then

$$\|T^*Ty - y\| = 0,$$

or $T^*Ty = y$ for all y in H. Applying T, we have

$$TT^*Ty = Ty.$$

Letting $Ty = x$, we have $TT^*x = x$ for all x in H. Thus $T^*T = TT^* = I$ and T is unitary: $T^* = T^{-1}$. $\square$

III.4.4. Theorem. *Let T be the Cayley transform of a self-adjoint operator A, defined on a Hilbert space H. Then*

$$A = i(I + T)(I - T)^{-1}.$$

Proof. Let y be in H. Then there exists $x \in D$ such that

$$y = (A + iI)x, \quad Ty = (A - iI)x.$$

Add and subtract to find

$$(I + T)y = Ax, \quad (I - T)y = 2ix.$$

This implies $(I - T)^{-1}$ exists and is bounded, and

$$A = i(I + T)(I - T)^{-1}.$$ $\square$

III.4.5. Theorem. *Let H be a Hilbert space. Let T be the Cayley transform of a self-adjoint operator A on H, and let*

$$T = \int_0^{2\pi} e^{i\theta} d\, E(\theta)$$

be the spectral decomposition of T. Then $E(\theta)$ is continuous at $\theta = 0$ and $\theta = 2\pi$.

Proof. If E is discontinuous either at 0 or 2π, then $\Delta E \neq 0$ at the appropriate point. Thus there exist $x, y \in H$, $y \neq 0$, such that $y = \Delta Ex$. Since E is a family of projections, $\Delta E y = y$. So 1 is an eigenvalue of T with eigenfunction y. But then $(I - T)^{-1}$ would not exist, giving a contradiction. $\square$

III.4.6. Theorem. *Let H be a Hilbert space. Let T be the Cayley transform of a self-adjoint operator A on H, and let $E(\theta)$, $0 \leq \theta \leq 2\pi$, be the spectral measure for T. Let $[0, 2\pi)$ be decomposed by $\{\theta_m\}$, determined by*

$$-\cot(\theta_m/2) = m, \qquad -\infty < m < \infty.$$

Then the projections $P_m = E(\theta_m) - E(\theta_{m-1})$ are pairwise orthogonal, commute with T and A, and satisfy $\sum_{m=-\infty}^{\infty} P_m = I$.

Proof. T commutes with each P_m and hence with A.

$$\sum_{m=-\infty}^{\infty} P_m = \lim_{\theta\to 2\pi} E(\theta) - \lim_{\theta\to 0} E(\theta) = I - 0 = I. \qquad \square$$

III.4.7. Theorem. *Under the notation of Theorem III.4.6, let $H_m = P_m H$. If $x \in H_m$, then*

$$Ax = \int_{m-1}^{m+} \lambda\, d\, E_A(\lambda)x,$$

where $\lambda = -\cot(\theta/2)$ and $d\,E_A(\lambda) = d\,E(-2\cos^{-1}\lambda)$.

Proof. If $x \in H_m$,

$$\begin{aligned}
Ax &= i(I+T)(I-T)^{-1}P_m x \\
&= \int_0^{2\pi} i(1+e^{i\theta})(1+e^i)^{-1} K_{(\theta_{m-1},\theta_m]}(\theta) d\,E(\theta)x \\
&= \int_{\theta_{m-1}}^{\theta_m+} i(1+e^{i\theta})(1-e^{i\theta})^{-1} d\,E(\theta)x.
\end{aligned}$$

(Here K stands for the characteristic function of the interval $(\theta_{m-1},\theta_m]$.) Thus

$$Ax = \int_{\theta_{m-1}}^{\theta_m+} -\cot(\theta/2)\, d\,E(\theta)x.$$

If we let $\lambda = -\cot(\theta/2)$ and $E_A = E(-2\cot^{-1}\lambda)$, then,

$$Ax = \int_{m-1}^{m+} \lambda\, d\,E_A(\lambda)x\ . \qquad \square$$

III.4.8. Theorem. *Let H be a Hilbert space. Let A be an unbounded-self-adjoint operator on H with domain D. There exists a collection of projection operators $\{E_A(\lambda),\ -\infty < \lambda < \infty\}$, strong limits of the Cayley transform of A, satisfying*

(1) $E_A(\lambda) \leq E_A(\mu)$ *when* $\mu \leq \nu$,

(2) $\displaystyle\lim_{\lambda\to-\infty} E_A(\lambda) = 0$,

(3) $\displaystyle\lim_{\lambda\to\infty} E_A(\lambda) = I$,

(4) $E_A(\lambda)$ *is continuous from above,*

(5) $Ax = \int_{-\infty}^{\infty} \lambda\, d\,E_A(\lambda)x$

   for all $x \in D$.

Proof. If $x \in D$, then

$$
\begin{aligned}
Ax &= A \sum_{-\infty}^{\infty} P_m\, x \\
&= \sum_{-\infty}^{\infty} A\, P_m\, x \\
&= \sum_{-\infty}^{\infty} \int_{m-1}^{m+} \lambda\, d\, E_A(\lambda) x \\
&= \int_{-\infty}^{\infty} \lambda\, d\, E_A(\lambda) x.
\end{aligned}
$$

The equality of A and its integral is strong equality. In the bounded case it was uniform.

The formula

$$
A = \int_{-\infty}^{\infty} \lambda\, d\, E_A(\lambda)
$$

is elegant, but it is very abstract. The purpose of Section 2 of the text is to see precisely what this looks like when A is a *differential* operator on an appropriate Hilbert (L^2) space. We must not only specify the form of A, but also find its domain D. This is not a trivial task. The result is equally as elegant as the abstract integral.

$$\square$$

References

[1]   N. I. Akhiezer and I. M. Glazman, **Theory of Linear Operators in Hilbert space, vol. I and II**, Frederick Ungar, New York, 1961.

[2]   A. M. Krall, **Applied Analysis**, D. Reidel, Dordrecht, 1986.

[3]   F. Riesz and B. Sz.-Nagy, **Functional Analysis**, Frederick Ungar, New York, 1955.

[4]   A. E. Taylor, **Introduction to Functional Analysis**, John Wiley and Sons, New York, 1958.

Chapter IV

Regular Linear Hamiltonian Systems

Examination of systems of differential equations began in the early 1900's with the work of G. D. Birkhoff and R. E. Langer (see [2] for example.), R. L. Wilder and L. Schlesinger. G. A. Bliss [3] in 1926 seems to have been the first to discuss regular, self-adjoint differential systems. Additional references to their works may be found in the papers of Birkhoff and Langer [2], and in the book [4] by Coddington and Levinson.

After World War II, W. T. Reid [7] continued the work of Bliss, and F. V. Atkinson wrote his classic book [1] on self-adjoint differential systems. It was in Atkinson's book that the linear Hamiltonian system format seems to have first appeared.

Further motivation to study Hamiltonian systms began in the 1970's with the books of B. M. Levitan and I. S. Sargsjan [5], [6], and the papers mentioned there which discussed Dirac systems, both regular and singular.

There are a number of very interesting ways in which linear Hamiltonian systems can arise. We mention three. First, every formally self-adjoint differential operator in one variable can be written as a Hamiltonian system [12]. Second, Dirac systems are special cases of Hamiltonian systems [5], [6]. Third, S-Hermitian systems are generalizations of Hamiltonian systems [10–12]. In turn, every S-Hermitian system can be written as a Hamiltonian system (Heinz Langer).

Of course, linear Hamiltonian systems are linearizations of more general (nonlinear) Hamiltonian systems.

IV.1 The Representation of Scalar Problems

Consider, for example, the formally self-adjoint differential equation of first order,

$$i[(q_0 y)' + q_0 y'] + p_0 y = \lambda w\, g + w\, f,$$

where $p_0, q_0 > 0$, $w > 0$ are real valued, Lebesgue measurable functions on an interval I of the real line. If we let $Y = (2q_0)^{\frac{1}{2}} y$, $J = i$, $A = w/2q_0$, $B = -p_0/2q_0$

and $F = (2q_0)^{\frac{1}{2}} f$, where A, B and AF are integrable, the equation becomes

$$iY' = [\lambda(w/2q_0) + (-p_0/2q_0)]Y + (w/2q_0)(2q_0)^{\frac{1}{2}} f,$$

or

$$JY' = [\lambda A + B]Y + AF,$$

where J is constant, $J^* = -J = J^{-1}$, $A = A^* \geq 0$, $B = B^*$.

The formally self-adjoint second order equation

$$-(p_0 y')' + i[(q_0 y)' + q_0 y'] + p_1 y = \lambda w\, y + w\, f$$

can likewise be transformed. Let

$$Y = \begin{pmatrix} y \\ p_0 y' - iq_0 y \end{pmatrix}, J = \begin{pmatrix} 0 & -1 \\ 1 & 0 \end{pmatrix}, A = \begin{pmatrix} w & 0 \\ 0 & 0 \end{pmatrix},$$

$$B = \begin{pmatrix} -p_1 + q_0^2/p_0 & -iq_0/p_0 \\ iq_0/p_0 & 1/p_0 \end{pmatrix} \text{ and } F = \begin{pmatrix} f \\ 0 \end{pmatrix}.$$

The second order equation becomes

$$\begin{pmatrix} 0 & -1 \\ 1 & 0 \end{pmatrix} Y' = \left[\lambda \begin{pmatrix} w & 0 \\ 0 & 0 \end{pmatrix} + \begin{pmatrix} p_1 + q_0^2/p_0 & -iq_0/p_0 \\ iq_0/p_0 & 1/p_0 \end{pmatrix} \right] Y$$
$$+ \begin{pmatrix} w & 0 \\ 0 & 0 \end{pmatrix} \begin{pmatrix} f \\ 0 \end{pmatrix},$$

or

$$JY' = [\lambda A + B]Y + AF,$$

where J is constant, $J^* = -J = J^{-1}$, $A = A^* \geq 0$, $B = B^*$. Conditions on the coefficients are similar to the first order equation. We assume A, B and AF are integrable.

If $q_0 = 0$, the system is a bit simpler:

$$\begin{pmatrix} 0 & -1 \\ 1 & 0 \end{pmatrix} \begin{pmatrix} y \\ p_0 y' \end{pmatrix}' = \left[\lambda \begin{pmatrix} w & 0 \\ 0 & 0 \end{pmatrix} + \begin{pmatrix} -p_1 & 0 \\ 0 & 1/p_0 \end{pmatrix} \right] \begin{pmatrix} y \\ p_0 y' \end{pmatrix} + \begin{pmatrix} w & 0 \\ 0 & 0 \end{pmatrix} \begin{pmatrix} f \\ 0 \end{pmatrix}.$$

Here lies the key to thinking about linear Hamiltonian systems of even order. Think of the first components of Y as y itself! Think of the second components as the derivative of y, even if the dimension of Y is greater than 2, even if A and B are not as simple as those above!

The formally self-adjoint third order equation has a similar representation. Consider

$$-[i(q_0(q_0 y')') + (p_0 y')'] + i[(q_1 y)' + q_1 y'] + p_1 y$$
$$= \lambda w\, y + w\, f.$$

If we replace q_0 and q_1 by $u_0 = q_0^2/2$ and $u_1 = q_1$, this can be written as

$$- i[(u_0 y')'' + (u_0 y'')'] - (p_0 y')'$$
$$+ i[(u_1 y)' + (u_1 y')] + p_1 y = \lambda w y + w f.$$

We let $Y = \begin{pmatrix} y \\ y^{[1]} \\ y^{[2]} \end{pmatrix}$, where, with $\theta = (1 + i)/\sqrt{2}$,

$$y^{[1]} = -\theta q_0 y',$$
$$y^{[2]} = i q_0 (q_0 y')' + p_0 y' - i q_1 y.$$

The scalar equation is equivalent to

$$\begin{pmatrix} 0 & 0 & -1 \\ 0 & i & 0 \\ 1 & 0 & 0 \end{pmatrix} Y' = \left[\lambda \begin{pmatrix} w & 0 & 0 \\ 0 & 0 & 0 \\ 0 & 0 & 0 \end{pmatrix} + \begin{pmatrix} -p_1 & \theta q_1/q_0 & 0 \\ \bar{\theta} q_1/q_0 & -p_0/q_0^2 - V\bar{\theta} q_0 & 0 \\ 0 & -1/\theta q_0 & 0 \end{pmatrix} \right] Y$$
$$+ \begin{pmatrix} w & 0 & 0 \\ 0 & 0 & 0 \\ 0 & 0 & 0 \end{pmatrix} Y,$$

or

$$J Y' = [\lambda A + B] Y + A F,$$

where J is constant, $J^* = -J = J^{-1}$, $A = A^* \geq 0$, $B = B^*$.

In fact, Walker [12] has showed that *any* formally self-adjoint differential equation can be so written. It is a remarkable paper. Since the details in general are so messy, we cite the source [12], but omit the details.

There are two additional examples of scalar equations we wish to exhibit in system format: those of fourth and sixth order with real coefficients. They will be needed later.

The formally self-adjoint, real, differential equation of fourth order is

$$(p_0 y'')'' - (p_1 y') + p_2 y = \lambda w y + f.$$

If we let $y_1 = y$, $y_2 = y'$, $y_3 = -(p_0 y'')' + p_1 y'$, $y_4 = p_0 y''$, then

$$\begin{pmatrix} 0 & 0 & -1 & 0 \\ 0 & 0 & 0 & -1 \\ 1 & 0 & 0 & 0 \\ 0 & 1 & 0 & 0 \end{pmatrix} \begin{pmatrix} y_1 \\ y_2 \\ y_3 \\ y_4 \end{pmatrix}' = \left[\lambda \begin{pmatrix} w & 0 & 0 & 0 \\ 0 & 0 & 0 & 0 \\ 0 & 0 & 0 & 0 \end{pmatrix} + \begin{pmatrix} -p_2 & 0 & 0 & 0 \\ 0 & -p_1 & 1 & 0 \\ 0 & 1 & 0 & 0 \\ 0 & 0 & 0 & 1/p_0 \end{pmatrix} \right]$$
$$\times \begin{pmatrix} y_1 \\ y_2 \\ y_3 \\ y_4 \end{pmatrix} + \begin{pmatrix} w & 0 & 0 & 0 \\ 0 & 0 & 0 & 0 \\ 0 & 0 & 0 & 0 \end{pmatrix} \begin{pmatrix} f \\ 0 \\ 0 \\ 0 \end{pmatrix}.$$

The formally self-adjoint, real, differential equation of sixth order is

$$-(p_0y''')''' + (p_1y'')'' - (p_2y')' + (p_3y) = w\,g + w\,f.$$

If we let $y_1 = y$, $y_2 = y'$, $y_3 = y''$, $y_4 = (p_0y''')'' - (p_1y'')' + (p_2y')$, $y_5 = -(p_0y''')' + (p_1y'')$ and $y_6 = (p_0y''')$, then

$$\begin{pmatrix} 0 & 0 & 0 & -1 & 0 & 0 \\ 0 & 0 & 0 & 0 & -1 & 0 \\ 0 & 0 & 0 & 0 & 0 & -1 \\ 1 & 0 & 0 & 0 & 0 & 0 \\ 0 & 1 & 0 & 0 & 0 & 0 \\ 0 & 0 & 1 & 0 & 0 & 0 \end{pmatrix} \begin{pmatrix} y_1 \\ y_2 \\ y_3 \\ y_4 \\ y_5 \\ y_6 \end{pmatrix}' =$$

$$\left[\lambda \begin{pmatrix} w & 0 & 0 & 0 & 0 & 0 \\ 0 & 0 & 0 & 0 & 0 & 0 \\ 0 & 0 & 0 & 0 & 0 & 0 \\ 0 & 0 & 0 & 0 & 0 & 0 \\ 0 & 0 & 0 & 0 & 0 & 0 \\ 0 & 0 & 0 & 0 & 0 & 0 \end{pmatrix} + \begin{pmatrix} -p_3 & 0 & 0 & 0 & 0 & 0 \\ 0 & -p_2 & 0 & 1 & 0 & 0 \\ 0 & 0 & -p_1 & 0 & 1 & 0 \\ 0 & 1 & 0 & 0 & 0 & 0 \\ 0 & 0 & 1 & 0 & 0 & 0 \\ 0 & 0 & 0 & 0 & 0 & 1/p_0 \end{pmatrix} \right] \begin{pmatrix} y_1 \\ y_2 \\ y_3 \\ y_4 \\ y_5 \\ y_6 \end{pmatrix}$$

$$+ \begin{pmatrix} w & 0 & 0 & 0 & 0 & 0 \\ 0 & 0 & 0 & 0 & 0 & 0 \\ 0 & 0 & 0 & 0 & 0 & 0 \\ 0 & 0 & 0 & 0 & 0 & 0 \\ 0 & 0 & 0 & 0 & 0 & 0 \\ 0 & 0 & 0 & 0 & 0 & 0 \end{pmatrix} \begin{pmatrix} f \\ 0 \\ 0 \\ 0 \\ 0 \\ 0 \end{pmatrix}. \qquad \text{(IV.1)}$$

IV.2 Dirac Systems

Consider the pair of linear differential equations [5], [6]

$$y_2' + p_{11}y_1 + p_{12}y_2 = \mu y_1,$$
$$-y_1' + p_{21}y_1 + p_{22}y_2 = \mu y_2,$$

where $p_{12} = p_{21}$ and p_{ij}, $j = 1, 2$, are integrable over a real interval I. Such a pair is called a "one-dimensional" Dirac system. It is, of course, two-dimensional.

If we let $B = \begin{pmatrix} 0 & 1 \\ -1 & 0 \end{pmatrix}$, $P = \begin{pmatrix} p_{11} & p_{12} \\ p_{21} & p_{22} \end{pmatrix}$, $I = \begin{pmatrix} 1 & 0 \\ 0 & 1 \end{pmatrix}$ and $y = \begin{pmatrix} y_1 \\ y_2 \end{pmatrix}$, then we have

$$B\,y' = (\mu I - P)y.$$

The matrix Dirac system has two canonical forms. Let

$$H = \begin{pmatrix} \cos\phi(x) & -\sin\phi(x) \\ \sin\phi(x) & \cos\phi(x) \end{pmatrix},$$

and $y = H z$. Further let the resulting coefficients of z not depending on λ be represented by $Q = \begin{pmatrix} q_{11} & q_{12} \\ q_{21} & q_{22} \end{pmatrix}$. We set $q_{12} = q_{21} = 0$. Hence

$$p_{12} \cos 2\phi(x) + \frac{1}{2}(p_{22} - p_{11}) \sin 2\phi(x) = 0\,,$$

and

$$\phi(x) = \frac{1}{2} \tan^{-1}\left[\frac{2p_{12}}{p_{11} - p_{22}}\right]\,.$$

Relabelling $Q = \begin{pmatrix} p & 0 \\ 0 & r \end{pmatrix}$, we arrive at

$$B\,z' = [\mu\,I - Q]z.$$

Finally, multiply by $-I$ and let $\mu = -\lambda$, $J = \begin{pmatrix} 0 & -1 \\ 1 & 0 \end{pmatrix}$ to find

$$J z' = [\lambda\,I + Q]z.$$

With $A = I$, $B = Q$ we have the equation

$$J z' = [\lambda\,A + B]z.$$

If a nonhomogeneous term were initially present, then there is also a nonhomogeneous term present above.

The second canonical form is found by requiring that the trace of q, $q_{11} + q_{22} = 0$. This implies

$$2\phi'(x) + (p_{11} + p_{22}) = 0,$$

and

$$\phi(x) = -\frac{1}{2}\int (p_{11} + p_{12})dx.$$

The result is that $Q = \begin{pmatrix} p & q \\ q & -p \end{pmatrix}$ and

$$J z' = [\lambda I + Q]z\,.$$

With $A = I$, $B = Q$, we have the equation

$$J z' = [\lambda A + B]z.$$

The case $p_{12} = p_{21} = 0$, $p_{11} = V(x) + m$, $p_{22} = V(x) - m$, where $V(x)$ is a potential function, where m is the mass of a particle, is called a stationary, one-dimensional, Dirac system in relativistic quantum mechanics [5], [6].

IV.3 S-Hermitian Systems

During the 1970's two German mathematicians F. W. Schafke and A. Schneider [9–11], began a study of what are called S-Hermitian systems: Let F_j, G_j, S_j, $j = 1, 2$, be integrable $n \times n$ matrices, I_n be the $n \times n$ identity matrix, for which there is a continuously differentiable $n \times n$ matrix H satisfying for all x in I, λ in C,

$$\begin{pmatrix} F_1 - \overline{\lambda}G_1 & F_2 - \overline{\lambda}G_2 \\ S_1 & S_2 \end{pmatrix}^* \begin{pmatrix} 0 & -I_n \\ I_n & 0 \end{pmatrix} \begin{pmatrix} F_1 - \lambda G_1 & F_2 - \lambda G_2 \\ S_1 & S_2 \end{pmatrix} = \begin{pmatrix} 0 & H \\ H & H' \end{pmatrix}.$$

We further assume that F_1 is invertible. Then

$$F_1 y' + F_2 y = \lambda [G_1 y' + G_2 y] \,,$$

or

$$F y = \lambda G y$$

is an S-Hermitian system over I.

It is possible to show there exists a positive definite matrix W such that $G_1 = W S_1$, $G_2 = W S_2$, or

$$G y = W S y.$$

The setting for the system is the Hilbert space generated by the inner product

$$\begin{aligned} \langle y, z \rangle &= \int_I (Sz)^*(Gy)dx, \\ &= \int_I (Sz)^* W (Sy)dx. \end{aligned}$$

As a special case let

$$\begin{aligned} F_1 &= J, & G_1 &= 0, & S_1 &= 0, \\ F_2 &= -B, & G_2 &= A, & S_2 &= I_n, \end{aligned}$$

where $J^* = -J = J^{-1}$, $A = A^* \geq 0$, $B = B^*$. Then $W = A$, $H = J$; the matrix equation is automatically satisfied; the differential equation is

$$J y' = [\lambda A + B]y,$$

a linear Hamiltonian system.

Remarkably, Heinz Langer, in a private communication to Reinhard Mennicken, has shown the converse: An S-Hermitian system can be written as a linear Hamiltonian system.

IV.4 Regular Linear Hamiltonian Systems

So far in this chapter we have been rather informal in attempting to motivate the study of linear Hamiltonian systems. We have shown that they arise in a number of ways, particularly as a generalization of scalar self-adjoint differential equations and also of Dirac systems. These two types of problems generate two different settings in the sense that the weight matrix A is singular in the second. Singularity of A causes a bit of difficulty in the definition of a differential operator, but it can be overcome.

IV.4.1. Definition. Let J be an $n \times n$ constant matrix satisfying $J^* = J^{-1} = -J$. Let A and B be integrable over a finite interval $[a, b]$, and assume that A and B satisfy $A = A^* \geq 0$, $B = B^*$. Let λ be a complex number. By a regular, linear Hamiltonian system we mean the differential system

$$JY' = [\lambda A + B]Y.$$

We assume that the system is definite over $[a, b]$. That is, if Y is a solution to the equation above, then

$$\int_a^b Y^* A Y \, dx = 0$$

if and only if $Y \equiv 0$. (Otherwise some strange things happen. We shall consider this in the discussion of spectral resolutions.)

There is a corresponding nonhomogeneous system. If F is appropriately defined, it is

$$JY' = [\lambda A + B]Y + AF.$$

IV.4.2. Definition. We denote by $L_A^2(a, b)$ the Hilbert space of n-dimensional vectors Y, Z, etc., generated by the inner product

$$\langle Y, Z \rangle_A = \int_a^b Z^* A Y \, dx$$

and norm

$$\|Y\| = \left(\int_a^b Y^* A Y \, dx \right)^{\frac{1}{2}}.$$

Note that if A is singular, we are definitely dealing with equivalence classes. As a matter of fact we are always dealing with equivalence classes because of the use of the Lebesgue integral, but, when A is singular, there may be some null elements Y in $L_A^2(a, b)$ for which $Y \neq 0$ a.e.

Note further that if A is zero except for $a_{11} = w$, then

$$\langle Y, Z \rangle_A = \int_a^b \bar{z}_{11} w \, y_{11} \, dx \, .$$

This is the inner product in $L^2(a, b; w)$, traditionally the setting for scalar differential operators.

In order to properly define differential operators in $L_A^2(a, b)$ we need to first introduce the minimal and maximal operators L_m and L_M. (The operators of interest are extensions of L_m, restrictions of L_M. The adjoint operators are likewise.)

IV.4.3. Definition. We denote by D_0 those elements Y in $L_A^2(a, b)$ satisfying

 (1) $\ell Y = JY' - BY = AF$ exists a.e. in (a, b) and F is in $L_A^2(a, b)$.

 (2) $Y(a) = 0$, $Y(b) = 0$.

We define the operator L_0 by setting $L_0 Y = F$ if and only if

$$JY' - BY = AF$$

for all Y in D_0.

Here is where the difficulty arises if A is singular. It is the possible singularity of A which causes the hyperhole in the definition of L_0. If A is nonsingular, we could much more easily say $L_0 Y = F$ if and only if

$$A^{-1}[JY' - BY] = F.$$

IV.4.4. Definition. We denote by D_M those elements Y in $L_A^2(a, b)$ satisfying

 1. $\ell y = JY' - BY - AF$ exists a.e. in (a, b) and F is in $L_A^2(a, b)$.

We define the operator L_M by setting $L_M Y = F$ if and only if

$$JY' - BY = AF$$

for all Y in D_M.

L_0 is traditionally called the minimal operator associated with ℓ when the interval is finite and the Hamiltonian system is regular (A and B are integrable over $[a, b]$). If the problem is singular, the domain D_0 must be modified slightly.

L_M is traditionally called the maximal operator associated with ℓ. In the singular case its definition is unchanged.

IV.4.5. Theorem. *In $L_A^2(a, b)$ the domain of L_0^* is D_M. $L_0^* = L_M$. The domain of L_M^* is D_0. $L_M^* = L_0$.*

The proof will seem a bit complicated but it is a generalization of the proof of the same theorem in the scalar, first order case where the differential operator can be written as $\ell y = (iy' - py)/w$. (See [8]).

Proof. Let Y be in D_0 and let Z be in the domain of L_0^*. Then

$$\langle L_0 Y, Z \rangle_A = \langle Y, L_0^* Z \rangle_A.$$

This is equivalent to

$$\int_a^b Z^*[JY' - BY]dx = \int_a^b [L_0^* Z]^* AY \, dx,$$

or

$$\int_a^b Z^* J Y' \, dx - \int_a^b [BZ + A(L_0^* Z)]^* Y \, dx = 0.$$

We integrate the second integral by parts, keeping in mind that $Y(a) = 0$, $Y(b) = 0$.

$$\int_a^b Z^* J Y' \, dx + \int_a^b \left[\int_a^x [BZ + A(L_0^* Z)] d\xi \right]^* Y' \, dx = 0,$$

or

$$\int_a^b \left[-JZ + \int_a^x [BZ + A(L_0^* Z)] d\xi \right] Y' \, dx = 0.$$

We must now find out which elements are orthogonal to Y' (in $L_I^2(a, b)$).

If the bracketed term is a constant C, then

$$\int_a^b C^* Y' \, dx = C^* Y \Big|_a^b = 0.$$

Constants are permitted. $\qquad\qquad\qquad\qquad\qquad\qquad\qquad\qquad\square$

In general let the bracketed term be W. By using the Gram–Schmidt procedure [8], we assume that

$$\langle W, C \rangle_I = \int_a^b C^* W \, dx = 0.$$

Since the constant may be arbitrary, we find $\int_a^b W \, dx = 0$.

Now let

$$\widetilde{Y} = \int_a^x W \, d\xi.$$

Then $\widetilde{Y}(a) = 0$, $\widetilde{Y}(b) = 0$. So $\widetilde{Y}'$ is an acceptable Y' (is in D_0). Thus $\widetilde{Y}'$ is orthogonal to W. But $\widetilde{Y}' = W$ in $L_I^2(a, b)$, and so $W = 0$.

We conclude that the bracketed term must be constant, and

$$-JZ + \int_a^x [BZ + A(L_0^* Z)] d\xi = C.$$

This says that Z is differentiable and

$$A L_0^* Z = J Z' - BZ.$$

Setting $L_0^* Z = G$, we see $L_0^* Z = G$ if and only if

$$J Z' - BZ = AG.$$

We have the form of L_0^*. Since no constraints were placed on Z at a or b, we find that L_0^* and L_M have the same form and that the domain of L_0^*: $D_{L_0^*} \subset D_M$. Thus $L_0^* \subset L_M$.

To show the reverse inclusion we use a simple form of Green's formula. Let Y be in D_0, Z be in D_M, and $L_0 Y = F$, $L_M Z = G$. Then

$$JY' - BY = AF,$$

and

$$JZ' - BZ = AG,$$

with F and G in $L_A^2(a, b)$. We compute

$$\langle L_0 Y, Z \rangle_A = \int_a^b Z^* A(L_0 Y)\, dx$$

$$= \int_a^b Z^* A F\, dx$$

$$= \int_a^b Z^* [JY' - BY]\, dx.$$

We integrate the term $Z^* J Y'$ by parts.

$$\langle L_0 Y, Z \rangle_A = Z^* J Y \big|_a^b + \int_a^b [JZ' - BZ]^* Y\, dx.$$

Since Y is in D_0, $Y(a) = 0$, $Y(b) = 0$, and

$$\langle L_0 Y, Z \rangle_A = \int_a^b [JZ' - BZ]^* Y\, dx$$

$$= \int_a^b G^* A Y\, dx$$

$$= \int_a^b (L_M Z)^* A Y\, dx$$

$$= \langle Y, L_M Z \rangle_A.$$

This shows that Z is in $D_{L_0^*}$, $D_M \subset D_{L_0^*}$, and $L_M Z = L_0^* Z : L_M \subset L_0^*$. Hence $L_0^* = L_M$.

To find L_M^*, we first note $L_0 \subset L_M$, that is, $D_0 \subset D_M$ and for Y in D_0, $L_0 Y = L_M Y$. Now let Z be in D_{M^*} and Y be in D_0, a subset of D_M. Then $\langle L_m Y, Z \rangle_A = \langle L_0 Y, Z \rangle_A$. A reapplication of the first part of the previous proof establishes that Z is in D_M and $L_M^* Z = L_M Z$. So $L_M^* \subset L_M$.

A reapplication of the Green's formula now establishes further that $Z(a) = 0$, $Z(b) = 0$, since, for general Y in D_M, $Y(a)$ and $Y(b)$ are arbitrary. Thus $D_{M^*} \subset D_0$ and $L_M^* \subset L_0$.

Finally checking via Green's formula establishes $L_0 \subset L_M^*$, and so $L_M^* = L_0$.

The regular, self-adjoint Hamiltonian operators we seek lie between L_0 and L_M. Specifically we seek a restriction D of D_M so that $D_0 \subset D \subset D_M$. Its restricted operator L then satisfies $L_0 \subset L \subset L_M$. Their adjoints then satisfy $L_M^* \subset L^* \subset L_0^*$, or $L_0 \subset L^* \subset L_M$ (an exercise). We wish to determine when D^*, the domain of L^*, which also satisfies $D_0 \subset D^* \subset D_M$, is the same as D. Then, since the *forms* of L and L^* already agree, we have $L = L^*$.

We introduce regular linear Hamiltonian boundary value problems.

IV.4.6. Definition. We denote by D those elements Y in $L_A^2(a, b)$ satisfying

(1) $\ell y = JY' - BY = AF$ exists a.e. and F is in $L_A^2(a, b)$.

(2) Let M and N be $m \times n$ matrices such that rank $(M : N) = m$. Then set

$$MY(a) + NY(b) = 0.$$

We define the operator L by setting $LY = F$ if and only if

$$JY' - BY = AF$$

for all Y in D.

We require some additional notation. Let P and Q be $(2n - m) \times n$ matrices, chosen so that rank $(P : Q) = 2n - m$ and $\begin{pmatrix} M & N \\ P & Q \end{pmatrix}$ is nonsingular. Let $\begin{pmatrix} \widetilde{M} & \widetilde{N} \\ \widetilde{P} & \widetilde{Q} \end{pmatrix}$ be chosen so that

$$\begin{pmatrix} \widetilde{M} & \widetilde{N} \\ \widetilde{P} & \widetilde{Q} \end{pmatrix} \begin{pmatrix} M & N \\ P & Q \end{pmatrix} = \begin{pmatrix} -J & 0 \\ 0 & J \end{pmatrix} .$$

Then

$$\widetilde{M}^* M + \widetilde{P}^* P = -J , \qquad \widetilde{N}^* M + \widetilde{Q}^* P = 0,$$
$$\widetilde{M}^* N + \widetilde{P}^* Q = 0 , \qquad \widetilde{N}^* N + \widetilde{Q}^* Q = J .$$

We can now develop the regular Green's formula.

IV.4.7. Theorem. *(Green's Formula). Let Y and Z be in D_M. Then*

$$\langle L_M Y, Z \rangle_A - \langle Y, L_M Z \rangle_A = [\widetilde{M} Z(a) + \widetilde{N} Z(b)]^* [MY(a) + NY(b)]$$
$$+ [\widetilde{P} Z(a) + \widetilde{Q} Z(b)]^* [PY(a) + QY(b)].$$

Proof. The left side is

$$\int_a^b [Z^* A(L_M Y) - (L_M Z)^* AY] dx.$$

Let $L_M Y = AF$, $L_M Z = AG$. Then the left side becomes

$$\int_a^b [Z^*(AF) - (AG)^* Y]\,dx$$

$$= \int_a^b [Z^*(JY' - BY) - (JZ' - BZ)^* Y]\,dx$$

$$= \int_a^b [Z^* JY' + Z'^* JY]\,dx$$

$$= Z^* JY \Big|_a^b,$$

since J is constant.

Now

$$Z^* JY \Big|_a^b = Z^*(b)\, JY(b) - Z^*(a) JY(a)$$

$$= (Z^*(a), Z^*(b)) \begin{pmatrix} -J & 0 \\ 0 & J \end{pmatrix} \begin{pmatrix} Y(a) \\ Y(b) \end{pmatrix}.$$

If we substitute for the middle matrix, we get

$$(Z(a)^*, Z(b)^*) \begin{pmatrix} \widetilde{M}^* & \widetilde{P}^* \\ \widetilde{N}^* & \widetilde{Q} \end{pmatrix} \begin{pmatrix} M & N \\ P & Q \end{pmatrix} \begin{pmatrix} Y(a) \\ Y(b) \end{pmatrix}$$

$$= \left[\begin{pmatrix} \widetilde{M} & \widetilde{N} \\ \widetilde{P} & \widetilde{Q} \end{pmatrix} \begin{pmatrix} Z(a) \\ Z(b) \end{pmatrix} \right]^* \left[\begin{pmatrix} M & N \\ P & Q \end{pmatrix} \begin{pmatrix} Y(a) \\ Y(b) \end{pmatrix} \right]$$

$$= \begin{pmatrix} \widetilde{M}\, Z(a) + \widetilde{N}\, Z(b) \\ \widetilde{P}\, Z(a) + \widetilde{Q}\, Z(b) \end{pmatrix}^* \begin{pmatrix} M\, Y(a) + N\, Y(b) \\ P\, Y(a) + Q\, Y(b) \end{pmatrix},$$

which when multiplied together gives the right side of Green's formula.

We can now precisely describe the adjoint of L, L^*, and its domain D^*. $\quad\square$

IV.4.8. Theorem. *The domain of L^*, D^*, consists of those elements Z in $L_A^2(a,b)$ satisfying*

    1. $\ell Z = JZ' - BZ = AG$ exists a.e. in (a,b) and G is in $L_A^2(a,b)$.

    2. $\widetilde{P}\, Z(a) + \widetilde{Q}\, Z(b) = 0$.

For Z in D^*, $L^* Z = G$ if and only if

$$JZ' - BZ = AG.$$

Proof. Since $L_0 \subset L \subset L_M$, we have $L_0 \subset L^* \subset L_M$ as well. The form of L^* is the same as that of L. Let Y be in D, Z be in D^* and apply Green's formula. The left side is zero, and the first term on the right, $M\, Y(a) + N\, Y(b)$ is also zero. But the term $P\, Y(a) + Q\, Y(b)$ is arbitrary, and this forces Z to satisfy $\widetilde{P}\, Z(a) + \widetilde{Q}\, Z(b) = 0$.

Conversely, if Z satisfies the criteria listed above, then Z is in the domain of the adjoint. Hence the theorem.

Both elements in D and elements in D^* also satisfy *parametric* boundary conditions. Recall that

$$M\,Y(a) + N\,Y(b) = 0,$$
$$P\,Y(a) + Q\,Y(b) = \psi,$$

where ψ is arbitrary. These equations are equivalent to

$$\begin{pmatrix} M & N \\ P & Q \end{pmatrix} \begin{pmatrix} Y(a) \\ Y(b) \end{pmatrix} = \begin{pmatrix} 0 \\ \psi \end{pmatrix}.$$

Multiply this by

$$\begin{pmatrix} J & 0 \\ 0 & -J \end{pmatrix} \begin{pmatrix} \widetilde{M}^* & \widetilde{P}^* \\ \widetilde{N}^* & \widetilde{Q} \end{pmatrix} \quad \text{to find} \quad \begin{pmatrix} Y(a) \\ Y(b) \end{pmatrix} = \begin{pmatrix} J\widetilde{P}^*\psi \\ -J\widetilde{Q}^*\psi \end{pmatrix},$$

or

$$Y(a) = J\widetilde{P}^*\psi, \; Y(b) = -J\widetilde{Q}^*\psi,$$

parametric boundary conditions.

Similar manipulations can be used to find parametric boundary conditions for D^*. Since

$$\widetilde{M}\,Z(a) + \widetilde{N}\,Z(b) = \phi,$$
$$\widetilde{P}\,Z(a) + \widetilde{Q}\,Z(b) = 0,$$

where ϕ is arbitrary, we have

$$(Z(a)^* \; Z(b)^*) \begin{pmatrix} \widetilde{M}^* & \widetilde{P}^* \\ \widetilde{N}^* & \widetilde{Q}^* \end{pmatrix} = (\phi^* \; 0).$$

Right multiply this by $\begin{pmatrix} M & N \\ P & Q \end{pmatrix} \begin{pmatrix} -J & 0 \\ 0 & J \end{pmatrix} = \begin{pmatrix} MJ & -NJ \\ PJ & -QJ \end{pmatrix}$. The result is

$$Z(a) = -JM^*\phi, \; Z(b) = JN^*\phi.$$

These can be used to develop a criterion, under which L is self-adjoint. $\qquad \square$

IV.4.9. Theorem. *The regular, linear Hamiltonian operator L is self-adjoint if and only if $m = n$ and $MJM^* = NJN^*$.*

Proof. If L is self-adjoint, then Z satisfies the D-boundary conditions. Thus

$$M[-JM^*\phi] + N[JN^*\phi] = 0, \quad \text{and} \quad [-MJM^* + NJN^*]\phi = 0.$$

Since ϕ is arbitrary, $-MJM^* + NJN^* = 0$.

Conversely if $MJM^* = NJN^*$, then $(-MJ \; NJ)\binom{M^*}{N^*} = 0$. This says that the columns of $\binom{M^*}{N^*}$ for n independent solutions to the equation $(-MJ \; NJ)X = 0$. But from equations derived earlier, $(-MJ \; NJ)\binom{\widetilde{P}^*}{\widetilde{Q}^*} = 0$ as well. Again we have a full complement of n solutions. Therefore there must be a constant, nonsingular matrix C^* such that $\left(\begin{smallmatrix}\widetilde{P}^*\\ \widetilde{Q}^*\end{smallmatrix}\right)C^* = \left(\begin{smallmatrix}M^*\\ N^*\end{smallmatrix}\right)$, or

$$(M\ N) = C(\widetilde{P}\ \widetilde{Q}).$$

This says that the boundary conditions

$$MY(a) + NY(b) = 0$$

and

$$\widetilde{P}Y(a) + \widetilde{Q}Y(b) = 0$$

are equivalent. Since the forms of L and L^* are the same, $L = L^*$. $\square$

IV.5   The Spectral Resolution of a Regular Linear Hamiltonian Operator

By spectral resolution here we mean eigenfunction expansion. We shall show indeed that the spectrum of a regular linear Hamiltonian operator is discrete, consisting only of real eigenvalues. Each eigenvalue is associated with at least one, at most n, eigenfunctions. The eigenfunctions completely span $L_A^2(a,b)$.

We consider the operator L of the previous section: Set in $L_A^2(a,b)$, $LY = F$ if and only if

$$JY' - BY = AF,$$

where Y is constrained in part by the self-adjointness criterion $MJM^* = NJN^*$.

IV.5.1. Lemma. *Let $\mathcal{Y}(x,\lambda)$ be a fundamental matrix for $JY' = [\lambda A + B]Y$ satisfying $\mathcal{Y}(a,\lambda) = I$. (That is, $\mathcal{Y}(x,\lambda)$ is an $n \times n$ matrix whose columns satisfy the differential equation. With $x = a$, $\mathcal{Y}(a,\lambda) = I$), then for all x, $\mathcal{Y}^*(x,\lambda)J\mathcal{Y}(x,\lambda) = J$.*

Proof. $\mathcal{Y}^*(x,\overline{\lambda})$ satisfies

$$-\mathcal{Y}^*(x,\overline{\lambda})'J = \mathcal{Y}^*(x,\overline{\lambda})[\lambda A + B],$$

while $\mathcal{Y}(x,\lambda)$ satisfies

$$J\mathcal{Y}(x,\lambda)' = [\lambda A + B]\mathcal{Y}(x,\lambda).$$

Right multiply the first by $\mathcal{Y}(x,\lambda)$, left multiply the second by $\mathcal{Y}^*(x,\overline{y})$, and subtract since J is constant

$$\mathcal{Y}^*(x,\overline{\lambda})'J\mathcal{Y}(x,\lambda) + \mathcal{Y}^*(x,\overline{\lambda})J\mathcal{Y}'(x,\lambda)$$
$$= [\mathcal{Y}^*(x,\overline{\lambda})J\mathcal{Y}(x,\lambda)]' = 0,$$

and $\mathcal{Y}^*(x,\overline{\lambda})J\mathcal{Y}(x,\lambda) = C$. If $x = a$, $\mathcal{Y}^* = \mathcal{Y} = I$, and so $C = J$. $\square$

We next attempt to solve the equation $(L - \lambda I)Y = F$ for Y. Y satisfies the nonhomogeneous differential equation

$$JY' = [\lambda A + B]Y + AF.$$

Since solutions to the homogeneous equation are given by $Y(x, \lambda) = \mathcal{Y}(x, \lambda)C$, for some constant C, we use variation of parameters. We have, with C now variable,

$$JY' = J\mathcal{Y}'C + J\mathcal{Y}C',$$

$$[\lambda A + B]Y = [\lambda A + B]\mathcal{Y}C.$$

Thus

$$\begin{aligned}
JY' = [\lambda A + B]Y &= \{J\mathcal{Y}' - [\lambda A + B]\mathcal{Y}\}C + J\mathcal{Y}C' \\
&= J\mathcal{Y}C' \\
&= AF.
\end{aligned}$$

Therefore

$$J\mathcal{Y}C' = AF.$$

Now from the Lemma IV. 5.1 $(J\mathcal{Y})^{-1} = -J\mathcal{Y}^*(x, \overline{\lambda})$, so

$$C' = -J\mathcal{Y}^*(x, \overline{\lambda})A(x)F(x).$$

Thus

$$Y = -\mathcal{Y}(x, \lambda) \int_a^x J\mathcal{Y}^*(\xi, \overline{\lambda})A(\xi)F(\xi)d\xi + \mathcal{Y}(x, \lambda)K.$$

But we must also impose the boundary condition $MY(a) + NY(b) = 0$. We see that

$$Y(a) = K,$$

$$Y(b) = -\mathcal{Y}(b, \lambda) \int_a^b J\mathcal{Y}^*(\xi, \overline{y})A(\xi)F(\xi)d\xi + \mathcal{Y}(b, \lambda)K.$$

These yield

$$Y(x) = -\mathcal{Y}(x, \lambda)[M + N\mathcal{Y}(b, \lambda)]^{-1}M \int_a^x J\mathcal{Y}^*(\xi, \overline{\lambda})A(\xi)F(\xi)d\xi$$

$$+ \mathcal{Y}(x, \lambda)[M + N\mathcal{Y}(b, \lambda)]^{-1}N \int_x^b J\mathcal{Y}^*(\xi, \overline{\lambda})A(\xi)F(\xi)d\xi,$$

which we rewrite as

$$Y(x) = \int_a^b G(\lambda, x, \xi)A(\xi)F(\xi)d\xi,$$

$$G(\lambda, x, \xi) = -\mathcal{Y}(x, \lambda)[M + N\mathcal{Y}(b, \lambda)]^{-1}M J\mathcal{Y}^*(\xi, \overline{y}), \quad a \le \xi \le x \le b$$

$$= \mathcal{Y}(x, \lambda)[M + N\mathcal{Y}(b, \lambda)]^{-1}N J\mathcal{Y}^*(\xi, \overline{\lambda}), \quad a \le x \le \xi \le b.$$

IV.5.2. Theorem. *$(L - \lambda I)^{-1}$ exists for all nonreal λ as a bounded operator. It exists also for all real λ for which $\det[M + N\mathcal{Y}(b, \lambda)] \neq 0$ as a bounded operator. The spectrum of L consists entirely of isolated eigenvalues, zeros of $\det[M + N\mathcal{Y}(b, \lambda)] = 0$.*

Proof. $(L - \lambda I)^{-1}$ is given by the formula

$$Y = \int_a^b G(\lambda, x, \xi) A(\xi) F(\xi) d\xi,$$

as long as $\det[M + N\mathcal{Y}(b, \lambda)] \neq 0$. Since L is self-adjoint it must exist for all complex λ. It clearly exists for all real λ except the zeros of $\det[M + NY(b, \lambda)] = 0$. Since the determinant is analytic in λ and is not identically zero, it can have only isolated zeros, which can accumulate only at $\pm\infty$.

In order to see that $(L - \lambda I)^{-1}$ is bounded, we let $f(\xi) = A(\xi)^{\frac{1}{2}} F(\xi)$ and

$$M(\lambda, x, \xi) = A(\xi)^{\frac{1}{2}} G(\lambda, x, \xi) A(x)^{\frac{1}{2}}.$$

(Recall $A \geqq 0$, and so it has a positive square root $A^{\frac{1}{2}}$.) Then

$$\|Y\|^2 = \int_a^b Y^*(x) A(x) Y(x) dx$$

$$= \int_a^b \left[\int_a^b f^*(\xi) M^*(\lambda, x, \xi) d\xi \right] \left[\int_a^b M(\lambda, x, \eta) f(\eta) d\eta \right] dx.$$

Apply Schwarz's inequality to both terms.

$$\|Y\|^2 \leqq \|M\|^2 \|F\|^2,$$

where (with $M = (M_{ij})$,

$$\|M\|^2 = \int_a^b \int_a^b \sum_{i=1}^n \sum_{j=1}^n |M_{ij}(\lambda, x, \xi)|^2 \, d\xi dx.$$

To see what happens when $\det[M + N\mathcal{Y}(b, \lambda)] = 0$, we note that if the zero λ_j is of order n_j, then the rank of $[M + N\mathcal{Y}(b, \lambda)]$ is $n - n_j$. There are n_j solutions of

$$[M + N\mathcal{Y}(b, \lambda)]C = 0,$$

$C = C_1, C_2, \ldots, C_j, \ldots, C_{n_j}$. For each nontrivial column C_j, $\mathcal{Y}(x, \lambda_j)C_j$ is an eigenfunction. $\qquad\square$

IV.5.3. Theorem. *Eigenfunctions associated with different eigenvalues are mutually orthogonal. For each eigenvalue λ_j, its eigenfunctions can be made mutually orthogonal.*

Proof. Let Y_1 be associated with λ_1, Y_2 be associated with λ_2. Then $LY_1 = \lambda_1 Y_1$, $LY_2 = \lambda_2 Y_2$. Green's formula ($Y = Y_1, Z = Y_2$) yields

$$\begin{aligned}
\lambda_1 \langle Y_1, Y_2 \rangle_A &= \langle \lambda_1 Y_1, Y_2 \rangle_A \\
&= \langle LY_1, Y_2 \rangle_A \\
&= \langle Y_1, LY_2 \rangle_A \\
&= \langle Y_1, \lambda_2 Y_2 \rangle_A \\
&= \lambda_2 \langle Y_1, Y_2 \rangle_A.
\end{aligned}$$

Since $\lambda_1 \neq \lambda_2$, $\langle Y_1, Y_2 \rangle = 0$.

Now let $Y_1, \ldots, Y_m$ be eigenfunctions associated with λ. We let $u_1 = Y_1/\|Y_1\|$. We then define inductively

$$V_k = Y_k - \sum_{i=1}^{k-1} U_i \langle Y_k, U_i \rangle_A$$

and

$$U_k = V_k/\|V_k\|.$$

It is a simple matter to check that U_k is orthogonal to $U_1, \ldots, U_{k-1}$. $\quad\square$

Thus the spectrum of L consists only of discrete eigenvalues $\{\lambda_j\}_{j=1}^{\infty}$ which can accumulate only at $\pm\infty$. Associated with each eigenvalue λ_j there is a collection of orthonormal eigenfunctions $Y_j(x) = \mathcal{Y}(x, \lambda_j)C_j$ which number at most n. We show these eigenfunctions form a complete set in $L_A^2(a,b)$.

We first note that we may assume that 0 is not an eigenvalue. This follows since we can perturb the eigenvalues at will. We write

$$JY' = [(\lambda - \lambda_0)A + (\lambda_0 A + B)]Y, \qquad \lambda_0 \text{ real.}$$

By redefining λ to replace $\lambda - \lambda_0$ and B to replace $\lambda_0 A + B$, we see that what was λ originally is now $\lambda - \lambda_0$. We choose λ_0 so that the new $\lambda = 0$ is not an eigenvalue if necessary.

Having done so, the solution to

$$JY' = BY + AF,$$
$$MY(a) + NY(b) = C,$$

is given by

$$Y = \int_a^b G(x,\xi)A(\xi)F(\xi)d\xi,$$

where $G(x,\xi) = G(0,x,\xi)$. We shorten the notation by setting

$$Y = \mathcal{G}F.$$

IV.5.4. Theorem. *Let $Y = \mathcal{G}F = L^{-1}F$. Then $\mathcal{G}$ is bounded. $\|\mathcal{G}\| = \sup\{|1/\lambda_j|;$ $\lambda_j \in \sigma(L)\}$.*

Proof. this follows from Theorem II.4.6, once it is realized that if $\{\lambda_j, U_j\}_{j=1}^{\infty}$ constitute the eigenvalue-eigenfunction pairs for L, then $\{1/\lambda_j, U_j\}_{j=1}^{\infty}$ constitute the eigenvalue-eigenfunction pairs for L^{-1}. Indeed, $LU_j = \lambda_j U_j$ if and only if $\mathcal{G}U_j = (1/\lambda_j)U_j$.

It is convenient to order the eigenvalues of $\mathcal{G}$, $\mu_j = 1/\lambda_j$, so that $|\mu_1| \geq |\mu_2| \cdots \geq |\mu_j| \geq \cdots$, where $\lim_{j \to \infty} |\mu_j| = 0$.

We define inductively $\{\mathcal{G}_n\}_{n=1}^{\infty}$ by

$$\mathcal{G}_1 F = \mathcal{G}F \ ,$$

$$\mathcal{G}_2 F(x) = \mathcal{G}F(x) - \mu_1 U_1(x) \int_a^b U_1^*(\xi)A(\xi)F(\xi)d\xi,$$

or

$$\mathcal{G}_2 F = \mathcal{G}F - \mu_1 U_1 \langle F, U_1 \rangle_A,$$

and

$$\mathcal{G}_n F = \mathcal{G}F - \sum_{j=1}^{n-1} \mu_j U_j \langle F, U_j \rangle_A. \qquad \square$$

IV.5.5. Theorem. $\|\mathcal{G}_n\| = |\mu_n|$, $n = 1, \ldots$ $\lim_{n \to \infty} \mathcal{G}_n = 0$.

Proof. We see that

$$\mathcal{G}_n U_k = \mathcal{G}U_k - \sum_{j=1}^{n-1} \mu_j U_j \langle U_k, U_J \rangle_A.$$

If $1 \leq k \leq n - 1$, this is $(\mu_k - \mu_k) = 0$. If $n \leq k < \infty$, this is μ_k. Further $\mathcal{G}_n$ is bounded and self-adjoint, just as $\mathcal{G}$ is. Thus

$$\|\mathcal{G}_n\| = \sup_{\substack{U \in L^2 \\ \|u\|=1}} |\langle \mathcal{G}_n U, U \rangle_A|$$

$$= \sup_{\substack{U \in L^2 \\ \|u\|=1 \\ U \neq U_1, \ldots, U_{n-1}}} |\langle \mathcal{G}_n U, U \rangle_A|$$

$$= |\mu_n| \ .$$

If this process stops with a finite n so that $\mathcal{G}_n = 0$, then for all F in $L_A^2(a, b)$,

$$\mathcal{G}F = \sum_{j=1}^{n-1} \mu_j U_j \langle F_0 U_j \rangle_A.$$

Applying L,

$$F = \sum_{j=1}^{n-1} U_j \langle F, U_j \rangle_A \ ,$$

which says that F is differentiable. Since there are F's which are not, the process cannot stop. Since $\lim_{n\to\infty} |\mu_n| = 0$, we find nonetheless $\lim_{n\to\infty} \mathcal{G}_n = 0$. (Note: this also shows the number of eigenvalues is really infinite.) $\qquad\square$

IV.5.6. Theorem. *(Spectral Resolutions). For all F in $L^2(a,b)$,*

$$F = \sum_{j=1}^{\infty} U_j \langle F, U_j \rangle_A \,,$$

$$\mathcal{G}\, F = \sum_{j=1}^{\infty} \mu_j\, U_j \langle F, U_j \rangle_A \,.$$

For all Y in D,

$$LY = \sum_{j=1}^{\infty} \lambda_j U_j \langle Y, U_j \rangle_A \,.$$

Proof. Since $\lim_{n\to\infty} \mathcal{G}_n = 0$, we have

$$\mathcal{G}\, F = \sum_{j=1}^{\infty} \mu_j\, U_j \langle F, U_j \rangle_A \,.$$

If we apply L to this, we have

$$F = \sum_{j=1}^{\infty} U_j \langle F, U_j \rangle_A \,.$$

If $LY = F$, then the coefficient

$$\begin{aligned}
\langle F, U_j \rangle_A &= \langle LY, U_j \rangle_A \\
&= \langle Y, LU_j \rangle_A \\
&= \lambda_j \langle Y, U_j \rangle_A \,.
\end{aligned}$$

Thus,

$$LY = \sum_{j=1}^{\infty} \lambda_j U_j \langle Y, U_j \rangle \,. \qquad\square$$

IV.5.7. Theorem. *There exist a collection of projection operators $\{E(\lambda)\}$ satisfying*

(1) $E(\mu) \leq E(\lambda)$ *when* $\mu < \lambda$,

(2) $\lim_{\lambda\to-\infty} E(\lambda) = 0$,

(3) $\lim_{\lambda\to\infty} E(\lambda) = I$,

(4) $E(\lambda)$ *is continuous from above,*

(5) *For all F in $L_A^2(a,b)$ and Y in D*

$$F = \int_{-\infty}^{\infty} dE(\lambda)F, \qquad \mathcal{G}F = \int_{-\infty}^{\infty} (1/\lambda) dE(\lambda)F, \qquad LY = \int_{-\infty}^{\infty} \lambda\, dE(\lambda)Y.$$

Proof. Let

$$P_j F = U_j \langle F, U_j \rangle_A.$$

P_j is a projection operator. If we define

$$E(\lambda)F = \sum_{\lambda_j \leqq \lambda} P_j F,$$

then $E(\lambda)$ generates a Stieltjes measure. The integrals are merely restatements of the series. $\qquad\qquad\qquad\qquad\qquad\qquad\qquad\qquad\qquad\qquad\qquad\qquad\qquad\qquad\Box$

IV.6   Example

Consider the differential operator $ly = -y'' + y$, defined on a domain D set in $L^2(0, 2\pi)$ which is in part restricted by the boundary conditions

$$y(0) = y(2\pi), \quad y'(0) = y'(2\pi).$$

If we let $y = y_1$, $y' = y_2$, we can recast the problem in system format. The differential equation

$$-y'' + y = \lambda y + f$$

becomes

$$\begin{pmatrix} 0 & -1 \\ 1 & 0 \end{pmatrix} \begin{pmatrix} y_1 \\ y_2 \end{pmatrix}' = \left[\lambda \begin{pmatrix} 1 & 0 \\ 0 & 0 \end{pmatrix} + \begin{pmatrix} -1 & 0 \\ 0 & 1 \end{pmatrix} \right] \begin{pmatrix} y_1 \\ y_2 \end{pmatrix} + \begin{pmatrix} 1 & 0 \\ 0 & 0 \end{pmatrix} \begin{pmatrix} f \\ 0 \end{pmatrix}.$$

The boundary conditions become

$$\begin{pmatrix} 1 & 0 \\ 0 & 1 \end{pmatrix} \begin{pmatrix} y_1 \\ y_2 \end{pmatrix} (0) + \begin{pmatrix} -1 & 0 \\ 0 & -1 \end{pmatrix} \begin{pmatrix} y_1 \\ y_2 \end{pmatrix} (2\pi) = \begin{pmatrix} 0 \\ 0 \end{pmatrix}.$$

Thus $J = \begin{pmatrix} 0 & -1 \\ 1 & 0 \end{pmatrix}$, $A = \begin{pmatrix} 1 & 0 \\ 0 & 0 \end{pmatrix}$, $B = \begin{pmatrix} -1 & 0 \\ 0 & 1 \end{pmatrix}$ $M = \begin{pmatrix} 1 & 0 \\ 0 & 1 \end{pmatrix}$ and $N = \begin{pmatrix} -1 & 0 \\ 0 & -1 \end{pmatrix}$. It is easy to see $M^* J M = N^* J N$. The self-adjoint differential operator in $L_A^2(0, 2\pi)$ is given by $L \begin{pmatrix} y_1 \\ y_2 \end{pmatrix} = \begin{pmatrix} f \\ 0 \end{pmatrix}$ if and only if

$$\begin{pmatrix} 0 & -1 \\ 1 & 0 \end{pmatrix} \begin{pmatrix} y_1 \\ y_2 \end{pmatrix}' - \begin{pmatrix} -1 & 0 \\ 0 & 1 \end{pmatrix} \begin{pmatrix} y_1 \\ y_2 \end{pmatrix} = \begin{pmatrix} 1 & 0 \\ 0 & 0 \end{pmatrix} \begin{pmatrix} f \\ 0 \end{pmatrix},$$

$\begin{pmatrix} y_1 \\ y_2 \end{pmatrix}$ satisfying the boundary condition.

$L - \lambda I$ has an inverse given by

$$\begin{pmatrix} y_1 \\ y_2 \end{pmatrix} = \begin{pmatrix} \cos s\,x & \frac{1}{s}\sin s\,x \\ -s\sin s\,x & \cos s\,x \end{pmatrix}$$

$$\times \frac{\begin{pmatrix} 1 - \cos 2\pi\,s & \frac{1}{s}\sin 2\,\pi\,s \\ -\sin 2\,\pi\,s & 1 - \cos 2\,\pi\,s \end{pmatrix}}{2 - 2\cos 2\,\pi\,s} \int_0^x \begin{pmatrix} \frac{1}{s}\sin s\,\xi \\ -\cos s\,\xi \end{pmatrix} f(\xi)d\xi$$

$$+ \begin{pmatrix} \cos s\,x & \frac{1}{s}\sin s\,x \\ -s\sin s\,x & \cos s\,x \end{pmatrix}$$

$$\times \frac{\begin{pmatrix} -1 + \cos 2\,\pi\,s & \frac{1}{s}\sin 2\,\pi\,s \\ -s\sin 2\,\pi\,s & -1 + \cos\cos 2\,\pi\,s \end{pmatrix}}{2 - 2\cos 2\pi s} \int_x^{2\pi} \begin{pmatrix} \frac{1}{s}\sin s\,\xi \\ -\cos s\,\xi \end{pmatrix} f(\xi)d\xi ,$$

where $\lambda - 1 = s^2$. It exists whenever $2 - 2\cos 2\pi s \neq 0$. When $2 - 2\cos 2\pi s = 0$, or $s = n$, an integer, it does not. For $s = \pm n$, $\lambda_n = 1 + n^2$ has two mutually orthogonal eigenfunctions

$$\begin{pmatrix} y_{n1} \\ y_{n2} \end{pmatrix} = \begin{pmatrix} e^{in\,x}/\sqrt{2\pi} \\ in\,e^{inx}/\sqrt{2\pi} \end{pmatrix}$$

and

$$\begin{pmatrix} y_{-n1} \\ y_{-n2} \end{pmatrix} = \begin{pmatrix} e^{-inx}/\sqrt{2\pi} \\ -ine^{-inx}/\sqrt{2\pi} \end{pmatrix}$$

when $n \neq 0$ and one, $\begin{pmatrix} y_{01} \\ y_{02} \end{pmatrix} = \begin{pmatrix} 1/\sqrt{2\pi} \\ 0 \end{pmatrix}$ when $n = 0$. The discrete spectral

resolution, or eigenfunction expansion, says that for arbitrary $\begin{pmatrix} f \\ * \end{pmatrix}$ in $L_A^2(0, 2\pi)$,

$$\begin{pmatrix} f \\ * \end{pmatrix} = \sum_{n=-0}^{\infty} \int_0^{2\pi} (\overline{y}_{n1}, \overline{y}_{n2}) \begin{pmatrix} 1 & 0 \\ 0 & 0 \end{pmatrix} \begin{pmatrix} f \\ * \end{pmatrix} d\xi \begin{pmatrix} y_{n1} \\ y_{n2} \end{pmatrix},$$

where the $*$ may be any function at all. Multiplied out, this says that f equals its Fourier series in $L^2(0, 2\pi)$:

$$f(x) = \sum_{n=-\infty}^{\infty} \int_0^{2\pi} \frac{e^{-in\xi}}{\sqrt{2\pi}} f(\xi)d\xi \frac{e^{inx}}{\sqrt{2\pi}} .$$

Again we make the point. *For even-dimensional systems, think of the first part of the vector Y as y, and the second part as y'.*

We also recommend that the reader verify in this example that the system theory is in fact equivalent to the classical L^2 theory. The real reason for the system format is that it is the same for second, fourth, sixth, in order boundary value problems.

References

[1] F. V. Atkinson, **Discrete and Continuous Boundary Problems**, Academic Press, NY, 1964.

[2] G. D. Birkhoff and R. E. Langer, *The boundary problems and developments associated with a system of ordinary linear differential equations of the first order*, Proc. Amer. Acad. Arts and Sci. **58** (1923), pp. 51–128.

[3] G. A. Bliss, *A boundary value problem for a system of ordinary linear differential equations of the first order*, Trans. Amer. Math. Soc. **28** (1926), pp. 561–589.

[4] E. A. Coddington and N. Levinson, **Theory of Ordinary Differential Equations**, McGraw-Hill, N. Y., 1955.

[5] B. M. Levitan and I. S. Sargsjan, **Introduction to Spectral Theory:/Self-Adjoint Ordinary Differential Operators**, Amer. Math. Soc., Providence, R. I., 1975.

[6] ——, **Sturm-Liouville and Dirac Operators**, Kluwer Academic Publ., Dordrecht, 1991.

[7] W. T. Reid, **Ordinary Differential Equations**, John Wiley and Sons, N. Y., 1971.

[8] F.Riesz and B. Sz.-Nagy, **Functional Analysis**, Fred. Ungar Pub., N. Y. 1955.

[9] F. W. Schafke and A. Schneider, *S-hermitesche Rand-Eigenwert Probleme, I*, Math. Ann. **162** (1965), pp. 9–26.

[10] ——, *S-hermitesche Rand-Eigenwert Probleme, II*, Math. Ann. **165** (1966), pp. 236–260.

[11] ——, *S-hermitesche Rand-Eigenwert Probleme, III* Math. Ann. **177** (1968), pp. 67–94.

[12] P. W. Walker, *A vector-matrix formulation for formally symmetric ordinary differential equations with applications to solutions of integrable square*, J. London Math. Soc. **9** (1974), pp. 151–159.

Chapter V

Atkinson's Theory for Singular Hamiltonian Systems of Even Dimension

When considering the linear Hamiltonian system

$$JY' = [\lambda A + B]Y \,,$$

over $L_A^2(a, b)$, if something goes wrong at either $x = a$ or $x = b$, the theory of the previous chapter may no longer be entirely valid. The problem is no longer regular and is called *singular*. Considerably more care must be used.

The first to consider singular systems appears to be F. Brauer [4], who extended the work begun by Bliss [3], following the techniques used in scalar problems (see [5]). F. V. Atkinson followed quickly with his book [2] in 1964, a treatise which has some very interesting and novel ideas in it, and is still well worth reading, even today.

Dirac systems on $[0, \infty)$, studied by the Russian school, are represented by the books of B. M. Levitan and I. S. Sargsjan [17], [18]. The first appeared in 1975; the second, a revision of sorts, appeared in 1991 in the West.

Likewise in the 1970's H. D. Niessen and A. Schneider [20–24] systematically attacked the problem of singular S-Hermitian systems. This was closely followed in 1981–1984 by the discussion in depth of singular linear Hamiltonian systems by D. B. Hinton and J. K. Shaw [6–13]. They approached the problem from a point of view that, while different from that of Niessen and Schneider, still generalized the fundamental work of H. Weyl, who first solved the scalar second order problem.

It was during this time, via several discussions with Professors Hinton and Shaw that the author become interested in the subject. His papers began with [16], [19], and culminated with two articles in 1989, [13] and [14].

For several reasons, both technical and because of applications, we now restrict the focus of our discussion to systems of *even* dimension $(2n)$ with *real* coefficients. As we continue, we shall point out where and why we make these assumptions. We remark that the works of Brauer, Atkinson, and Niessen and

Schneider, Hinton and Shaw do not always make these assumptions. All permitted complex coefficients, while only Hinton and Shaw restricted their attention to systems of even dimension. We do so here so we may push farther than they. In particular we wish to derive details conerning the spectral resolution, details the others do not reach.

Hence we consider over an interval (a, b) the linear Hamiltonian system

$$JY' = [\lambda A + B]Y + AF$$

of dimension $2n$. In other words we examine the system

$$\begin{pmatrix} 0 & -I \\ I & 0 \end{pmatrix} \begin{pmatrix} Y_1 \\ Y_2 \end{pmatrix}' = \left[\lambda \begin{pmatrix} A_{11} & A_{12} \\ A_{21} & A_{22} \end{pmatrix} + \begin{pmatrix} B_{11} & B_{12} \\ B_{21} & B_{22} \end{pmatrix} \right] \begin{pmatrix} Y_1 \\ Y_2 \end{pmatrix} + \begin{pmatrix} A_{11} & A_{12} \\ A_{21} & A_{22} \end{pmatrix} \begin{pmatrix} F_1 \\ F_2 \end{pmatrix} ,$$

where Y_1, Y_2 and F_2 are n-dimensional vectors, $A_{i,j}$ and $B_{i,j}$, $i, j = 1, 2$, are $n \times n$ matrices, and I is the n-dimensional identity.

We assume that the matrices $A = (A_{i,j})$, $B = (B_{i,j})$ and the vector $F = (F_i)$ are locally integrable in (a, b). As in the regular case, we suppose that A and B satisfy $A = A^* \geq 0$, $B = B^*$. Likewise we assume that the homogeneous system is definite. I.e. over compact subintervals $[a', b']$ of (a, b), if Y is a solution of

$$JY' = [\lambda A + B]Y ,$$

then

$$\int_{a'}^{b'} Y^* A Y \, dx = 0$$

if and only if $Y \equiv 0$.

V.1   Singular Linear Hamiltonian Systems

The differences between regular systems and singular systems are more far reaching than the definition (to follow) indicates. In laymen's terms, let us say that regular systems behave beautifully, while for singular systems "all - - - - breaks loose."

V.1.1. Definition. Let J be a constant $2n \times 2n$ matrix, $J = \begin{pmatrix} 0 & -I \\ I & 0 \end{pmatrix}$, let A and B be locally integrable $2n \times 2n$ matrices over (a, b) satisfying $A = A^* \geq 0$, $B = B^*$. Let λ be a complex number. If either A or B is not integrable near $x = a$, or if $a = -\infty$, then $x = a$ is a singular point. If either A or B is not integrable near $x = b$ or if $b = \infty$, then b is a singular point.

If $x = a$ or $x = b$ is a singular point, then the linear Hamiltonian system

$$JY' = [\lambda A + B]Y$$

is singular.

Hence anywhere within (a, b), solutions are well defined (by Picard iterations), but as x approaches a or as x approaches b, whichever is singular, limiting values may not exist. Existence in L_A^2, in particular, may fail.

V.2   Existence of Solutions in $\mathbf{L^2_A}(\mathbf{a}, \mathbf{b})$

Let us consider the real, homogeneous, definite linear Hamiltonian system of order $2n$,

$$JY' = [\lambda A + B]Y \,,$$

where $J = \begin{pmatrix} 0 & -I \\ I & 0 \end{pmatrix}$, and where the real symmetric matrices $A \geq 0$ and B are integrable over every subinterval $[a', b']$ of (a, b). We choose a value c between a' and b', so $a < a' < c < b' < b$, and consider separately $L^2_A(a, c)$ and $L^2_A(c, b)$. These spaces will be later welded together to form $L^2_A(a, b)$.

The reason we make such considerations is to find the building blocks, the L^2_A solution needed for further work.

V.2.1. Theorem. *Let* $a < c < b$, *and* $Im\ \lambda \neq 0$. *Then*

$$JY' = [\lambda A + B]Y$$

has m_a *linearly independent solutions in* $L^2_A(a, c)$ *and* m_b *linearly independent solutions in* $L^2_A(c, b)$, $n \leq m_a, m_b \leq 2n$.

This proof seems to be due to Atkinson [2]. He refers to Akhiezer and Glazman's book [1] in which it is not. It is relatively easy to prove. Unfortunately it does not give us enough information.

Proof. Assume that $Im\ \lambda > 0$. (Only minor modifications are needed when $Im\ \lambda < 0$.)

Since

$$JY' = [\lambda A + B]Y \,,$$

we find, taking complex conjugate transposes, that

$$Y^{*'} J = -Y^*[\bar{\lambda} A + B] \,.$$

(Recall that $J^* = -J$.) Premultiplying the first by Y^*, postmultiplying the second by Y, and then adding, we find

$$Y^* JY' + Y^{*'} JY = (\lambda - \bar{\lambda})Y^* AY \,.$$

Integrate from c to b',

$$Y^* JY \Big|_c^{b'} = 2i\mathrm{Im}\ \lambda \int_c^{b'} Y^* AY \, dx \,.$$

Hence

$$\int_c^{b'} Y^* AY \, dx = (1/2\mathrm{Im}\ \lambda)[Y^*(J/i)Y]_c^{b'} \,.$$

If $\mathcal{Y}(x,\lambda)$ is a fundamental matrix satisfying $\mathcal{Y}(c,\lambda) = I$, then $Y = \mathcal{Y}(x,\lambda)C$, and

$$\int_c^b Y^* A Y \, dx = (1/2\mathrm{Im}\ \lambda)[C^* \mathcal{Y}^*(J/i)\mathcal{Y}C(b') - C^*(J/i)C].$$

We now examine the matrix $\mathcal{Y}^*(J/i)\mathcal{Y}$. It is symmetric. When $x = c$, it is reduced to (J/i), which has n eigenvalues equal to 1 and n equal to -1. Since it is never singular, when $x = b'$, it still has n positive and n negative eigenvalues, and each possesses an eigenvector which is linearly independent of all the others. Let those associated with negative eigenvalues be denoted by $C_1,\ldots,C_n$, and assume they have norm 1. If C is one of them, then

$$(1/2\mathrm{Im}\ \lambda)\{C^* Y^*(J/i)YC\}(b') < 0\,.$$

So

$$\int_c^{b'} Y^* A Y \, dx \leqq -C^*(J/i)C/(2\mathrm{Im}\ \lambda)\,.$$

Since (J/i) is a unitary matrix,

$$\int_c^{b'} Y^* A Y \, dx \leq 1/(2\mathrm{Im}\ \lambda)\,.$$

We would now like to simply let b' approach b. Unfortunately, the eigenvalues $C_1,\ldots,C_n$ depend upon b', so we must first establish that $C_1,\ldots,C_n$ have limits.

Since $C_1,\ldots,C_n$ all lie on the unit sphere in R^{2n}, which is compact, we can extract subsequences of $C_1(b'),\ldots,C_n(b')$ which converge. Since for all b' they are mutually orthogonal, so must be their limiting vectors, $K_1,\ldots,K_n$ as well, and are therefore also linearly independent.

Through these subsequences we pass to the limit b, finding

$$\int_c^b Y_j^* A Y_j \, dx \leq 1/(2\mathrm{Im}\ \lambda)\,,$$

where $Y_j = \mathcal{Y}(x,\lambda)K_j$, $j = 1,\ldots,n$. These are all linearly independent and are in $L_A^2(c,b)$.

For any λ, $\mathrm{Im}\ \lambda \neq 0$, there may be additional solutions in $L_A^2(c,b)$. We denote the total by m_b, $n \leq m_b \leq 2n$.

Minor modifications are also needed to examine solutions in $L_A^2(a,c)$. We leave those to the reader. Since there may be additional solutions in $L_A^2(a,c)$, we denote the total by m_a, $n \leq m_a \leq 2n$.

One may think that the numbers m_a and m_b may vary with λ. So long as $\mathrm{Im}\ \lambda \neq 0$, this is not the case. The proof depends upon the existence of a Green's function, which we will derive later. $\square$

V.2.2. Theorem. *Let $a < c < b$, and Im $\lambda \neq 0$. Then the number of solutions of*

$$JY' = [\lambda A + B]Y ,$$

m_a *in* $L_A^2(a,c)$, *and* m_b *in* $L_A^2(c,b)$, *does not vary with* λ.

Proof. We assume specifically that there is a matrix function $G(\lambda, x, \xi)$ which generates solutions of

$$JY' = [\lambda A + B]Y + AF$$

lying in $L_A^2(a,c)$ or $L_A^2(c,b)$ when F is in the same space and Im $\lambda \neq 0$, via

$$Y(x) = \int_a^c G(\lambda, x, \xi) A(\xi) F(\xi) d\xi$$

or

$$Y(x) = \int_c^b G(\lambda, x, \xi) A(\xi) F(\xi) d\xi .$$

We also assume that

$$\int Y^* AY \, dx \leq |1/\mathrm{Im}\ \lambda| \int F^* AF \, dx,$$

where integration is over the appropriate integral.

   We shall verify such a function G exists later in Section VI.5.

   Consider (c,b). Let $s = \min_{\mathrm{Im}\ \lambda \neq 0} \{m_b\}$. Since $s \geq n$, it achieves a minimum at some point $\lambda = \mu$. We can assume that Im $\mu > 0$ as well, since m_b is integer valued. Choose λ such that $|\lambda - \mu| < |\mathrm{Im}\ \mu|$.

   Let $\{Y_q\}$, $1 \leq q \leq s$, be the solutions of

$$JY' = [\mu A + B]Y$$

in $L_A^2(c,b)$. Let $\{Y_q\}$, $s + 1 \leq q \leq 2n$, complete the collection of independent solutions.

   Let $\{Z_p\}$, $1 \leq p \leq S = m_b(\lambda)$, be the solutions of

$$JZ' = [\lambda A + B]Z$$

in $L_A^2(c,b)$. Let $\{Z_p\}$, $S + 1 \leq p \leq 2n$, complete the collection of independent solutions. Assume $S > s$.

   Write

$$JZ' = [\mu A + B]Z + (\lambda - \mu)AZ .$$

Then

$$\tilde{Z}_p = (\lambda - \mu) \int_c^b G(\mu, x, \xi) A(\xi) Z_p(\xi) d\xi ,$$

$1 \leq p \leq S$, satisfies

$$J\widetilde{Z}_p' - [\mu A + B][\widetilde{Z}_p = (\lambda - \mu)AZ_p$$

and is in $L_A^2(c, b)$. Consequently there are constants $\{\alpha_{pq}\}$, $1 \leq p \leq S$, $1 \leq q \leq s$, such that

$$Z_p = \sum_{q=1}^{2n} \alpha_{pq}Y_q + \widetilde{Z}_p.$$

Since Z_p, $\{Y_q\}_{q=1}^S$ and $\widetilde{Z}_p$ are all in $L_A^2(c, b)$, and no combination of $\{Y_q\}_{q=s+1}^{2n}$ can be, $\alpha_{pq} = 0$, if $s + 1 \leq q \leq 2n$, $1 \leq p \leq S$. Hence

$$Z_p = \sum_{q=1}^{s} \alpha_{pq}Y_q + \widetilde{Z}_p, \qquad 1 \leq p \leq S.$$

Since $S > s$, there exist nontrivial solutions $\{\beta_p\}_{p=1}^S$, to

$$(\beta_1, \ldots, \beta_S) \begin{bmatrix} \alpha_{11} & \cdots & \alpha_{1s} \\ \vdots & & \vdots \\ \alpha_{S1} & \cdots & \alpha_{Ss} \end{bmatrix} = (0, \ldots, 0).$$

(There are s equations, S unknowns.) Define ϕ by

$$\phi = \sum_{p=1}^{S} \beta_p Z_p.$$

Then ϕ is not identically 0, and

$$\begin{aligned}
\phi &= \sum_{p=1}^{S} \beta_p Z_p \\
&= \sum_{p=1}^{S} \beta_p \left[\sum_{q=1}^{s} \alpha_{pq}Y_q + \widetilde{Z}_p \right] \\
&= \sum_{q=1}^{s} \left[\sum_{p=1}^{S} \beta_p \alpha_{pq} \right] Y_q + \sum_{p=1}^{S} \beta_p \widetilde{Z}_p \\
&= 0 + \sum_{p=1}^{S} \beta_p \widetilde{Z}_p \\
&= \sum_{p=1}^{S} (\lambda - \mu) \int_c^b G(\mu, x, t)A(t)\beta_p Z_p(t)dt \\
&= (\lambda - \mu) \int_c^b G(\mu, x, t)A(t)\phi(t)dt.
\end{aligned}$$

This implies that

$$\int_c^b \phi^* A\phi\, dt \leqq \frac{|\lambda - \mu|}{|\text{Im } \mu|} \int_c^b \phi^* A\phi\, dt\,.$$

This is impossible. And so m_b does not vary with λ, Im $\lambda > 0$. Conjugation shows the same holds for Im $\lambda < 0$, when A and B are real.

The interval (a, c) may be discussed in a similar manner. $\qquad\square$

V.3 Boundary Conditions

In order to define a differential operator which is self-adjoint on $L_A^2(a, b)$, we need to develop the idea of singular boundary conditions. It will not be too surprising that regular boundary conditions ultimately turn out to be special cases of the singular conditions.

Recall the definition of the maximal operator (Def. IV.4.4). It remains the same for singular problems. The minimal operator must be redefined. We defer this until later.

V.3.1. Theorem. *Let Y_j be a solution of*

$$JY' = [\bar{\lambda}A + B]Y\,, \qquad Im \ \lambda \neq 0\,.$$

Then for all Y in D_M , the domain of the maximal operator,

$$B_{aj}(Y) = \lim_{x \to a} Y_j^* JY$$

exists if and only if Y_j is in $L_A^2(a, c)$.

$$B_{bj}(Y) = \lim_{x \to b} Y_j^* JY$$

exists if and only if Y_j is in $L_A^2(c, b)$.

Proof. Consider (c, b). Let Y be in D_M, so that $JY' - BY = AF$, where F is $L_A^2(c, b)$. Further let Y_j be a solution of $JY' = [\bar{\lambda}A + B]Y$, Im $\lambda \neq 0$. Then

$$Y_j^* JY' = Y_j^* BY + Y_j^* AF$$

and

$$Y_j^{*\prime} JY = -Y_j^* BY - \lambda Y_j^* AY\,.$$

Adding, we have

$$(Y_j^* JY)' = Y_j^* A[F - \lambda Y]\,.$$

Integrating from c to x gives

$$Y_j^* JY(x) = Y_j^* JY(c) + \int_c^x Y_j^* A[F - \lambda Y]dx\,.$$

If Y_j is in $L^2_A(c, b)$, then the limit as x approaches b exists, and $B_{bj}(Y)$ exists as well.

Conversely, suppose that $\lim_{x \to b} Y_j^* JY(x)$ exists for all Y in D_M. Suppose that G is arbitrary in $L^2_A(c, b)$. Then using the Green's function (to be derived later), we solve

$$JY' - [\lambda A + B]Y = AG$$

in $L^2_A(c, b)$. Setting $F = G + \lambda Y$, we find

$$JY' - BY = AF.$$

That is, for every G in $L^2_A(c, b)$ there is a Y in D_M such that $JY' - BY = AF$. Thus $G = F - \lambda Y$ can be arbitrary, and

$$\lim_{x \to b} \int_c^x Y_j^* AG\, dt$$

defines a linear functional whose domain is all of $L^2_A(c, b)$.

Define T_{nj} by setting

$$T_{nj}(G) = \int_c^{b-[(b-c)/n]} Y_j^* AG\, dt\,,$$

if b is finite, or by setting

$$T_{nj}(G) = \int_c^n Y_j^* AG\, dt\,,$$

if b is ∞. By assumption T_{nj} exists for all G in $L^2_A(c, b)$ and converges to

$$T_j(G) = \int_c^b Y^* AG\, dt$$

as n approaches ∞. By the Hahn–Banach theorem, T_j is bounded. By the Riesz representation theorem, Y_j is in $L^2_A(c, b)$.

The interval (a, c) is similar.

It is tempting to call the forms B_{aj} and B_{bj} boundary forms, and indeed some are. But, sometimes, some are annihilators. That is, sometimes $B_{aj}(Y)$ or $B_{bj}(Y)$ will be zero for *all* Y in D_M. Technically they are not boundary forms, and, so, we will have to be more careful to distinguish between genuine boundary forms and annihilators. We cannot do so yet.

In the case where, say b is regular, then all Y_j's are in $L^2_A(c, b)$. If they are represented by a fundamental matrix y, we can write

$$B_b(Y) = \lim_{x \to b} y^* JY = y^*(b, \overline{\lambda}) JY(b)\,,$$

$2n$ boundary forms.

Likewise, if a is regular, then

$$B_a(Y) = \lim_{x \to a} \mathcal{Y}^* J Y = \mathcal{Y}^*(a, \overline{\lambda}) J Y(a)$$

represents $2n$ boundary forms.

If these are combined into

$$M\mathcal{Y}^*(a, \overline{\lambda})^{-1} J^{-1} B_a(Y) + N\mathcal{Y}^*(b, \overline{\lambda})^{-1} J^{-1} B_b(Y) = 0\,,$$

where M and N are $m \times 2n$ matrices for which $(M : N)$ has rank m, inserting the values for $B_a(Y)$ and $B_b(Y)$, we find this is equivalent to

$$M Y(a) + N Y(b) = 0\,,$$

the classical representation of regular boundary conditions. $\qquad\square$

V.4 A Preliminary Green's Formula

In anticipation of stronger results to come we present a preliminary version of the singular Green's formula.

V.4.1. Lemma. *Let $\mathcal{Y}(x, \lambda)$ be a fundamental matrix for*

$$JY' = [\lambda A + B]Y$$

satisfying $\mathcal{Y}(c, \lambda) = I$, $a < c < b$. Let

$$\mathcal{Y}^t(x, \lambda) = \mathcal{Y}^*(x, \overline{\lambda}).$$

Then

$$\mathcal{Y}^t J \mathcal{Y} \equiv J,$$

and

$$J\mathcal{Y}J\mathcal{Y}^t J = -J\,.$$

Proof. $\mathcal{Y}^t$ satisfies

$$\mathcal{Y}^{t'} J = -\mathcal{Y}^t[\lambda A + B]$$

and $\mathcal{Y}$ satisfies

$$J\mathcal{Y}' = [\lambda A + B]\mathcal{Y}\,.$$

Multiply the $\mathcal{Y}^t$-equation on the right by $\mathcal{Y}$; multiply the $\mathcal{Y}$-equation on the left by $\mathcal{Y}^t$. Add to find

$$(\mathcal{Y}^t J \mathcal{Y})' = 0.$$

Thus $\mathcal{Y}^t J \mathcal{Y} = C$. Letting $x = c$, we find $C = J$.

To show the second, note that

$$\mathcal{Y}^t J \mathcal{Y} = J \qquad \text{implies} \qquad (J\mathcal{Y}^t)(J\mathcal{Y}) = -I\,,$$

since $J^* = -J = J^{-1}$. This shows that the two terms in parentheses commute, so reverse their order and right multiply by J. $\qquad\square$

V.4.2. Theorem. *Assume that at both a and b,*

$$JY' = [\lambda A + B]Y$$

has $2n$ solutions in $L_A^2(a,c)$ and $L_A^2(c,b)$. Define the boundary forms B_a and B_b by using solutions with $\lambda = 0$.[1] Then for all Y, Z in D_M

$$\langle L_M Y, Z\rangle_A - \langle Y_0 L_M Z\rangle_A$$

$$= \sum_{j=1}^{n} \overline{B_{b(n+j)}(Z)} B_{bj}(Y) + \sum_{j=1}^{n} \overline{B_j(Z)} B_{b(n+j)}(Y)$$

$$- \sum_{j=1}^{n} \overline{B_{a(n+j)}(Z)} B_{aj}(Y) - \sum_{j=1}^{n} \overline{B_{aj}(Z)} B_{b(n+j)}(Y).$$

Proof. Note that

$$\lim_{x\to a} Z^* JY = \lim_{x\to a} [\mathcal{Y}^* JZ]^* J[\mathcal{Y}^* JY],$$

$$\lim_{x\to b} Z^* JY = \lim_{x\to a} [\mathcal{Y}^* JZ]^* J[\mathcal{Y}^* JY].$$

All the limits exist. Now let

$$\lim_{x\to a}[\mathcal{Y}^t JY] = \begin{bmatrix} B_{a1}(Y) \\ \vdots \\ B_{a2n}(Y) \end{bmatrix}, \lim_{x\to a}[\mathcal{Y}^t JZ] = \begin{bmatrix} B_{a1}(Z) \\ \vdots \\ B_{a2n}(Z) \end{bmatrix},$$

$$\lim_{x\to b}[\mathcal{Y}^t JY] = \begin{bmatrix} B_{b1}(Y) \\ \vdots \\ B_{b2n}(Y) \end{bmatrix}, \lim_{x\to b}[\mathcal{Y}^t JZ] = \begin{bmatrix} B_{a1}(Z) \\ \vdots \\ B_{a2n}(Z) \end{bmatrix},$$

and set $L_M Y = F$, $L_M Z = G$ if and only if $JY' - BY = AF$, $JZ' - BZ = AG$. Then for $a < a' < c < b' < b$,

$$\langle L_M Y, Z\rangle_A - \langle Y, L_M Z\rangle_A = \lim_{\substack{a'\to a \\ b'\to b}} \int_{a'}^{b'} \{Z^* A(L_M Y) - (L_M Z)^* AY\} dx$$

$$= \lim_{\substack{a'\to a \\ b'\to b}} \int_{a'}^{b'} \{Z^* AF - G^* AY\} dx$$

$$= \lim_{\substack{a'\to a \\ b'\to b}} \int_{a'}^{b'} \{Z^*[JY' - (L_M Z)^* AY\} dx$$

$$= \lim_{\substack{a'\to a \\ b'\to b}} \int_{a'}^{b'} \{Z^* JY' + Z^{*'} JY\} dx = \lim_{\substack{a'\to a \\ b'\to b}} Z^* JY \Big|_{a'}^{b'}.$$

At this point we merely substitute the appropriate expressions for the limits.

---

[1] When *all* solutions are in $L_A^2(a,b)$ for Im $\lambda \neq 0$ it is easy to show [5] that all solutions are in $L_A^2(a,b)$ for *all* λ.

A further modification of the boundary forms is also possible. As before, let

$$B_a(Y) = \lim_{x \to a} [\mathcal{Y}^t J Y], \qquad B_a(Z) = \lim_{x \to a} [\mathcal{Y}^t J Z],$$
$$B_b(Y) = \lim_{x \to b} [\mathcal{Y}^t J Y], \qquad B_b(Z) = \lim_{x \to b} [\mathcal{Y}^t J Z].$$

Then

$$\langle L_M Y, Z \rangle_A - \langle Y, L_M Z \rangle_A$$
$$= B_b(Z)^* J B_b(Y) - B_a(Z)^* J B_A(Z)$$
$$= (B_a(Z)^* B_b(Z)^*) \begin{pmatrix} -J & 0 \\ 0 & J \end{pmatrix} \begin{pmatrix} B_a(Y) \\ B_b(Y) \end{pmatrix}.$$

We can now insert coefficient matrices

$$\begin{pmatrix} M & N \\ P & Q \end{pmatrix}, \begin{pmatrix} \widetilde{M} & \widetilde{N} \\ \widetilde{P} & \widetilde{Q} \end{pmatrix}$$

in the middle, just as was done in the regular case in Theorem IV.4.4. Here the B-terms replace the evaluations of Y and Z at $x = a$ and $x = b$, but the results are the same. If

$$M B_a(Y) + N B_b(Y) = 0,$$

then the adjoint boundary condition is

$$\widetilde{P} B_a(Z) + \widetilde{Q} B_b(Z) = 0.$$

There are also parametric boundary conditions,

$$B_a(Y) = J \widetilde{P}^* \psi,$$
$$B_b(Y) = -J \widetilde{Q}^* \psi,$$
$$B_a(Z) = -J M^* \phi,$$
$$B_b(Z) = -J N^* \phi.$$

Any operator L, defined using the $M - N$ boundary condition, is self-adjoint if and only if

$$M J N^* = N J N^*.$$

While there is a Green's formula of a similar nature when the numbers of solutions of $JY' = [\lambda A + B]Y$ which are in $L_A^2(a,c)$ and $L_A^2(c,b)$, m_a and m_b, respectively are less than $2n$, the actual limiting forms are more complicated. If a solution Y_j is not in L_A^2, then its form $B_{jn}(Y)$ may become infinite as x approaches a or b. Its coefficient, therefore, must always approach zero, i.e., be an annihilator. Exactly how this is done, however, is not yet clear. $\qquad \square$

References

[1] N. I. Akhiezer and I. M. Glazman, **Theory of Linear Operators in Hilbert Space, vols. I and II**, Frederick Ungar Pub. Co., New York, 1963.

[2] F. V. Atkinson, **Discrete and Continuous Boundary Problems**, Academic Press, NY, 1964.

[3] G. A. Bliss, *A boundary value problem for a system of ordinary linear differential equations of the first order*, Trans. Amer. Math. Soc. **28** (1926), pp. 561–589.

[4] F. Brauer, *Spectral theorem for linear systems of differential equations*, Pacific J. Math. **10** (1960), pp. 17–34.

[5] E. A. Coddington and N. Levinson, **Theory of Ordinary Differential Equations**, McGraw-Hill, N. Y., 1955.

[6] D. B. Hinton and J. K. Shaw, *Titchmarsh-Weyl theory for Hamiltonian systems*, Spectral Theory Diff. Ops, North Holland, I. W. Knowles and R. E Lewis, eds., 1981, pp. 219–230

[7] ______, *On Titchmarsh-Weyl $M(\lambda)$-functions for linear Hamiltonian systems*, J. Diff. Eq. **40** (1981), pp. 316–342.

[8] ______, *On the spectrum of a singular Hamiltonian system*, Quaes. Math. **5** (1982), pp. 29–81.

[9] ______, *Titchmarsh's λ-dependent boundary conditions for Hamiltonian systems*, Lect. Notes in Math. (Springer-Verlag) **964** (1982), pp. 318–326.

[10] ______, *Well-posed boundary value problems for Hamiltonian systems of limit point or limit circle type*, Lect. Notes in Math. (Springer-Verlag) **964** (1982), pp. 614–631.

[11] ______, *Parameterization of the $M(\lambda)$ function for a Hamiltonian system of limit circle type*, Proc. Roy. Soc. Edin. **93** (1983), pp. 349–360.

[12] ______, *Hamiltonian systems of limit point or limit circle type with both ends singular*, J. Diff. Eq. **50** (1983), pp. 444–464.

[13] ______, *On boundary value problems for Hamiltonian systems with two singular points*, SIAM J. Math. Anal. **15** (1984), pp. 272–286.

[14] A. M. Krall, *$M(\lambda)$ theory for singular Hamiltonian systems with one singular point*, SIAM J. Math. **20** (1989), pp. 644–700.

[15] ______, *$M(\lambda)$ theory for singular Hamiltonian systems with two singular points*, SIAM J. Math. **20** (1989), pp. 701–715.

[16] A. M. Krall, B. D. Hinton and J. K. Shaw, *Boundary conditions for differential systems in intermediate situations*, Proc. Conf. Diff. Eq. Alabama-Birmingham (1983), pp. 301–305.

[17] B. M. Levitan and I. S. Sargsjan, **Introduction to Spectral Theory:/Self-Adjoint Ordinary Differential Operators**, Amer. Math. Soc., Providence, R. I., 1975.

[18] ______, **Sturm-Liouville and Dirac Operators**, Kluwer Academic Publ., Dordrecht, 1991.

[19] L. L. Littlejohn and A. M. Krall, *Orthogonal polynomials and higher order singular Sturm-Liouville systems*, Acta. Applicandae Math. **17** (1989), pp. 99–170

[20] H. D. Niessen, *Singuläre S-hermitesche Rand-Eigenwert Probleme*, Manuscripta Math. **3** (1970), pp. 35–68.

[21] ______, *Zum verallgemeinerten zweiten Weylschen Satz*, Archiv der Math. **22** (1971), pp. 648–656.

[22] ______, *Greensche Matrix und die Formel von Titchmarch-Kodaira für singuläre S-hermitesche Eigenwert Probleme*, J. f. d. reine ang. Math. **261** (1972), pp. 164–193.

[23] H. D. Niessen and A. Schneider, *Integral-Transformationen zu singulären S-hermiteschen*, Manuscripta Math. **5** (1971), pp. 133–145.

[24] A. Schneider, *Untersuchungen über singuläre reele S-hermitesche Differentialgleichungssysteme im Normalfall*, Math. Zeit. **107** (1968), pp. 271–296.

Chapter VI

The Niessen Approach to Singular Hamiltonian Systems

In 1910, Herman Weyl [7] took a major step in opening up the study of singular boundary value problems. He showed that if a second order problem, singular at one end, is restricted to a regular interval, then each regular, separated boundary condition imposed near a singular end is in a 1-to-1 correspondence with a point on a circle in the complex plane. That is, each such boundary condition corresponds to a different point on the circle, and every point on the circle corresponds to a different boundary condition.

As the regular point moves nearer to the singular, the circles, associated with different boundary points, contract. As the regular point approaches the singular, the limit of these circles is either a circle or a point.

Finally for each point on the limit circle, or for the limit point, there is a square integrable solution of the original homogeneous differential equation with a complex eigenparameter over an interval containing the singular boundary point.

A remarkable work, it has been further developed by M. H. Stone [5], E. C. Titchmarsh [6], E. C. Coddington and N. Levinson [1] in their famous books.

Since the study of linear Hamiltonian systems began, there have been two different, but connected extensions of Weyl's theory. The first, by H. D. Niessen [2–4], is the subject of this chapter. The second, by D. B. Hinton and J. K. Shaw, follows.

Niessen's ingenious idea was to examine the matrix

$$\mathcal{A}(x) = [1/2 \operatorname{Im} \lambda] \mathcal{Y}^*(x, \lambda)(J/i) \mathcal{Y}(x, \lambda),$$

where $\mathcal{Y}(x, \lambda)$ is a fundamental solution of

$$JY' = [\lambda A + B]Y,$$
$$\operatorname{Im} \lambda \neq 0,$$

satisfying $\mathcal{Y}(c, \lambda) = I$, as x approaches a. He shows that its eigenvalues begin with n at 1 and at -1. As x approaches the lower limit a (with Im $\lambda > 0$), the eigenvalues decrease, but since the matrix $\mathcal{A}$ is never singular, n eigenvalues remain greater than or equal to zero. Of the other n, some may have finite limits (the same number as those that have positive limits), while some approach $-\infty$ (the same number that approach 0). Finite limiting eigenvalues generate $L_A^2(a, c)$ solutions, while those approaching $-\infty$ do not. Finally, the positive and negative but finite eigenvalue pairs generate circles, closely related to Weyl's circle. The other pairs generate points.

We call the two extremes the limit circle and limit point cases. This not only corresponds to the Weyl notation, but to that of Hinton and Shaw. We call the intermediate case, where there are m_a solutions in $L_A^2(a, c)$, $n < m_a < 2n$, the limit m_a case.

We note when $\mathcal{Y}(x, \lambda)$ is a fundamental matrix for $JY' = [\lambda A + B]Y$, Im $\lambda \neq 0$, $\mathcal{Y}(c, \lambda) = I$, $a < c < b$, that

$$\mathcal{Y}^*(x, \lambda)(J/i)\mathcal{Y}(x, \lambda)\bigg|_\alpha^\beta = 2\mathrm{Im}\ \lambda \int_\alpha^\beta \mathcal{Y}^*(x, \lambda)A(x)\mathcal{Y}(x, \lambda)dx\,.$$

Since $A(x) \geqq 0$, if Im $\lambda > 0$, we find that $\mathcal{Y}^*(x, \lambda)(J/i)\mathcal{Y}(x, \lambda)$ is monotone nondecreasing.

VI.1   Boundary Values of Hermitian Forms

We assume that $\mathcal{A}(x)$ is a monotone nondecreasing transformation of $[a, b]$ into the Hermitian $2n \times 2n$ matrices as x approaches a and that $\mathcal{A}(x)$ has eigenvalues

$$\mu_1(x) \geqq \mu_2(x) \geqq \cdots \geqq \mu_{2n}(x)\,.$$

Then $\mu_j(x)$, $j = 1, \ldots, 2n$, is monotone nondecreasing as x approaches a.

VI.1.1. Definition.

(1) Let $\mu_j = \lim_{x \to a} \mu_j(x)$, $j = 1, \ldots, 2n$. $-\infty \leqq \mu_j \leqq M$, $j = 1, \ldots, 2n$.

(2) Let $v_1(x), v_2(x), \ldots, v_{2n}(x)$ be a complete orthonormal set of eigenvectors for $\mathcal{A}(x)$, with $v_j(x)$ corresponding to the eigenvalue $\mu_j(x)$, $j = 1, \ldots, 2n$. $\|v_j(x)\| = 1$, $j = 1, \ldots, 2n$.

(3) Let $\lim_{x \to a} v_j(x) = v_j$. (The limits exist through subsequences.)

(4) Let $v(x) = (v_1(x), \ldots, v_{2n}(x))$, $V = (v_1, \ldots, v_{2n})$.

(5) Let r equal the number of eigenvalues μ_j that are greater than $-\infty$, i.e., finite.

VI.1.2. Lemma. *Let*

$$
I_r = \begin{pmatrix}
1 & & & & & & 0 \\
& \ddots & & & & & \\
& & 1 & & & & \\
& & & 0 & & & \\
& & & & \ddots & & \\
0 & & & & & & 0
\end{pmatrix},
$$

with r 1's. Then

(1) $\lim\limits_{x \to a} I_r^* V^* \mathcal{A}(x) V(x) I_r = diag\ (\mu_1, \ldots, \mu_r, 0, \ldots, 0)$.

(2) $\lim\limits_{x \to a} f^* \mathcal{A}(x) f = \sum\limits_{j=1}^{r} \mu_j |(f, v_j)|^2$, *where $\langle \cdot, \cdot \rangle$ denotes the inner product in $\mathbb{C}^{2n}$,
and f is in the span of $v_1, \ldots, v_r$.*

(3) *f is in the span of $v_1, \ldots, v_r$ if and only if $\lim\limits_{x \to a} f^* \mathcal{A}(x) f$ exists.*

(4) *r is the maximum of the dimensions of the spaces R that are linear and
$\lim\limits_{x \to a} f^* \mathcal{A}(x) f$ exists for all f in R.*

(5) *The number of eigenvalues $\mu_j \geqq 0$ is the maximum of the dimensions of all
spaces R that are linear and satisfy $\lim\limits_{x \to a} f^* \mathcal{A}(x) f \geqq 0$ for all f in R.*

(6) *The number of eigenvalues μ_j, $0 \geqq \mu_j > -\infty$, is the maximum of the dimensions of all spaces R that are linear and satisfy $\lim\limits_{x \to a} f^* \mathcal{A}(x) f \leqq 0$ for all f in
R.*

Proof. 1. Let

$$
N(x) = \begin{pmatrix}
\mu_1(x) & & & \\
& \ddots & & \\
& & \ddots & \\
& & & \mu_{2n}(x)
\end{pmatrix},\
M(x) = \begin{pmatrix}
\mu_1(x) & & & & & \\
& \ddots & & & & \\
& & \mu_r(x) & & & \\
& & & 0 & & \\
& & & & \ddots & \\
& & & & & 0
\end{pmatrix},
$$

$M = \lim\limits_{x \to a} M(x)$. Then because $\mathcal{A}(x)$ is monotone and near a, $N(x) \leqq M(x)$, and

$$
\begin{aligned}
M &= \lim\limits_{x \to a} M(x) = \lim\limits_{x \to a} I_r N(x) I_r = \lim\limits_{x \to a} I_r V^*(x) \mathcal{A}(x) V(x) I_r \\
&\leqq \limsup\limits_{y \to a} \lim\limits_{x \to a} I_r V^*(x) \mathcal{A}(y) V(x) I_r \leqq \limsup\limits_{y \to a} I_r V^* \mathcal{A}(y) V I_r \\
&\leqq \limsup\limits_{y \to a} I_r V^* V(y) N(y) V^*(y) V I_r \leqq \limsup\limits_{y \to a} I_r V^* V(y) M(y) V(y) V I_r \\
&= I_r V^* V M V^* V I_r = M .
\end{aligned}
$$

So $\lim\limits_{x \to a} I_r V^* \mathcal{A}(x) V I_r = M$.

2. All f in the span of $v_1, \dots, v_r$ have the form

$$f = \sum_{j=1}^{r} v_j c_j = VC,$$

where $C = (c_1, \dots, c_r, 0, \dots, 0)^T = I_r C$. So

$$\lim_{x \to a} f^* \mathcal{A}(x) f = \lim_{x \to a} C^* V^* \mathcal{A}(x) VC = \lim_{x \to a} C^* I_r V^* \mathcal{A}(x) V I_r C = C^* MC.$$

Since $c_j = (f, v_j) = v_j^* f$, we find $C = V^* f$, and

$$\lim_{x \to a} f^* \mathcal{A}(x) f = \sum_{j=1}^{r} \mu_j |v_j^* f|^2.$$

3. Let $\lim_{x \to a} f^* \mathcal{A}(x) f$ exist. Since

$$f^* \mathcal{A}(x) f = \sum_{j=1}^{2n} \mu_j(x) |v_j^* f|^2,$$

we must have $\lim_{x \to a} v_j^*(x) f = 0$ for $j > r$. Thus $v_j^* f = 0$ for $j > r$ and f is in the span of $v_1, \dots, v_r$.

4. For f in R, the span of $v_1, \dots, v_r$, $\lim_{x \to a} f^* \mathcal{A}(x) f$ exists, so the maximum is at least r. Likewise if $\lim_{x \to a} f^* \mathcal{A}(x) f$ exists, then f is in the span of $v_1, \dots, v_r$, and so r is at least as large as the maximum.

5. Let p equal the number of eigenvalues μ_j which are greater than or equal to 0. Let R be the space spanned by $v_1, \dots, v_p$. Then $\dim R = p$, and if f is in R,

$$\lim_{x \to a} f^* \mathcal{A}(x) f = \sum_{j=1}^{p} \mu_j |v_j^* f|^2 \geqq 0.$$

p is less than or equal to the maximum.

If p is smaller, than there is a space R with dimension greater than p, such that if f is in R, $\lim f^* \mathcal{A}(x) f \geqq 0$. Choose f_0 in R such that f_0 is orthogonal to $v_1, \dots, v_p$. For this f_0,

$$\lim_{x \to a} f_0^* \mathcal{A}(x) f_0 = \sum_{j=p+1}^{r} \mu_j |v_j^* f_0|^2 < 0.$$

This is a contradiction.

6. This is similar to 5. $\qquad\qquad\square$

VI.2　The Eigenvalues of $\mathcal{A}(x,\lambda) = [1/2\mathbf{Im}\ \lambda]\mathcal{Y}^*(x,\lambda)(J/i)\mathcal{Y}(x,\lambda)$

We assume that a is singular, $a < c < b$ and examine the interval (a,c). We also assume that Im $\lambda > 0$. Im $\lambda < 0$ merely reverses the monotonic nature of $\mathcal{A}$. We note that since

$$\mathcal{Y}^*(x,\overline{\lambda})(J/i)\mathcal{Y}(x,\lambda) = (J/i)\,,$$

(J/i) is Hermitian and $(J/i)^2 = I$, that

$$\mathcal{Y}^*(x,\lambda)(J/i)\mathcal{Y}(x,\overline{\lambda}) = (J/i)$$

as well, and

$$\mathcal{Y}(x,\overline{\lambda}) = (J/i)\mathcal{Y}^*(x,\lambda)^{-1}(J/i)\,.$$

Thus

$$\mathcal{A}(x,\overline{\lambda}) = [-1/(2\mathrm{Im}\ \lambda)^2](J/i)\mathcal{A}(x,\lambda)^{-1}(J/i)$$

and

$$\mathcal{A}(x,\overline{\lambda}) - \rho I = -\rho(J/i)\mathcal{A}(x,\lambda)^{-1}[\mathcal{A}(x,\lambda) + [1/\rho(2\mathrm{Im}\ \lambda)^2]I](J/i)\,.$$

We have established the following.

VI.2.1. Theorem. *ρ is an eigenvalue of $\mathcal{A}(x,\overline{\lambda})$ if and only if $-1/\rho(2Im\ \lambda)^2$ is an eigenvalue of $\mathcal{A}(x,\lambda)$. Let $\mu_{1,\lambda}(x) \geq \cdots \geq \mu_{2n,\lambda}(x)$ be the eigenvalues of $\mathcal{A}(x,\lambda)$. We find for Im $\lambda > 0$,*

$$\mu_{j,\overline{\lambda}}(x) = -1/(2Im\ \lambda)^2\mu_{j+n,\lambda}(x), \qquad j = 1,\ldots,n\,,$$
$$\mu_{j,\overline{\lambda}}(x) = -1/(2Im\ \lambda)^2\mu_{j-n,\lambda}(x), \qquad j = n+1,\ldots,2n\,.$$

We recall that $\mathcal{A}(x,\lambda)$ is never singular, and that, when $x = c$, $\mathcal{A}(c,\lambda) = (J/i)$. Hence $\mathcal{A}(x,\lambda)$ has n positive and n negative eigenvalues for all x. Since the eigenvalues behave in a monotone fashion, let

$$\mu_{j,\lambda} = \lim_{x\to a}\mu_{j,\lambda}(x)\,, \qquad j = 1,\ldots,2n\,.$$

If $m_{(a,\lambda)}$ denotes the number of eigenvalues $\mu_{j,\lambda}$ which are greater than $-\infty$ (finite), then since n eigenvalues of $\mathcal{A}(x,\lambda)$ are greater than 0, $m_{(a,\lambda)} \geq n$.

The formulas of the previous theorem show that $\mu_{j,\lambda} = 0$ if and only if $\mu_{j+n,\overline{\lambda}} = -\infty$. This implies $j \leq n$, and $j+n > m_{(a,\overline{\lambda})}$, or

$$m_{(a,\overline{\lambda})} < j \leq n\,.$$

More explicitly,

$$\infty > \mu_1 \geq \cdots \geq \mu_{(m_{a,\overline{\lambda}}-n)} > 0\,,$$
$$\mu_{(m_{a,\overline{\lambda}}-n+1)} = \cdots = \mu_n = 0\,,$$
$$0 > \mu_{n+1} \geq \cdots \geq \mu_{(m_{a,\overline{\lambda}})} > -\infty\,,$$
$$\mu_{(m_{a,\overline{\lambda}}+1)} = \cdots = \mu_{2n} = -\infty\,.$$

Therefore
$$\overline{\mathcal{A}(x,\lambda)} = [1/2Im(\lambda)]\overline{\mathcal{Y}^*(x,\lambda)(J/i)\mathcal{Y}(x,\lambda)}$$
$$= [1/2Im(\lambda)]\mathcal{Y}^*(x,\overline{\lambda})(J/i)\mathcal{Y}(x,\overline{\lambda})$$
$$= \mathcal{A}(x,\overline{\lambda}).$$

Hence $\mu_{j,\lambda}(x) = \overline{\mu_{j,\lambda}(x)} = \mu_{j,\overline{\lambda}}(x)$. Therefore $m_{(a,\lambda)} = m_{(a,\overline{\lambda})}$ and $\mu_{j,\lambda} = 0$ if and only if $\mu_{j+n,\lambda} = -\infty$, $j = (m_{(a,\lambda)-n+1}, \ldots, n)$. For $j = 1, \ldots, m_{(a,\lambda)-n}$,

$$\mu_{j,\lambda} = -1/2(2Im\lambda)\mu_{j+n,\lambda}.$$

VI.3   Generalization of the Second Weyl Theorem

We set
$$[Y,Z](a) = \lim_{x\to a} Z^*(x)JY(x).$$

Here Y and Z are $2n$-dimensional vectors. We have the following.

VI.3.1. Theorem. *For $Im\,(\lambda) \neq 0$, there is a decomposition of the solution space of*
$$JY' = (\lambda A + B)Y$$

into n two-dimensional subspaces N_j, $j = 1, \ldots, n$, such that for all j, k:

(1) $\dim N_j \cap L_A^2(a,c) \geqq 1$.

(2) *If* $\dim N_j \cap L_A^2(a,c) = 1$, *then* $[Y,Y](a) = 0$ *for Y in* $N_j \cap L_A^2(a,c)$.

(3) *If* $\dim N_j \cap L_A^2(a,c) = 2$, *then there is a basis Y_j, Z_j in N_j and a circle*

$$K_j = \{\alpha : |\alpha| = \rho_j\},$$

with $\rho_j > 0$, so that for each

$$Y = Y_j + \alpha Z_j, \qquad \alpha \quad on \ K_j,$$

we have $[Y,Y](a) = 0$. Further it is impossible for $\alpha \neq \beta$, to have

$$Y = Y_j + \alpha Z_j, \qquad Z = Y_j + \beta Z_j$$

and $[Y,Z](a) = 0$.

(4) $[Y,Z](a) = 0$ *for all Y in $N_j \cap L_A^2(a,c)$ and Z in $N_k \cap L_A^2(a,c)$, $j \neq k$.*

Proof. 1. Let $v_{1,\lambda}(x), \ldots, v_{2n,\lambda}(x)$ be an orthonormal system of eigenvectors for $\mathcal{A}(x,\lambda)$ and let $\mu_{1,\lambda}(x), \ldots, \mu_{2n,\lambda}(x)$ be the corresponding eigenvalues. Choose $x_k \to a$ so that these converge (or diverge to $-\infty$).

Let
$$V_\lambda = (v_{1,\lambda}, \ldots, v_{2n,\lambda}),$$

let
$$Y_j(x, \lambda) = \mathcal{Y}(x, \lambda) v_{j,\lambda},$$

and let
$$Z_j(x, \lambda) = \mathcal{Y}(x, \lambda) v_{n+j,\lambda},$$

$j = 1, \ldots, n$. Then define N_j as the two-dimensional linear space spanned by Y_j and Z_j, $j = 1, \ldots, n$. Since $Y_1, \ldots, Y_n$ are in $L_A^2(a, c)$, the first statement is proved.

2. If $\dim N_j \cap L_A^2(a, c) = 1$, then $j + n > m_{a,\lambda}$. Thus $\mu_{j,\lambda} = 0$ and

$$[Y_j, Y_j](a) = 2i\text{Im }(\lambda) \lim_{x \to a} v_{j,\lambda}^* \mathcal{A}(x, \lambda) v_j = 2i\text{Im }(\lambda) \bullet \mu_{j,\lambda} = 0 \, .$$

3. If $\dim N_j \cap L_A^2(a, c) = 2$, then $j + n \leqq m_{a,\lambda}$, and $\mu_{j,\lambda} > 0$, $0 > \mu_{n+j,\lambda} > -\infty$. Set

$$\rho_j = \sqrt{|\mu_{j,\lambda}|/|\mu_{n+j,\lambda}|} = \sqrt{-\mu_{j,\lambda}/\mu_{n+j,\lambda}} > 0 \, .$$

Let $|\alpha| = \rho_j$ and $Y = Y_j + \alpha Z_j$. Then

$$\begin{aligned}
[Y, Y](a) &= 2i\text{Im }(\lambda) \lim_{x \to a} [(v_{j,\lambda} + \alpha v_{n+j,\lambda}) * \mathcal{A}(x, \lambda)(v_{j,\lambda} + \alpha v_{n+j,\lambda}) \\
&= 2i\text{Im }(\lambda)(\mu_{j,\lambda} + |\alpha|^2 \mu_{n+j,\lambda}) \\
&= 0 \, .
\end{aligned}$$

Further

$$\begin{aligned}
[Y, Z](a) &= 2i\text{Im } \lambda \lim_{x \to a} (v_{j,\lambda} + \beta v_{n+j,\lambda}) * \mathcal{A}(x, \lambda)(v_{j,\lambda} + \alpha v_{n+j,\lambda}) \\
&= 2i\text{Im } \lambda(\mu_{j,\lambda} + \beta * \alpha \mu_{n+j,\lambda}) \, .
\end{aligned}$$

Since both α and β lie on K_j, $\alpha = \rho_j e^{i\theta}$ and $\beta = \rho_j e^{i\theta}$. Consequently, if $[Y, Z](a) = 0$, then $\mu_{j,\lambda} + \beta * \alpha \mu_{n+j,\lambda} = 0$, and $\theta - \phi = 2n\pi$. Thus $\alpha = \beta$, a contradiction.

4. Here

$$[Y, Z](a) = 2i\text{Im } \lambda \lim_{x \to a} v_{k,\lambda}^* \mathcal{A}(x, \lambda) v_{j,\lambda}$$

$$= 0 \, ,$$

since the eigenvectors are mutually orthogonal.   □

If $n = 1$, so that the vectors Y, Z, etc., are two-dimensional, then the number of subspaces, N_j is 1. There is at least one solution in $L_A^2(a, c)$. If there is exactly one, then $[Y, Y](a) = 0$. If there are two, then $Y = Y_1 + \alpha Z_1$, α on K_1, satisfies $[Y, Y](a) = 0$. If different α's are chosen, then their corresponding Y and Z fail to satisfy $[Y, Z](a) = 0$. This is essentially Weyl's second theorem.

We comment in closing this section that as x approaches b, the same analysis applies. There are n two-dimensional subspaces which span the space of solutions.

Intersected with $L_A^2(c, b)$, they contain at least one solution, with a total of $m_{(b,\lambda)}$ in all. The difference between the limits at a and at b is that the eigenvalues of $\mathcal{A}(x, \lambda)$ increase as x approaches b, with $2n - m_{(b,\lambda)}$ approaching ∞.

The situation for Im $\lambda < 0$ is exactly the reverse of the case Im $\lambda > 0$. As x approaches a, some eigenvalues of $\mathcal{A}(x, \lambda)$ approach ∞; as x approaches b, some approach $-\infty$.

Further, since the number of L_A^2 solutions is independent of λ, we can drop that subscript in m_a and m_b.

VI.4   Singular Boundary Value Problems

We are now in a position to establish that certain differential operators whose domains are restricted by singular boundary conditions are self-adjoint. We do this by using a Green's function, since as indicated earlier, Green's formula is useful only in the limit-$2n$ or limit circle case. Verification is through Niessen's decomposition of the solution subspaces.

Since we will use it extensively, we note again that if $\mathcal{Y}(x, \lambda)$ is a fundamental matrix for

$$Jy' = (\lambda A + B)Y$$

and $\mathcal{Y}(c, \lambda) = I$, $a < c < b$, then

$$\mathcal{Y}(x, \overline{\lambda})J\mathcal{Y}(x, \lambda) \equiv J.$$

Now on $[c, b)$, there exist m_b solutions of

$$JY' = (\overline{\lambda} A + B)Y, \qquad \text{Im } \lambda \neq 0,$$

in $L_A^2(c, b)$, $n \leq m_b \leq 2n$. We choose n of these, one from each of Niessen's subspaces, and call them $U_1, \ldots, U_n$. We let

$$U = (U_1, \ldots, U_n, 0, \ldots, 0), = \mathcal{Y}R,$$

where $R = (R_1, \ldots, R_n, 0, \ldots, 0)$.

On (a, c) there exist m_a solutions in $L_A^2(a, c)$, $n \leq m_a \leq 2n$. We choose n of these, one from each of Niessen's subspaces, and call them $V_1, \ldots, V_n$. We let

$$V = (0, \ldots, 0, V_1, \ldots, V_n) = \mathcal{Y}S,$$

where $S = (0, \ldots, 0, S_1, \ldots, S_n)$.

We assume that

$$(U + V) = (U_1, \ldots, U_n, V_1, \ldots, V_n) = \mathcal{Y}(R + S)$$

is nonsingular. This assures that no solution in $L_A^2(a, b)$, if any exists, is in the space spanned by $\{U_1, \ldots, U_n\}$ and also in the space spanned by $\{V_1, \ldots, V_n\}$. Hence $U + V$ is a fundamental matrix. $(R + S)$ is invertible.

VI.4.1. Definition. We denote by D_L those elements Y in $L_A^2(a,b)$ satisfying:

(1) $\ell Y = JY' - BY = AF$ exists a.e., and F is in $L_A^2(a,b)$.

(2) $\lim\limits_{x\to a} V^*(x,\overline{\lambda})JY(x) = 0.$

(3) $\lim\limits_{x\to a} U^*(x,\overline{\lambda})JY(x) = 0.$

We define the operator L by setting $LY = F$ if and only if

$$JY' - BY = AF$$

for all Y in D_L.

Please note that $\overline{\lambda}$ in U and V is fixed with Im $\lambda \neq 0$. We shall remove this restriction shortly.

VI.5 The Green's Function

We attempt to solve for Im $\lambda \neq 0$, $(L - \lambda I)Y = F$. That is, we solve

$$JY' - BY - \lambda AY = AF\,,$$

$$\lim\limits_{x\to a} V^*(x,\overline{\lambda}JY(x) = 0\,, \quad \lim\limits_{x\to b} U^*(x,\overline{\lambda}JY(x) = 0\,.$$

Letting $Y(x) = \mathcal{Y}(x,\lambda)C(x)$, variation of parameters yields $J\mathcal{Y}C' = AF$. Since $\mathcal{Y}^t J\mathcal{Y} \equiv J$, where $\mathcal{Y}^t = \mathcal{Y}^*(x,\overline{\lambda})$, we find $J\mathcal{Y}^{-1} = -J\mathcal{Y}^{-1}$, and $C' = -J\mathcal{Y}^t AF$. Thus

$$C = -\int_c^x J\mathcal{Y}^t AF\,dt + K\,,$$

and

$$Y(x) = -\mathcal{Y}(x,\lambda)\int_c^x J\mathcal{Y}^*(t,\overline{\lambda})A(t)F(t)\,dt + \mathcal{Y}(x,\lambda)K\,.$$

We impose the boundary condition at b. Since

$$U^*(x,\overline{\lambda})JY(x) = -R^t\mathcal{Y}^t J\mathcal{Y}\int_c^x J\mathcal{Y}^t AF\,dt + R^t\mathcal{Y}^t J\mathcal{Y}K\,,$$

we find after simplification that

$$U^*(x,\overline{\lambda})JY = \int_c^x R^t\mathcal{Y}^t AF\,dt + R^t JK\,.$$

As x approaches b, the integral converges, since $\mathcal{Y}(x,\overline{\lambda})R(\overline{\lambda})$ is in $L_A^2(c,b)$. Hence

$$\int_c^b R^t\mathcal{Y}^t AF\,dt + R^t JK = 0\,.$$

Likewise, as x approaches a,

$$\int_c^a S^t \mathcal{Y}^t AF \, dt + S^t JK = 0 \,.$$

Adding, we find that

$$K = J(R^t + S^t)^{-1} \left[\int_c^b R^t \mathcal{Y}^t AF \, dt + \int_c^a S^t \mathcal{Y}^t AF \, dt \right] ,$$

and

$$Y(x) = \mathcal{Y}(x, \lambda) J(R^*(\overline{\lambda}) + S^*(\overline{\lambda}))^{-1} \int_x^b R^*(\overline{\lambda}) \mathcal{Y}^*(t, \overline{\lambda}) A(t) F(t) \, dt$$
$$- \mathcal{Y}(x, \lambda) J(R^*(\overline{\lambda}) + S^*(\overline{\lambda}))^{-1} \int_a^x S^*(\overline{\lambda}) \mathcal{Y}^*(t, \overline{\lambda}) A9t) F(t) \, dt \,.$$

VI.5.1. Theorem. *Let $\mathrm{Im}\, \lambda \neq 0$; then $(L - \lambda I)Y = F$ has a unique solution given by*

$$Y(x) = \int_a^b G(\lambda, x, t) A(t) F(t) \, dt \,,$$

where

$$G(\lambda, x, t) = \mathcal{Y}(x, \lambda) J(R^*(\overline{\lambda}) + S^*(\overline{\lambda}))^{-1} R^*(\overline{\lambda}) \mathcal{Y}^*(t, \overline{\lambda}), \quad a \leq x \leq t \leq b$$
$$= \mathcal{Y}(x, \lambda) J(R^*(\overline{\lambda}) + S^*(\overline{\lambda}))^{-1} S^*(\overline{\lambda}) \mathcal{Y}^*(t, \overline{\lambda}), \quad a \leq t \leq x \leq b \,.$$

We can verify that the boundary conditions are satisfied.

$$R^t \mathcal{Y}^t JY = R^t \mathcal{Y}^t J\mathcal{Y} J(R^t + S^t)^{-1}$$
$$\times \int_x^b R^t \mathcal{Y}^t AF \, dt - R^t \mathcal{Y}^t J\mathcal{Y} J(R^t + S^t)^{-1} \int_a^x S^t \mathcal{Y}^t AF \, dt$$
$$= -\int_x^b R^t(R^t + S^t)^{-1} R^t \mathcal{Y}^t AF \, dt + \int_a^x R^t(R^t + S^t)^{-1} S^t \mathcal{Y}^t AF \, dt \,.$$

Recall that $R = \begin{pmatrix} R_1 & 0 \\ R_2 & 0 \end{pmatrix}, S = \begin{pmatrix} 0 & S_1 \\ 0 & S_2 \end{pmatrix}$. It is easy to show that

$$R^t(R^t + S^t)^{-1} = \begin{pmatrix} I & 0 \\ 0 & 0 \end{pmatrix}, S^t(R^t + S^t)^{-1} = \begin{pmatrix} 0 & 0 \\ 0 & I \end{pmatrix} ,$$

and hence

$$R^t(R^t + S^t)^{-1} R^t = R^t \,, \quad R^t(R^t + S^t)^{-1} S^t = 0 \,,$$
$$S^t(R^t + S^t)^{-1} R^t = 0 \,, \quad S^t(R^t + S^t)^{-1} S^t = S^t \,.$$

Therefore

$$R^t \mathcal{Y}^t JY = -\int_x^b R^t \mathcal{Y}^t AF dt.$$

As x approaches b, the integral vanishes.

As x approaches a, we find that

$$S^t \mathcal{Y}^t JY = \int_a^x S^t \mathcal{Y}^t AF dt,$$

which also vanishes.

VI.6 Self-Adjointness

It is not obvious from the previous sections that the boundary value problem is self-adjoint. Indeed, in general it is not. It is not obvious that the Green's function has the correct symmetry either. We consider a shrunken interval and its corresponding regular problem. Then the reduced interval is allowed to swell.

Consider the boundary value problem in which F vanishes outside (a', b'), $a < a' < b' < b$. Then

$$Y = \mathcal{Y}J(R^t + S^t)^{-1} \int_x^{b'} R^t \mathcal{Y}^t AF dt - \mathcal{Y}J(R^t + S^t)^{-1} \int_{a'}^x S^t \mathcal{Y}^t AF dt$$

satisfies
$$JY' - BY - \lambda AY = AF,$$
$$\lim_{x \to a'} S^t \mathcal{Y}^t JY(x) = 0, \qquad \lim_{x \to b'} R^t \mathcal{Y}^t JY(x) = 0.$$

This is a REGULAR Sturm-Liouville problem. The boundary conditions may be written as
$$MY(x) + NY(b) = 0,$$

where
$$M = S^t \mathcal{Y}^t(a')J, \qquad N = R^t \mathcal{Y}^t(b')Y.$$

Note $M = \begin{pmatrix} 0 & 0 \\ M_1 & M_2 \end{pmatrix}$, $N = \begin{pmatrix} N_1 & N_2 \\ 0 & 0 \end{pmatrix}$.

Now the REGULAR problem is self-adjoint if and only if

$$MJM^* = NJN^*.$$

If we compute MJM^* and NJN^*, we find

$$MJM^* = (\mathcal{Y}(a')S)^* J\mathcal{Y}(a')S.$$

Letting

$$\mathcal{Y}(a')S = (0, \ldots, 0, V_1, \ldots, V_n),$$

then

$$\overline{MJM^*} = \begin{pmatrix} 0 & 0 \\ 0 & [V_i, V_j](a') \end{pmatrix},$$

where

$$[V_i, V_j](a') = V_j^* J V_i(a').$$

This is 0, provided V_i, V_j come from different Niessen subspaces, and are chosen according to Section VI.3.

Likewise

$$\overline{NJN^*} = \begin{pmatrix} [U_i, U_j](b') & 0 \\ 0 & 0 \end{pmatrix},$$

where

$$[U_i, U_j](b') = U_j^* J U_i(b').$$

This is again 0, provided U_i, U_j come from different Niessen subspaces according to Section VI.3.

The Green's function for the (a', b') self-adjoint problem satisfies

$$G_{a',b'}(\lambda, x, t) = G^*_{a'b'}(\overline{\lambda}, t, x).$$

As $(a', b') \to (a, b)$, we find that

$$G(\lambda, x, t) = G^*(\overline{\lambda}, t, x).$$

So, if we let $[L - \lambda I]Y = F$ and $[L^* - \overline{\lambda} I]Z = G$,

$$\langle [L - \lambda I]^{-1} F, G \rangle = \int_a^b G^*(x) A(x) \left[\int_a^b G(\lambda, x, t) A(t) F9t) dt \right] dx$$

$$= \int_a^b \left[\int_a^b G^*(\lambda, x, t) A(x) G(x) dx \right]^* A(t) F(t) dt$$

$$= \int_a^b \left[\int_a^b G^*(\lambda, t, x) A(t) G(t) dt \right]^* A(x) F(x) dx$$

$$= \int_a^b \left[\int_a^b G(\overline{\lambda}, x, t) A(t) G(t) dt \right]^* A(x) F(x) dx$$

$$= \langle F, [L - \overline{\lambda}]^{-1} G \rangle$$

$$= \langle [L^* - \lambda]^{-1} F, G \rangle.$$

In other words

$$[L^* - \lambda I]^{-1} = [L - \lambda I]^{-1}.$$

Taking inverses,

$$L^* - \lambda I = L - \lambda I,$$

or $L^* = L$.

VI.6.1. Theorem. *Let the nonzero columns of $\mathcal{Y}(x,\lambda)R(\lambda)$ and $\mathcal{Y}(x,\lambda)S(\lambda)$ be chosen from different Niessen subspaces according to Section IV.5.d. Then the singular Sturm–Liouville problem*

$$JY' = (\lambda A + B)Y + AF,$$

$$\lim_{x \to a} S^*(\overline{\lambda})\mathcal{Y}^*(x,\overline{\lambda})JY(x) = 0, \qquad \lim_{x \to b} R^*(\overline{\lambda})\mathcal{Y}^*(x,\overline{\lambda})JY(x) = 0,$$

is self-adjoint.

Note that if any of the Niessen circles at a' or b' remains a circle as a' approaches a or b' approaches b, then the solution from the subspace associated with it must be determined by a point on the Niessen circle. If the circle contracts to a point, then the solution associated with it is mandated.

VI.6.2. Theorem. *Let $Im \ \lambda \neq 0$. Then the resolvent operator $[L - \lambda I]^{-1}$ satisfies*

$$\|[L - \lambda I]^{-1}\| \leqq 1/|Im \ \lambda|\,,$$

or, equivalently, if $[L - \lambda I]Y = F$, then

$$\|Y\|_A \leqq (1/|Im \ \lambda|)\|F\|_A\,.$$

Proof.

$$\langle Y, F \rangle - \langle F, Y \rangle = \langle Y, [L - \lambda I]Y \rangle - \langle [L - \lambda I]Y, Y \rangle = (\lambda - \overline{\lambda})\langle Y, Y \rangle\,.$$

Applying Schwarz's inequality, we find

$$2|Im \ \lambda| \ \|Y\|_A^2 \leqq 2\|Y\|_A\|F\|_A\,,$$

which gives the result.

As a corollary to the symmetry of the Green's function, we find that

$$S^*(\overline{\lambda})JS(\lambda) = R^*(\overline{\lambda})JR(\lambda) = 0\,.$$

This implies that the Green's function may be written in a more symmetric fashion as

$$G(\lambda, x, t) = \mathcal{Y}(x,\lambda)S(\lambda)[R(\lambda) + S(\lambda)]^{-1}J[R^*(\overline{\lambda}) + S^*(\overline{\lambda})]^{-1}R^*(\overline{\lambda})\mathcal{Y}^*(t,\overline{\lambda})\,,$$
$$a \leqq x \leqq t \leqq b\,,$$
$$= -\mathcal{Y}(x,\lambda)R(\lambda)[R(\lambda) + S(\lambda)]^{-1}J[R^*(\overline{\lambda}) + S^*(\overline{\lambda})]^{-1}S^*(\overline{\lambda})\mathcal{Y}^*(t,\overline{\lambda})\,,$$
$$a \leqq t \leqq x \leqq b\,. \qquad \square$$

VI.7 Modification of the Boundary Conditions

We now show that the choice of λ in generating boundary conditions is largely optional as long as $\operatorname{Im}\lambda \neq 0$, and in the limit circle case (all solutions square integrable), λ may even be real. We denote the fixed, chosen λ of the previous sections by λ_0.

We focus our attention on the boundary condition at b and restrict our attention to $L_A^2(c,b)$, $a < c < b$.

VI.7.1. Lemma. *Let $U_{j0}(x) = \mathcal{Y}(x,\lambda_0)R_j(\lambda_0)$, where $R_j(\lambda_0)$ is one of the nonzero columns of $R(\lambda_0)$. Then*

$$\lim_{x \to b} R^*(\overline{\lambda}_0)\mathcal{Y}^*(x,\overline{\lambda}_0)JU_{j0}(x) = 0\,.$$

Proof.

$$R^*(\overline{\lambda}_0)y^*(x,\overline{\lambda}_0)J\mathcal{Y}(x,\lambda_0)R_j(\lambda_0) = R^*(\overline{\lambda}_0)JR_j(\lambda_0)\,.$$

This is one of the components of $R_0^*JR_0$, which is 0.

The elements $U_{j0}(x)$, chosen earlier, satisfy the boundary conditions at b.

Consider now the equation

$$U_j(x,\lambda) = (\lambda - \lambda_0)\int_c^b G(\lambda_0,x,t)A(t)U_j(t,\lambda)dt + U_{j0}(x)$$

where, here, G is the Green's function over the interval (c,b) rather than (a,b). If $|\lambda - \lambda_0| < |\operatorname{Im}\lambda_0|^{-1}$, this is well known to have a solution in $L_A^2(c,b)$ given by a Neumann series. Further, the solution can be continued analytically throughout the half plane containing λ_0. By using the reflection principle, it can be extended to the other half plane not containing λ_0. In all cases $U_j(x,\lambda)$

$$JU_j' = (\lambda A + B)U_j\,,$$

satisfies the integral equation, and is in $L_A^2(c,b)$. $\qquad\qquad\square$

VI.7.2. Definition. *We call such extensions $U_j(x,\lambda)$ the analytic extension of $U_{j0}(x)$, $j = 1,\ldots,n$.*

VI.7.3. Lemma. *For all λ, $\operatorname{Im}\lambda \neq 0$,*

$$\lim_{x \to b} R^*(\overline{\lambda}_0)\mathcal{Y}^*(x,\overline{\lambda}_0)JU_j(x,\lambda) = 0\,.$$

Proof. If λ_0 and λ lie in the same half plane, then

$$R_0^t\mathcal{Y}_0^tJU_j = (\lambda - \lambda_0)\left[R_0^t\mathcal{Y}_0^tJ\mathcal{Y}_0J(R_0^t + S_0^t)^{-1}\int_x^b R_0^t\mathcal{Y}_j^t AU_0 dt\right]$$

$$= R_0^t\mathcal{Y}_0^tJ\mathcal{Y}_0J(R_0^t + S_0^t)^{-1}\int_c^x S_0^t\mathcal{Y}_0^t AU_j dt + R_0^t\mathcal{Y}_0^tJ\mathcal{Y}_0R_{0j}$$

where $\mathcal{Y}_0 = \mathcal{Y}(x, \lambda_0)$, $\mathcal{Y}_0^t = \mathcal{Y}^*(x, \overline{\lambda}_0)$, $R_0 = R(\lambda_0)$, $R_0^t = R^*(\overline{\lambda}_0)$, $S_0 = S(\lambda_0)$, $S_0^t = S^*(\overline{\lambda}_0)$. Recalling that $\mathcal{Y}_0^t J \mathcal{Y}_0 = J$, $J^2 = -I$, $R_0^t(R_0^t + S_0^t)^{-1} S_0^t = 0$, $R_0^t J R_0 = 0$, this reduces to

$$R_0^t \mathcal{Y}_0^t J U_j = -(\lambda - \lambda_0) \int_x^b R_0^t \mathcal{Y}_0^t A U_j \, dt \,,$$

which vanishes as x approaches b.

If λ_0 and λ are not in the same half plane, we modify $U_j(x, \lambda)$ so that it smoothly vanishes near c. So modified, we can now say that $U_j(x, \lambda)$ is in D_L, restricted to (c, b). If elements $U_{k0}(x) = \mathcal{Y}(x, \lambda_0) R_k(\lambda_0)$ are also modified, then they, too, will be in D_L, restricted to (c, b). Hence

$$0 = \langle LU_j, U_{k0} \rangle - \langle U_j, LU_{k0} \rangle = \lim_{x \to b} U_{k0}^*(x) J U_j(x, \lambda),$$

since the lim at a is 0. Taking complex conjugates,

$$\lim_{x \to b} R_k^*(\overline{\lambda}_0) \mathcal{Y}^*(x, \overline{\lambda}_0) J U_j(x, \overline{\lambda}) = 0 \,.$$

This is equivalent to the statement in the Lemma for $\overline{\lambda}$. $\qquad\square$

VI.7.4. Theorem. *Let Y be in D_L. Then*

$$\lim_{x \to b} R^*(\lambda) \mathcal{Y}^*(x, \lambda) J Y(x) = 0 \,, \qquad \lim_{x \to a} S^*(\lambda) \mathcal{Y}^*(x, \lambda) J Y(x) = 0 \,,$$

for all λ, Im $\lambda \neq 0$, and in the limit circle case even for real λ.

Proof. Suitably altered, the columns of $\mathcal{Y}(x, \lambda) R(\lambda)$ are in D_L. Consequently, if $U_j(x, \lambda) = \mathcal{Y}(x, \lambda) R_j(\lambda)$ is one of these, then

$$0 = \langle LY, U_j \rangle - \langle Y < LU_J \rangle = \lim_{x \to b} U_j^*(x, \lambda) J Y(x) \,.$$

Or

$$\lim_{x \to b} R^*(\lambda) \mathcal{Y}^*(x, \lambda) J Y(x) = 0 \,.$$

At a, the situation is similar. $\qquad\square$

In many cases, especially for equations of second order, λ is chosen to be $i = \sqrt{-1}$. For second order problems, however, λ can be chosen to be real if a boundary condition is not automatically satisfied. In this instance (the limit circle case) $\lambda = 0$ is often a simpler choice.

The restriction of λ to a specific value is a convenience, not a necessity.

VI.8 Other Boundary Conditions

On occasion people have attempted to use other elements, Z, of D_M to generate boundary conditions through $\lim_{x \to a} Z^* JY$ or $\lim_{x \to b} Z^* JY$. We can show that this gives nothing new beyond those which are solution generated.

Let Y and Z be in D_M. Let $L_M Z = G$. Then for any λ,

$$Z = \mathcal{Y} J (R^t + S^t)^{-1} \int_x^b R^t \mathcal{Y}^t AG dt - \mathcal{Y} J (R^t + S^t)^{-1} \int_a^x S^t \mathcal{Y}^t AG dt + \mathcal{Y} C_1,$$

where $\mathcal{Y} C_1$ represents the solutions in $L_A^2(c, b)$. Thus

$$Y^* J Z = Y^* J \mathcal{Y} J (R^t + S^t)^{-1}$$
$$\times \int_x^b R^t \mathcal{Y}^t AG dt - Y^* J \mathcal{Y} J (R^t + S^t)^{-1} \int_a^x S^t \mathcal{Y}^t AG dt + Y^* J \mathcal{Y} C_1,$$

and

$$\lim_{x \to b} Y^* J Z = - \lim_{x \to b} Y^* J \mathcal{Y} \left[J (R^t + S^t)^{-1} \int_a^b S^t \mathcal{Y}^t AG dt + C_1 \right].$$

Thus

$$\lim_{x \to b} Z^* JY = \lim_{x \to b} C_b \mathcal{Y}^* JY.$$

The right side represents solution generated boundary conditions, where C_b is the term in the brackets.

At a,

$$\lim_{x \to b} Z^* JY = \lim_{x \to a} C_a \mathcal{Y}^* JY.$$

as well.

VI.9 The Limit Point Case

The limit point case, in which there are exactly n solutions in $L_A^2(a, c)$ or $L_A^2(c, b)$ has already attracted considerable attention [1], [17–22]. We show that the boundary conditions are satisfied by all elements in D_M when this occurs. In some sense, then, no boundary condition needs to be imposed. They are all annihilators.

VI.9.1. Theorem. *Let the limit point case hold at a or b. Let Y be in D_M. Then*

$$\lim_{x \to a} S^* (\overline{\lambda}) \mathcal{Y}^* (x, \overline{\lambda}) JY(x) = 0,$$

or

$$\lim_{x \to b} R^* (\overline{\lambda}) \mathcal{Y}^* (x, \overline{\lambda}) JY(x) = 0.$$

Proof. Let Y be modified smoothly so that it vanishes near a. Let $LY = F$. Assume that the limit point case holds at b, so that $U(x, \lambda) = \mathcal{Y}(x, \lambda)R(\lambda)$ is uniquely determined as the matrix of solutions in $L_A^2(c, b)$. Then

$$Y = \mathcal{Y}J(R^t + S^t)^{-1} \int_x^b R^t \mathcal{Y}^t AF dt - \mathcal{Y}J(R^t + S^t)^{-1} \int_a^x S^t \mathcal{Y}^t AF dt + \mathcal{Y}RC$$

for some suitable choice of S and C. It is a trivial computation to show that $\lim_{x \to b} R^t \mathcal{Y}^t JY = 0$. $\qquad\square$

We can also show the following:

VI.9.2. Theorem. *Let the limit point case hold at a or b. Let Y, Z be in D_M. Then*

$$\lim_{x \to a} Z^*(x)JY(x) = 0 \,,$$

or

$$\lim_{x \to b} Z^*(x)JY(x) = 0.$$

Proof. Consider the limit point case at b. Assume that Y and Z have been modified to vanish near a, and that suitable conditions have been imposed at a to define a self-adjoint problem. Since Y and Z satisfy the b-boundary condition, they are in the domain D_L. Hence

$$0 = \langle LY, Z \rangle - \langle Y, LZ \rangle - \lim_{x \to b} Z^* JY \,.$$

Note that this section only required the existence of the Green's function. The Niessen circle theory can be bypassed. $\qquad\square$

VI.10   The Limit m Case

The reader may wonder why we essentially impose n boundary conditions at each end, a $2n$ total, when the deficiency index theory says that fewer, in fact, are needed. The answer to this is that the boundary conditions imposed by one-dimensional Niessen subspaces are actually *annihilators*, and are automatically satisfied by all elements Y in D_M. Hence if there are m L_A^2 solutions at an end point, $m \geq n$, then $2(m - n)$ lie in two-dimensional subspaces, and $2n - m$ lie in one-dimensional subspaces. We are actually imposing $n - (2n - m) = m - n$ constraints.

VI.10.1 Theorem. *Let b be in the limit m case. Let Y be in D_M. Let $\mathcal{Y}R_j$ be a solution of*

$$JY' = (\lambda A + B)Y \,, \qquad Im\ \lambda \neq 0 \,,$$

with A and B real Hermitian, $A > 0$, lying in a one-dimensional Niessen subspace. Then $\lim_{x \to b}(\mathcal{Y}R_j)^* JY(x) = 0$.

Proof. Using the Green's function we write

$$Y = \mathcal{Y}J(R^t + S^t)^{-1} \int_x^b R^t \mathcal{Y}^t AF dt - \mathcal{Y}J(R^t + S^t)^{-1} \int_a^x R^t \mathcal{Y}^t AF dt$$

$$+ \sum_{j=1}^n \mathcal{Y}R_j c_j + \sum_{j=1}^{m-n} \mathcal{Y}S_j c_j,$$

where $(L_m - \lambda)Y = F$, where $\{R_j\}_{j=1}^n$ are the nonzero columns of R, $\{S_j\}_{j=1}^{m-n}$ are the nonzero columns of S which generate solutions $\mathcal{Y}S_j$ in $L_A^2(c,b)$. We arrange their order so that

$$N_j = \{\mathcal{Y}R_j, \mathcal{Y}S_j\}, \qquad j = 1, \ldots, n,$$

and

$$N_j \cap L_A^2(c,b) = \{\mathcal{Y}R_j, \mathcal{Y}S_j\}, \qquad j = 1, \ldots, m-n,$$
$$N_j \cap L_A^2(c,b) = [\mathcal{Y}R_j], \qquad j = m-n+1, \ldots, n.$$

Then, if $\lim_{x \to b}(\mathcal{Y}R_j)^* JY(x)$ is computed, the terms involving the integrals have limit 0 automatically. So do the terms in the series by the second Weyl theorem. $\square$

VI.11   The Limit Circle Case

We have now nine possible situations which can occur. At each end we may have the limit point case, and intermediate limit m case, or the limit circle case. The intermediate cases have already been discussed, and, as we have seen, if the limit point case holds, boundary terms at that end automatically vanish.

We would like to make some observations about the limit circle case. Here, again, much of the previous technical computation can be bypassed. Since there is an amazing similarity between limit circle and regular problems, we shall proceed informally and merely indicate the possibilities without formal proof.

If the limit circle case holds at a, while the limit point case holds at b, then boundary conditions are required only at a. Green's formula is Y, Z in D_M

$$\langle L_M Y, Z \rangle - \langle Y, L_M Z \rangle = -\lim_{x \to a} Z^* JY.$$

Inserting

$$-J\mathcal{Y}J\mathcal{Y}^* J = J$$

on the right side,

$$\langle L_M Y, Z \rangle - \langle Y, L_M Z \rangle = -\lim_{x \to a}(\mathcal{Y}^* JZ)^* J \lim_{x \to a}(\mathcal{Y}^* JY).$$

Recall that the limits are $B_a(Z)$ and $B_a(Y)$.

Let $\alpha, \beta, \gamma, \delta$ be $n \times n$ matrices such that $\begin{pmatrix} \alpha & \beta \\ \gamma & \delta \end{pmatrix}$ is nonsingular, and let $\begin{pmatrix} -\varepsilon^* & -\eta^* \\ \zeta^* & \theta^* \end{pmatrix}$ be its inverse. Then

$$-\begin{pmatrix} -\varepsilon^* & -\eta^* \\ \zeta^* & \theta^* \end{pmatrix} \begin{pmatrix} \alpha & \beta \\ \gamma & \delta \end{pmatrix} = J,$$

and

$$\langle L_M Y, Z \rangle - \langle Y, L_M Z \rangle = -[(\zeta\varepsilon)B_a(Z)]^*[(\alpha\beta)B_a(Y)] - [(\theta\eta)B_a(Z)]^*[(\gamma\delta)B_a(Y)].$$

If $(\alpha\ \beta)B_a(Y) = 0$ is used as a boundary condition to define an operator $\mathcal{L}$, then in specifying $\mathcal{L}^*$, the adjoint condition $(\theta\ \eta)B_a(Z) = 0$ is used.

The other expressions are arbitrary. Consequently, if

$$\begin{pmatrix} \alpha & \beta \\ \gamma & \delta \end{pmatrix} B_a(Y) = \begin{pmatrix} 0 \\ \psi \end{pmatrix}, \qquad \begin{pmatrix} \zeta & \varepsilon \\ \theta & \eta \end{pmatrix} B_a(Z) = \begin{pmatrix} \phi & 0 \end{pmatrix},$$

then

$$B_a(Y) = \begin{pmatrix} -\eta^* & \psi \\ \theta^* & \psi \end{pmatrix}, \qquad B_a(Z) = \begin{pmatrix} -\beta^* & \phi \\ \alpha^* & \phi \end{pmatrix}$$

are equivalent parametric boundary conditions for

$$(\alpha\ \beta)B_a(Y) = 0, \qquad (\theta\ \eta)B_a(Z) = 0.$$

The operator $\mathcal{L}$, defined in part by $(\alpha\ \beta)B_a(Y) = 0$, is self-adjoint if and only if

$$(\alpha\ \beta)B_a(Z) = (\alpha\ \beta)\begin{pmatrix} -\beta^* & \phi \\ \alpha^* & \phi \end{pmatrix} = (-\alpha\beta^* + \beta\alpha^*)\phi = 0,$$

i.e.,

$$-\alpha\beta^* + \beta\alpha^* = 0.$$

Finally, if the limit circle case holds at both a and b, Green's formula is (see Section V.4)

$$\langle L_M Y, Z \rangle - \langle Y, L_m Z \rangle = \lim_{x \to b}(\mathcal{Y}^* JZ)^* - J\lim_{x \to b}(\mathcal{Y}^* JZ)^* J \lim_{x \to a}(\mathcal{Y}^* JY)$$

with $B_b(Z), B_b(Y), B_a(Z), B_a(Y)$ denoting the limits,

$$\langle L_M Y, Z \rangle - \langle Y, L_M Z \rangle = B_b(Z)^* JB_b(Y) - B_a(Z)^* JB_a(Y).$$

At this point we can again call on the theory of regular operators. If we replace $Y(a)$ by $B_a(Y)$, $Y(b)$ by $B_b(Y)$, etc., then the computation of Section V.4 carries over here as well:

If the boundary value problem

$$JY' = (\lambda A + B)Y + AF, \qquad MB_a(Y) + NB_b(Y) = 0$$

is considered, its adjoint is given by

$$JZ' = (\overline{\lambda}A + B)Z + AG\,, \qquad \widetilde{P}B_a(Z) + \widetilde{Q}B_b(Z) = 0\,,$$

where $M, N, \widetilde{P}, \widetilde{Q}$ are given in Section IV.4.

There are parametric boundary conditions,

$$B_a(Y) = J\widetilde{P}^*\psi,; B_b(Y) = -J\widetilde{Q}\psi\,,$$

and adjoint conditions

$$B_a(Z) = -JM^*\phi\,, \qquad B_b(Z) = JN^*\phi\,.$$

Self-adjointness occurs if and only if

$$MJM^* = NJN^*\,.$$

VI.12   Comments Concerning the Spectral Resolution

In both Chapter V and Chapter VI we have derived self-adjoint operators under singular conditions. Such operators, of course, have spectral resolutions which follow fairly quickly from the regular eigenfunction expansion. This was indeed done by F. Brauer (see the reference in Chapter V). Such resolutions, however, turn out to be rather "abstract," and so at this point we prefer to defer such a derivation until more details can be given. With Hinton and Shaw's $M(\lambda)$ theory this can be done. Further, the preliminary results derived there, via the Helly theorems, fit the current cases as well.

References

[1]   E. A. Coddington and N. Levinson, **Theory of Ordinary Differential Equations** McGraw-Hill, N. Y., 1955.

[2]   H. D. Niessen, *Singuläre S-hermitesche Rand-Eigenwert Probleme*, Manuscripta Math. **3** (1970), pp. 35–68.

[3]   ______, *Zum verallgemeinerten zweiten Weylschen Satz*, Archiv der Math. **22** (1971), pp. 648–656.

[4]   ______, *Greensche Matrix and die Formel von Titchmarch-Kodaira für singuläre S-hermitesche Eigenwert Probleme*, J. f. d. reine arg. Math. **261** (1972), pp. 164–193.

[5]   M. H. Stone, **Linear Transformations in Hilbert Space and Their Applications to Analysis**, Amer. Math. Soc. Providence R. I., 1932.

[6]   E. C. Titchmarsh, **Eigenfunction Expansions**, Oxford Univ. Press, Oxford, 1962.

[7]   H. Weyl, *Über gewöhnliche Differentialgleichungen mit Singularitäten und die zugehörigen Entwicklungen Willkürlicher Funktionen*, Math. Ann. **68** (1910), pp. 220–269.

Chapter VII

Hinton and Shaw's Extension of Weyl's M(λ) Theory to Systems

D. B. Hinton and J. K. Shaw have developed an extension of the Weyl theory which is a bit different from that of Chapter VI, and which proves to be ultimately much more useful in deriving the spectral resolution for self-adjoint systems.

Here, just as in the previous chapters, we restrict our attention to *real* matrices A and B in order to handle not only the extreme, so-called, limit point and limit circle cases, but also the interesting intermediate case as well.

Hinton and Shaw's approach is in some sense more closely allied with Weyl's methods. It yields results which are difficult to picture (the limit circle is a rather strange ellipsoid), but, nonetheless, gives more information about the spectral resolution of the self-adjoint system operators.

Because of a number of interesting applications, Fourier sine and cosine transforms just to name two, we consider in this chapter a problem which is regular at the left end a, and singular at the right end b. We shall impose regular, self-adjoint, separated boundary conditions at a and at b', $a < b' < b$. Then let b' approach b.

The following chapter considers the case of two singular ends.

VII.1   Notations and Definitions

In order to ensure that elements in the domain of the maximal operator (to be defined) are dense in $L_A^2(a, b)$, we assume that if $JY' - BY = AF$ and $AY = 0$, then $Y = 0$. Most of what follows goes through without this assumption, but without it, some expressions must be restricted to subspaces, instead of holding on all of $L_A^2(a, b)$. Earlier works [Atkinson 1, Hinton and Shaw 12–19] made this assumption with $F = O$. We shall show the existence of m solutions of $JY' = (\lambda A + B)Y$ in $L_A^2(a, b)$, $n \leq m \leq 2n$, when Im $\lambda \neq 0$.

We impose a regular, self-adjoint boundary condition at a,

$$(\alpha_1 \alpha_2)Y(a) = 0\,,$$

where α_1, α_2 are $n \times n$ matrices satisfying rank $(\alpha_1, \alpha_2) = n$ and

$$\alpha_1 \alpha_1^* + \alpha_2 \alpha_2^* = I\,,$$
$$\alpha_1 \alpha_2^* - \alpha_2 \alpha_1^* = 0\,.$$

Self-adjointness is assured, since when written as

$$MY(a) = 0, \quad \text{where} \quad M = \begin{pmatrix} \alpha_1 & \alpha_2 \\ 0 & 0 \end{pmatrix},$$

$$MJM^* = \begin{pmatrix} \alpha_1 & \alpha_2 \\ 0 & 0 \end{pmatrix} \begin{pmatrix} 0 & -I_n \\ I_n & 0 \end{pmatrix} \begin{pmatrix} \alpha_1^* & 0 \\ \alpha_2^* & 0 \end{pmatrix} = 0,$$

the requirement for self-adjointness under separated conditions.

It is no imposition to ensure that $\alpha_1 \alpha_1^* + \alpha_2 \alpha_2^* = I_n$, for if the rank$(\alpha_1 \alpha_2) = n$, then rank $\begin{pmatrix} \alpha_1^* \\ \alpha_2^* \end{pmatrix} = n$, and the rank of

$$\alpha_1 \alpha_1^* + \alpha_2 \alpha_2^* = (\alpha_1 \alpha_2) \begin{pmatrix} \alpha_1^* \\ \alpha_2^* \end{pmatrix}$$

is n. It is nonsingular and positive. Hence if the sum does not equal I_n, replace α_1 and α_2 by $(\alpha_1 \alpha_1^* + \alpha_1 \alpha_2^*)^{1/2} \alpha_1$ and $(\alpha_1 \alpha_1^* + \alpha_2 \alpha_2^*)^{1/2} \alpha_2$.

Note that since

$$\begin{pmatrix} \alpha_1 & \alpha_2 \\ -\alpha_2 & \alpha_1 \end{pmatrix} \begin{pmatrix} \alpha_1^* & -\alpha_2^* \\ \alpha_2^* & \alpha_1^* \end{pmatrix} = \begin{pmatrix} I_n & 0 \\ 0 & I_n \end{pmatrix},$$

we have

$$\begin{pmatrix} \alpha_1^* & -\alpha_2^* \\ \alpha_2^* & \alpha_1^* \end{pmatrix} \begin{pmatrix} \alpha_1 & \alpha_2^* \\ -\alpha_2^* & \alpha_1^* \end{pmatrix} = \begin{pmatrix} I_n & 0 \\ 0 & I_n \end{pmatrix},$$

as well. This implies

$$\alpha_1^* \alpha_1 + \alpha_2^* \alpha_2 = I_n\,,$$
$$\alpha_1^* \alpha_2 - \alpha_2^* \alpha_1 = 0\,.$$

Next let b' be in (a, b) and impose the regular, self-adjoint boundary condition

$$(\beta_1 \beta_2)Y(b') = 0\,,$$

where

$$\beta_1 \beta_1^* + \beta_2 \beta_2^* = I_n\,,$$
$$\beta_1 \beta_2^* - \beta_2 \beta_1^* = 0\,.$$

If the boundary condition is written as $NY(b') = 0$, where $N = \begin{pmatrix} 0 & 0 \\ \beta_1 & \beta_2 \end{pmatrix}$, then

$$NJN^* = \begin{pmatrix} 0 & 0 \\ \beta_1 & \beta_2 \end{pmatrix} \begin{pmatrix} 0 & -I_n \\ I_n & 0 \end{pmatrix} \begin{pmatrix} 0 & \beta_1^* \\ 0 & \beta_2^* \end{pmatrix} = 0,$$

again the requirement for self-adjointness under separated conditions.

The differential equation $JY' = (\lambda A + B)Y$, together with the two boundary conditions defines a regular, self-adjoint Sturm–Liouville problem.

Finally, let $E = \begin{pmatrix} \alpha_1^* & -\alpha_2^* \\ \alpha_2^* & \alpha_1^* \end{pmatrix}$ (note that $E^{-1} = E^*$), and let $\mathcal{Y}$ be the fundamental matrix for $JY' = (\lambda A + B)Y$, satisfying $\mathcal{Y}(a) = E$.

If $\mathcal{Y}$ is partitioned into $\mathcal{Y} = (\theta, \phi) = \begin{pmatrix} \theta_1 & \phi_1 \\ \theta_2 & \phi_2 \end{pmatrix}$, then at $x = a$, $(\alpha_1 \alpha_2)\theta(a) = I_n$, and $(\alpha_1 \alpha_2)\phi(a) = 0$. $\phi = \begin{pmatrix} \phi_1 \\ \phi_2 \end{pmatrix}$ satisfies the boundary condition at a. We cite [Hinton and Shaw, 12–19] as a further reference to this notation.

VII.2 The M(λ) Matrix

Let Im $\lambda \neq 0$, and impose on

$$\chi_{b'} = \mathcal{Y} \begin{pmatrix} I \\ M(b') \end{pmatrix}$$

the boundary condition $(\beta_1 \beta_2)\chi_{b'}(b') = 0$. Hence, with $x = b'$,

$$(\beta_1 \beta_2) \begin{pmatrix} \theta_1 & \phi_1 \\ \theta_2 & \phi_2 \end{pmatrix} \begin{pmatrix} I \\ M(b') \end{pmatrix} = 0.$$

This yields

$$(\beta_1 \theta_1(b') + \beta_2 \theta_2(b')) + (\beta_1 \phi_1(b') + \beta_2 \phi_2(b'))M(b') = 0,$$

and

$$M(b') = -(\beta_1 \phi_1(b') + \beta_2 \phi_2(b'))^{-1}(\beta_1 \theta_1(b') + \beta_2 \theta_2(b')).$$

The inverse must exist. Otherwise λ, which is complex, would be an eigenvalue of the self-adjoint boundary value problem on $[a, b']$.

Since $(\beta_1, \beta_2)\chi_{b'}(b') = 0$,

$$\chi(b') = \begin{pmatrix} 0 & -I_n \\ I_n & 0 \end{pmatrix} \begin{pmatrix} \beta_1^* \\ \beta_2^* \end{pmatrix} C,$$

for

$$(\beta_1 \beta_2) \begin{pmatrix} 0 & -I_n \\ I_n & 0 \end{pmatrix} \begin{pmatrix} \beta_1^* \\ \beta_2^* \end{pmatrix} C = 0.$$

This in turn implies that

$$\chi_{b'}^*(b')J\chi_{b'}(b') = 0,$$

or

$$(I_n M(b')^*)\mathcal{Y}(b')^* J\mathcal{Y}(b') \begin{pmatrix} I_n \\ M(b') \end{pmatrix} = 0.$$

Conversely, if for some M,

$$(I_n M^*)\mathcal{Y}(b')^* J\mathcal{Y}(b') \begin{pmatrix} I_n \\ M \end{pmatrix} = 0,$$

define

$$(\beta_1\beta_2) = (I_n, M^*)\mathcal{Y}(b')^* J.$$

Then $\mathrm{rank}(\beta_1\beta_2) = n$,

$$(\beta_1\beta_2)\mathcal{Y}(b') \begin{pmatrix} I_n \\ M \end{pmatrix} = 0,$$

and

$$-\beta_1\beta_2^* + \beta_2\beta_1^* = (\beta_1, \beta_2) \begin{pmatrix} 0 & -I_n \\ I_n & 0 \end{pmatrix} \begin{pmatrix} \beta_1 \\ \beta_2 \end{pmatrix}$$
$$= (IM^*)\mathcal{Y}(b')^* J^3 \mathcal{Y}(b') \begin{pmatrix} I \\ M \end{pmatrix}$$
$$= 0.$$

Further

$$M = -(\beta_1\phi_1(b') + \beta_2\phi_2(b'))^{-1}(\beta_1\theta_1(b') + \beta_2\phi_2(b')),$$

and

$$\beta_1\beta_1^* + \beta_2\beta_2^* = (\beta_1\beta_2) \begin{pmatrix} \beta_1^* \\ \beta_2^* \end{pmatrix}$$
$$= (I_n M^*)\mathcal{Y}(b')^* \mathcal{Y}(b') \begin{pmatrix} I_n \\ M \end{pmatrix}$$
$$> 0,$$

and so with minor admustments like those of the previous section, we can have
$\beta_1\beta_1^* + \beta_2\beta_2^* = I_n$.

We summarize.

VII.2.1. Theorem. *Let β_1, β_2 satisfy*

$$\beta_1\beta_1^* + \beta_2\beta_2^* = I_n,$$
$$\beta_1\beta_2^* - \beta_2\beta_1^* = 0.$$

Let $\chi_{b'} = \mathcal{Y} \begin{pmatrix} I \\ M(b') \end{pmatrix}$, and suppose

$$(\beta_1\beta_2)\chi_{b'}(b') = 0.$$

Then

$$M(b') = -(\beta_1\phi_1(b') + \beta_2\phi_2(b'))^{-1}(\beta_1\theta_1(b') + \beta_2\theta_2(b'))),$$

and

$$\chi_{b'}^*(b')J\chi_{b'}(b') = 0.$$

Conversely, if for some M, $\chi_{b'} = \mathcal{Y}\begin{pmatrix} I_n \\ M \end{pmatrix}$ *satisfies*

$$\chi_{b'}(b')^*J\chi_{b'}(b') = 0,$$

then there exist β_1, β_2 *satisfying*

$$\beta_1\beta_1^* + \beta_2\beta_2^* = I_n,$$
$$\beta_1\beta_2^* - \beta_2\beta_1^* = 0,$$

such that $(\beta_1\beta_2)\chi_{b'}(b') = 0,$ *and*

$$M = -(\beta_1\phi_1(b') + \beta_2\phi_2(b'))^{-1}(\beta_1\theta_1(b') + \beta_2\theta_2(b')).$$

Again we cite [Hinton and Shaw 12–19].

VII.3   M Circles

The M circle equation is

$$\pm(I_n M^*)\mathcal{Y}(b')^*(J/i)\mathcal{Y}(b')\begin{pmatrix} I_n \\ M \end{pmatrix} = 0,$$

where for convenience, we have divided by i, and $(+)$ holds when Im $\lambda > 0$, $(-)$ holds when Im $\lambda < 0$.

Let

$$\begin{pmatrix} \mathbf{A} & \mathbf{B}^* \\ \mathbf{B} & \mathbf{D} \end{pmatrix} = \begin{cases} \mathcal{Y}(b')^*(J/i)\mathcal{Y}(b'), & \text{Im } \lambda > 0, \\ -\mathcal{Y}(b')^*(J/i)\mathcal{Y}(b'), & \text{Im } \lambda < 0, \end{cases}$$

and let

$$\mathbf{E}(M) = (IM^*)\begin{pmatrix} \mathbf{A} & \mathbf{B}^* \\ \mathbf{B} & \mathbf{D} \end{pmatrix}\begin{pmatrix} I \\ M \end{pmatrix}.$$

VII.3.1. Theorem. $\mathbf{E}(M) = 0$ *if and only if* $M = M_\beta$ *for some* $\beta = (\beta_1, \beta_2),$ *where*

$$M_\beta = -(\beta_1\phi_1(b') + \beta_2\phi_2(b'))^{-1}(\beta_1\theta_1(b') + \beta_2\theta_2(b')).$$

This is a restatement of Theorem VII.2.1.

Expanding, we find

$$\begin{aligned} \mathbf{E}(M) &= (M + \mathbf{D}^{-1}\mathbf{B})^*\mathbf{D}(M + \mathbf{D}^{-1}\mathbf{B}) + \mathbf{A} - \mathbf{B}^*\mathbf{D}^{-1}\mathbf{B} \\ &= (M - C)^*R_1^{-2}(M - C) - R_2^2 \\ &= 0, \end{aligned}$$

where $C = -\mathbf{D}^{-1}\mathbf{B}$, $R_1 = \mathbf{D}^{-1/2}$, $R_2 = (\mathbf{B}^*\mathbf{D}^{-1}\mathbf{B} - \mathbf{A})^{1/2}.$

VII.3.2. Lemma. $\mathbf{D} > 0$.

Proof.

$$\begin{pmatrix} \mathbf{A} & \mathbf{B}^* \\ \mathbf{B} & \mathbf{D} \end{pmatrix} = \pm \begin{pmatrix} \theta_1^* & \theta_2^* \\ \phi_1^* & \phi_2^* \end{pmatrix} \begin{pmatrix} 0 & iI_n \\ -iI_n & 0 \end{pmatrix} \begin{pmatrix} \theta_1 & \phi_1 \\ \theta_2 & \phi_2 \end{pmatrix} = \pm \begin{pmatrix} -i\theta^* J\theta & -i\theta^* J\phi \\ -i\phi^* J\theta & -i\phi^* J\phi \end{pmatrix},$$

where $\phi = \begin{pmatrix} \theta_1 \\ \theta_2 \end{pmatrix}$, $\phi = \begin{pmatrix} \phi_1 \\ \phi_2 \end{pmatrix}$. So $\mathbf{D} = \pm\phi^*(b')(J/i)\phi(b')$. Now manipulation of the differential equation

$$J\phi' = (\lambda A + B)\phi$$

yields

$$\phi^*(J/i)\phi\Big|_a^{b'} = 2\mathrm{Im}\,\lambda \int_a^{b'} \phi^* A\phi\,dt.$$

The limit at $x = a$ is 0. Hence

$$\phi(b')^*(J/i)\phi(b') > 0,$$

when $\mathrm{Im}\,\lambda > 0$, and

$$\phi(b')^*(J/i)\phi(b') < 0,$$

when $\mathrm{Im}\,\lambda < 0$. In either case $\mathbf{D} > 0$. $\square$

VII.3.3. Lemma. $\mathbf{B}^*\mathbf{D}\mathbf{B} - \mathbf{A} = \overline{\mathbf{D}}^{-1} > 0$, *where* $\overline{\mathbf{D}}^{-1} = \mathbf{D}^{-1}(\overline{\lambda})$.

Proof. Note that for λ and $\overline{\lambda}$

$$\mathcal{Y}(x, \overline{\lambda})^* J\mathcal{Y}(x, \lambda) = J$$

for all x. Hence

$$-J\mathcal{Y}(x, \overline{\lambda})^* J\mathcal{Y}(x, \lambda) = I,$$

and

$$(J\mathcal{Y}(x, \lambda))(-J\mathcal{Y}(x, \overline{\lambda})^*) = I$$

as well. Multiplying by J yields

$$\mathcal{Y}(x, \lambda))J\mathcal{Y}(x, \overline{\lambda})^* = J$$

for all x. As a result

$$\begin{aligned}
J &= \mathcal{Y}(x, \lambda)^* J\mathcal{Y}(x, \overline{\lambda}) \\
&= \mathcal{Y}(x, \lambda)^*[-J\mathcal{Y}(x, \lambda)J\mathcal{Y}(x, \overline{\lambda})^* J]\mathcal{Y}(x, \overline{\lambda}) \\
&= -[\mathcal{Y}(x, \lambda)^*(J/i)\mathcal{Y}(x, \lambda)]J[-\mathcal{Y}(x, \overline{\lambda})^*(J/i)\mathcal{Y}(x, \overline{\lambda})].
\end{aligned}$$

Or

$$\begin{pmatrix} 0 & -I_n \\ I_n & 0 \end{pmatrix} = -\begin{pmatrix} \mathbf{A} & \mathbf{B}^* \\ \mathbf{B} & \mathbf{D} \end{pmatrix} \begin{pmatrix} 0 & -I_n \\ I_n & 0 \end{pmatrix} \begin{pmatrix} \overline{\mathbf{A}} & \overline{\mathbf{B}}^* \\ \overline{\mathbf{B}} & \overline{\mathbf{D}} \end{pmatrix},$$

where $-$ indicates that $\overline{\lambda}$ replaces λ. (Remember that there is a sign change in the matrix when λ replaces $\overline{\lambda}$.) Hence

$$0 = \mathbf{A}\overline{\mathbf{B}} - \mathbf{B}^*\overline{\mathbf{A}}, \quad -I_n = \mathbf{A}\overline{\mathbf{D}} - \mathbf{B}^*\overline{\mathbf{B}},$$
$$I_n = \mathbf{B}\overline{\mathbf{B}} - \mathbf{D}\overline{\mathbf{A}}, \qquad 0 = \mathbf{B}\overline{\mathbf{D}} - \mathbf{D}\overline{\mathbf{B}}^*.$$

The last yields

$$\overline{\mathbf{B}}^*\overline{\mathbf{D}}^{-1} = \mathbf{D}^{-1}\mathbf{B}.$$

The second shows

$$\overline{\mathbf{D}}^{-1} = -\mathbf{A} + \mathbf{B}^*[\overline{\mathbf{B}}^*\overline{\mathbf{D}}^{-1}]$$
$$= -\mathbf{A} + \mathbf{B}^*\mathbf{D}^{-1}\mathbf{B}. \qquad \square$$

VII.3.4. Corollary. $R_2 = \overline{R}_1$.

Note that since the coefficient matrices A and B are real, $\mathcal{Y}(x, \overline{\lambda}) = \overline{\mathcal{Y}(x, \lambda)}$ and so also for $\mathbf{A}, \mathbf{B}, \mathbf{D}$. Hence $R_2 = \overline{R}_1$.

VII.3.5. Theorem. *As b' increases, $\mathbf{D}$ decreases, R_1 decreases, and R_2 decreases.*

Proof. Note that

$$\mathbf{D} = 2|\mathrm{Im}\ \lambda| \int_a^{b'} \phi^* A \phi \, dt.$$

The results are then immediate. $\qquad \square$

VII.3.6. Theorem. $\lim\limits_{b' \to b} R_1(b', \lambda) = R_0(\lambda),\ \lim\limits_{b' \to b} R_2(b', \lambda) = R_0(\overline{\lambda}) = \overline{R}_0$ *exist.* $R_0 \geq 0;\ \overline{R}_0 \geq 0$.

VII.3.7. Theorem. *As b' approaches b, the circles $\mathbf{E}(M) = 0$ are nested;* $\lim\limits_{b' \to b} C(b', \lambda) = C_0$ *exists.*

Proof. The interior of the circle $\mathbf{E}(M) = 0$ is given by $\mathbf{E}(M) \leq 0$, or by

$$\pm(IM^*)\mathcal{Y}(b')^*(J/i)\mathcal{Y}(b')\begin{pmatrix} I \\ M \end{pmatrix} \leq 0.$$

Using the differential equation $JY' = (\lambda A + B)Y$, we find

$$\mathbf{E}(M) = 2|\mathrm{Im}\ \lambda| \int_a^{b'} \chi_{b'}^* A \chi_{b'} \, dt \pm (M^* - M)/i.$$

Now, if M is in the circle at $b'' > b'$, then $\mathbf{E}(M) \leq 0$ at b''. At b', $\mathbf{E}(M)$ is certainly smaller, and so M is in the circle at b' as well. The circles are nested. $\qquad \square$

To show that the centers converge, we need to solve the circle equation

$$(M - C)^* R_1^{-2}(M - C) = \overline{R}_1^{-2}.$$

(Recall that $\mathbf{D}$, R_1 and $\overline{R}_1$ are self-adjoint matrices.) This is equivalent to

$$[R_1^{-1}(M - C)\overline{R}_1^{-1}]^*[R_1^{-1}(M - C)\overline{R}_1^{-1}] = I.$$

Therefore the bracketed term

$$R_1^{-1}(M - C)\overline{R}_1^{-1} = U,$$

a unitary matrix, and

$$M = C + R_1 U \overline{R}_1.$$

As U varies over the $n \times n$ unit sphere, M varies over a "circle" with center C. We shall have more to say about the range of M later.

Now let C_1 be the center at b', C_2 be the center at b''. If

$$M_1 = C_1 + R_1(b')U_1\overline{R}_1(b'),$$

and

$$M_2 = C_2 + R_1(b'')U_2\overline{R}_2(b''),$$

then M_2 lies in the b' circle as well, and

$$M_2 = C_1 + R_1(b')V_1\overline{R}_1(b'),$$

where V_1 is a contraction. Thus

$$C_1 - C_2 = R_1(b'')U_2\overline{R}_1(b'') - R_1(b')V_1\overline{R}_1(b').$$

Consider the mapping defined by the equations for M_2 above, defining V_1 in terms of U_2: $V_1 = F(U_2)$. It is a continuous transformation of the unit sphere into itself. It therefore has a unique fixed point U. Letting U_2 and V_1 be replaced by U, we find

$$\begin{aligned}
\|C_1 - C_2\| &= \|R_1(b'')U\overline{R}_1(b'') - R_1(b')U\overline{R}_1(b')\| \\
&\leq \|R_1(b'')U\overline{R}_1(b'') - R_1(b'')U\overline{R}_1(b')\| \\
&\quad + \|R_1(b'')U\overline{R}_1(b') - R_1(b')U\overline{R}_1(b')\| \\
&\leq \|R_1(b'')\|\,\|\overline{R}_1(b'') - \overline{R}_1(b')\| \\
&\quad + \|R_1(b'') - R_1(b')\|\,\|\overline{R}_1(b')\|.
\end{aligned}$$

As b' and b'' approach b, R_1 and $\overline{R}_1$ have limits. The centers then form a Cauchy sequence and converge.

A computation shows

$$B = \pm \left[2\mathrm{Im}\ \lambda \int_a^{b'} \phi^* A\theta dt - iI_n \right].$$

So at b', the center

$$C = -D^{-1}B = - \left[2\mathrm{Im}\ \lambda \int_a^{b'} \phi^* A\phi dt \right]^{-1} \left[2\mathrm{Im}\ \lambda \int_a^{b'} \phi^* A\theta dt - iI_n \right].$$

As we have seen, as b' approaches b, this has a limit C_0.

The limiting "circle" equation may not exist because both R_0 and $\overline{R}_0$ may have rank less than n and may be singular. Nonetheless

$$M = C_0 + R_0 U \overline{R}_0$$

is perfectly well defined. As U varies over the unit circle in $n \times n$ space, the limit "circle" or "point" $\mathbf{E}_0(M)$ is covered.

VII.4 Square Integrable Solutions

As a consequence of the previous section we have the following theorem.

VII.4.1. Theorem. *Let M be a point inside $\mathbf{E}_0(M) \le 0$. Let $\chi = \theta + \phi M$. Then χ is in $L_A^2(a,b)$.*

Proof.

$$2|\mathrm{Im}\ \lambda| \int_a^{b'} \chi^* A\chi dt \pm [M^* - M]/i = \mathbf{E}(M) \le 0.$$

As a result

$$0 \le \int_a^{b'} \chi^* A\chi dt \le [M - M^*]/2i\mathrm{Im}\ \lambda.$$

The upper bound is fixed, so let $b' \to b$.

The number of square integrable solutions depends upon the flexibility of M. Remember that

$$M = C_0 + R_0 U \overline{R}_0,$$

where R_0 and $\overline{R}_0$ are decreasing matrices, and U is either unitary or a contraction.

$\square$

Lemma. *Let rank $\overline{R}_0 = r$, let $S(U) = R_0 U \overline{R}_0$ where U is unitary. Then*

$$rank\ S(U) \le r.$$

Proof. This is well known. $\square$

VII.4.2. Theorem. *Under the conditions of Theorem VII.4.1,* $\sup\limits_{u} rank\ S(U) = r$.

Proof. The rank of $S(U)$ is the dimension of the image of $R_0 U \overline{R}_0$ acting on C^n. Let $\overline{R}_0$ acting on C^n be the subspace W of dimension r. Further note that there is a subspace X of C^n such that $\dim R_0 X = \operatorname{rank} R_0 = r$. Since $\dim X \geq \dim W$, there is a unitary matrix $U : W \to X$, injectively. Then $U(W)$ is a subspace of X of dimension r, and $\dim(R_0 U(W)) = \dim W = r$.

$$\operatorname{rank} R_0 U \overline{R}_0 = r. \qquad \qquad \Box$$

VII.4.3. Theorem. *Let* $m = n + r$. *Let* $Im\ \lambda \neq 0$. *Then there exist* m *solutions of* $JY' = (\lambda A + B)Y$ *in* $L_A^2(a,b)$, $n \leq m \leq 2n$.

Proof. $\theta + \phi C_0$ is made up of n solutions in $L_A^2(a,b)$. As U varies, $\phi(R_0 U \overline{R}_0)$ gives an additional $m - n$ solutions which are independent of the others. The number is the same for $Im\ \lambda > 0$ or $Im\ \lambda < 0$ by the reflection principle.

By way of example consider the following chart:

n	rank R_0	rank $\overline{R}_0$	m	Case name
1	1	1	2	limit-circle
1	0	0	1	limit-point
2	2	2	4	limit-circle
2	1	1	3	intermediate
2	0	0	2	limit-point
3	3	3	6	limit-circle
3	2	2	5	intermediate
3	1	1	4	intermediate
3	0	0	3	limit-point
etc.				

These cases actually occur in practice for scalar problems. Question: Is it possible for rank $R_0 \neq$ rank $\overline{R}_0$ when A and B are complex?

In closing this section we include a rather interesting theorem concerning the eigenvalues of $\mathbf{D}(b', \lambda)$ as b' approaches b. $\Box$

VII.4.4. Theorem. *Let* $\mu_1(b') \leq \cdots \leq \mu_n(b')$ *be the eigenvalues of* $\mathbf{D}(b', \lambda)$. *Let there be* m, $n \leq m \leq 2n$, *solutions of* $JY' = (\lambda A + B)Y$, $Im\ \lambda \neq 0$, *in* $L_A^2(a,b)$. *Then* $\mu_1(b'), \ldots, \mu_{m-n}(b')$ *remain finite and* $\mu_{m-n+1}(b'), \ldots, \mu_{2n}(b')$ *approach* ∞ *as* b' *approaches* b. *(See also [Hinton and Shaw, 19].)*

Proof. Suppose $\mu(b') < B$ for all b'. Let $v_{b'}$ be a unit eigenvector of $\mathbf{D}(b', \lambda)$. Setting $\chi_{b'} = \phi v_{b'}$ yields

$$2i\mathrm{Im}\,\lambda \int_a^{b'} \chi_{b'}^* A \chi_{b'}\, dx = v_{b'}^* \phi^* J \phi v_{b'} \Big|_a^{b'},$$

$$= i\,\mathrm{sgn}\,(\mathrm{Im}\,\lambda) \cdot \mu(b')\,.$$

Thus

$$\int_a^{b'} \chi_{b'}^* A \chi_{b'}\, dx = \mu(b')/|\mathrm{Im}\,\lambda|,$$

$$\leq B/|\mathrm{Im}\,\lambda|.$$

Choosing a subsequence of $(v_{b'})$'s that converge, we find a solution $\chi = \phi v$ in $L_A^2(a, b)$. Since there are only m L_A^2 solutions and $\chi = \theta + \phi M$ comprises n of these, there can only be $m - n$ such χ's and only $m - n$ finite μ's.

Theorem VII.4.4 should be compared with Niessen's theorem [25], which considers eigenvalues of

$$\begin{pmatrix} A & B^* \\ B & D \end{pmatrix}.$$

$\qquad\square$

VII.5 Singular Boundary Conditions

Unlike regular boundary points, such as b', where a regular boundary condition $(\beta_1 \beta_2) Y(b') = 0$ is imposed, singular boundary points, such as b, require a more careful approach. We define a singular boundary value that is valid on the domain of the maximal operator.

VII.5.1. Definition. We denote by D_M those elements Y in $L_A^2(a, b)$ satisfying

$$(1) \qquad lY = JY' - BY = AF \text{ exists a.e. for some } F \text{ in } L_A^2(a, b).$$

We define the maximal operator L_M by setting $L_M Y = F$ for all Y in D_M.

The following is a restatement of Theorem V.3.1. It is repeated here for the reader who is interested in jumping in without reading earlier material.

VII.5.2. Theorem. *Let Y_j be a solution of*

$$JY_j' = (\bar{\lambda}_0 A + B)Y_j\,, \quad Im\,\lambda_0 \neq 0\,.$$

Then for all Y in D_M,

$$B_{bj}(Y) = \lim_{x \to b} Y_j^* J Y$$

exists if and only if Y_j is in $L_A^2(a, b)$.

Proof. Manipulation of

$$JY' - BY = AF$$

and

$$JY_j' - BY_j = \overline{\lambda} AY_j$$

yields

$$(Y_j^* JY)' = Y_j^* A[F - \lambda Y].$$

Integrating, we get

$$(Y_j^* JY)(x) = Y_j^* JY(a) + \int_a^x Y_j^* A[F - \lambda Y]dx.$$

If Y_j is in $L_A^2(a,b)$, then as x approaches b, the integral on the right converges, and

$$\lim_{x \to b} Y_j^* JY(x)$$

exists.

Conversely, if the integral on the right converges for all Y, F in $L_A^2(a,b)$, the Hahn–Banach theorem states that it generates a bounded operator. The Riesz representation theorem then affirms that Y_j is in $L_A^2(a,b)$. $\square$

We now assume that λ_0, Im $\lambda_0 \neq 0$, is fixed. We cannot stress enough that λ_0 is not permitted to vary.

VII.5.3. Definition. Let Im $\lambda_0 \neq 0$. Let $M(\overline{\lambda}) = \overline{C}_0 + \overline{R}_0 U R_0$ be on the limit-circle. Let $\chi(x, \overline{\lambda}_0) = \theta(x, \overline{\lambda}_0) M(\overline{\lambda}_0)$ satisfy $JY' = (\lambda_0 A + B)Y$ and be in $L_A^2(a,b)$. We define the boundary value $B_{\lambda_0}(Y)$ by setting

$$B_{\lambda_0}(Y) = \lim_{x \to b} \chi(x, \overline{\lambda}_0)^* JY(x)$$

for all Y in D_M.

Note that $B_{\lambda_0}(Y)$ is explicitly written as

$$B_{\lambda_0}(Y) = \lim_{x \to b} (I_n M(\overline{\lambda}_0)^*) \mathcal{Y}(x, \overline{\lambda}_0)^* JY(x).$$

Note also that $\overline{\lambda}_0$ is used in the definition of $B_{\lambda_0}(Y)$. This is a convenience only. We shall remove the requirement later.

VII.6   The Differential Operator L

We assume that the number of solutions $JY = (\lambda A + BY)$ in either half plane Im $\lambda > 0$ and Im $\lambda < 0$ is m, as given by Theorems VII.4.3 and VII.4.4. Using boundary conditions at both a and b, we can now define a self-adjoint differential operator on $L_A^2(a,b)$.

VII.6.1. Definition. We denote by D those elements Y in $L_A^2(a,b)$ satisfying

(1) $lY = JY' - BY = AF$ exists a.e. for some F in $L_A^2(a,b)$,

(2) $(\alpha_1 \alpha_2) Y(a) = 0$, and

(3) $B_{\lambda_0}(Y) = 0$.

VII.6.2. Definition. We define the operator L by setting $LY = F$ if and only if $lY = JY' - BY = AF$ for all Y in D.

The inverse of $L - \lambda_0 I$ can now be calculated easily. If we solve

$$JY' = (\lambda_0 A + B)Y + AF$$

by variation of parameters, the substitution of $Y = \mathcal{Y}C$, where $\mathcal{Y}$ is the fundamental matrix, yields

$$J\mathcal{Y}C = AF,$$

so

$$C' = -J\mathcal{Y}^t AF,$$

where we have written $\mathcal{Y}^t(x, \lambda_0)$ for $\mathcal{Y}(x, \overline{\lambda_0})^*$, the transpose. Integrating, we find

$$Y = -\mathcal{Y}(x, \lambda_0) \int_a^x J\mathcal{Y}^t(\xi, \lambda_0) A(\xi) F(\xi) d\xi + \mathcal{Y}(x, \lambda_0) K,$$

where K is constant.

Since $\begin{pmatrix} \alpha_1 & \alpha_2 \\ 0 & 0 \end{pmatrix} Y(a) = 0$,

$$0 + \begin{pmatrix} \alpha_1 & \alpha_2 \\ 0 & 0 \end{pmatrix} \begin{pmatrix} \alpha_1^* & -\alpha_2^* \\ \alpha_2^* & \alpha_1^* \end{pmatrix} K = 0,$$

or

$$\begin{pmatrix} I & 0 \\ 0 & 0 \end{pmatrix} K = 0.$$

Further,

$$\begin{pmatrix} 0 & 0 \\ I_n & M^t \end{pmatrix} \mathcal{Y}^t(x, \lambda_0) JY(x)$$

$$= -\begin{pmatrix} 0 & 0 \\ I_n & M^t \end{pmatrix} \mathcal{Y}^t(x, \lambda_0 (J\mathcal{Y}(x, \lambda) \int_a^x J\mathcal{Y}^t(\xi, \lambda_0) A(\xi) F(\xi) d\xi$$

$$+ \begin{pmatrix} 0 & 0 \\ I_n & M^t \end{pmatrix} \mathcal{Y}^t(x, \lambda_0) J\mathcal{Y}(x, \lambda_0) K.$$

Since $\mathcal{Y}^t(x, \lambda_0) J\mathcal{Y}(x, \lambda_0) \equiv J$ for all x,

$$\begin{pmatrix} 0 & 0 \\ I_n & M^t \end{pmatrix} \mathcal{Y}^t(x, \lambda_0) JY(x)_0 = \begin{pmatrix} 0 & 0 \\ I_n & M^t \end{pmatrix} \int_a^x \mathcal{Y}^t(\xi, \lambda_0) A(t) F(t) dt$$

$$+ \begin{pmatrix} 0 & 0 \\ I_n & M^t \end{pmatrix} \begin{pmatrix} 0 & -I_n \\ I_n & 0 \end{pmatrix} K.$$

Letting $x \to b$, we get

$$0 = \int_a^b \begin{pmatrix} 0 & 0 \\ I_n & M^t \end{pmatrix} \mathcal{Y}^t(\xi, \lambda_0) A(\xi) F(\xi) d\xi + \begin{pmatrix} 0 & 0 \\ M_t & -I^n \end{pmatrix} K.$$

Adding

$$0 = \begin{pmatrix} I_n & 0 \\ 0 & 0 \end{pmatrix} K$$

to this, we have

$$0 = \int_a^b \begin{pmatrix} 0 & 0 \\ I_n & M^t \end{pmatrix} \mathcal{Y}^t(\xi, \lambda_0) A(\xi) F(\xi) d\xi + \begin{pmatrix} I_n & 0 \\ M_t & -I^n \end{pmatrix} K.$$

The coefficient of K is its own inverse. Hence

$$K = \int_a^b \begin{pmatrix} 0 & 0 \\ I_n & M^t \end{pmatrix} \mathcal{Y}^t(\xi, \lambda_0) A(\xi) F(\xi) d\xi,$$

and

$$Y = \mathcal{Y}(x, \lambda_0) \int_a^x \begin{pmatrix} 0 & I_n \\ 0 & M^t \end{pmatrix} \mathcal{Y}^t(\xi, \lambda_0) A(\xi) F(\xi) d\xi$$

$$+ \mathcal{Y}(x, \lambda_0) \int_x^b \begin{pmatrix} 0 & 0 \\ I_n & M^t \end{pmatrix} \mathcal{Y}^t(\xi, \lambda_0) A(\xi) F(\xi) d\xi.$$

At this point we note that the Green's function, the kernel of the integral operator, is the limit of Green's functions of regular Sturm–Liouville problems. Since regular problems satisfy

$$\mathbf{G}(\lambda_0, x, \xi) = \mathbf{G}^*(\overline{\lambda}_0, \xi, x),$$

so does the present Green's function. Comparison of the two sides gives a definition of $M(\lambda)$ in terms of $M(\overline{\lambda})$,

$$M(\lambda_0) = M^t = M^*(\overline{\lambda}_0),$$

and $\chi(x, \lambda_0) = \theta(x, \lambda_0) + \phi(x, \lambda_0) M(\lambda_0)$ is in $L_A^2(a, b)$. Now from the first integral, the terms

$$\mathcal{Y}(x, \lambda_0) \begin{pmatrix} 0 & I_n \\ 0 & M^t \end{pmatrix} \mathcal{Y}^*(\xi, \lambda_0) = \chi(x, \lambda_0) \phi^t(\xi, \lambda_0).$$

From the second, recalling that $M^t = M$, we have

$$\mathcal{Y}(x, \lambda_0) \begin{pmatrix} 0 & 0 \\ I_n & M^t \end{pmatrix} \mathcal{Y}^t(\xi, \lambda_0) = \phi(x, \lambda_0) \chi^t(\xi, \lambda_0).$$

Therefore

$$Y(x) = \chi(x, \lambda_0) \int_a^x \phi(\xi, \overline{\lambda}_0)^* A(\xi) F(\xi) d\xi$$

$$+ \phi(x, \lambda_0) \int_x^b \chi(\xi, \overline{\lambda}_0)^* A(\xi) F(\xi) d\xi,$$

or

$$Y = \int_a^b \mathbf{G}(\lambda_0, x, \xi) A(\xi) F(\xi) d\xi \,,$$

where

$$\mathbf{G}(\lambda_0, x, \xi) = \chi(x, \lambda_0) \phi(\xi, \overline{\lambda}_0)^* \,, \qquad a \leq \xi \leq x \leq b \,,$$
$$= \phi(x, \lambda_0 (\chi(\xi, \overline{\lambda}_0)^* \,, \qquad a \leq x \leq \xi \leq b \,.$$

VII.6.3. Theorem. *λ_0, used in defining the boundary condition B_{λ_0}, is in the resolvent of L,*

$$(L - \lambda_0 I)^{-1} F(x) = \int_0^b \mathbf{G}(\lambda_0, x, \xi) A(\xi) F(\xi) d\xi \,,$$

where $\mathbf{G}$ is given above.

VII.6.4. Theorem. *L is self-adjoint.*

Proof. Let $(L - \lambda_0 I) {=} F$, $(L^* - \overline{\lambda}_0 I) Z = G$. Then

$$\langle (L - \lambda_0 I)^{-1} F, G \rangle = \int_a^b G^*(x) A(x) \int_a^b \mathbf{G}(\lambda_0, x, \xi) A(\xi) F(\xi) d\xi dx$$
$$= \int_a^b \left[\int_a^b \mathbf{G}^*(\lambda_0, x, \xi) A(x) G(x) dx \right]^* A(\xi) F(\xi) d\xi$$
$$= \int_a^b [\mathbf{G}^*(\lambda_0, \xi, x) A(\xi) G(\xi) d(\xi]^* A(x) F(x) dx$$
$$= \int_a^b \left[\int_a^b \mathbf{G}(\overline{\lambda}_0, x, \xi) A(\xi) G(\xi) d\xi \right]^* A(x) F(x) dx$$
$$= \langle F, (L - \overline{\lambda}_0 I)^{-1} G \rangle \,,$$

since $\mathbf{G}(\lambda_0, x, \xi) = \mathbf{G}(\overline{\lambda}_0, \xi, x)^*$. But

$$\langle (L - \lambda_0 I)^{-1} F, G \rangle = \langle F, (L^* - \overline{\lambda}_0 I) G \rangle \,,$$

so

$$(L - \overline{\lambda}_0 I)^{-1} = (L^* - \overline{\lambda}_0 I)^{-1} \,.$$

From inverses, $L - \overline{\lambda}_0 I = L^* - \overline{\lambda}_0 I$. Cancelling $\overline{\lambda}_0 I$, we get $L = L^*$.

We remark that when $M(\overline{\lambda}_0)$ is on the limit-circle, then

$$\lim_{x \to b} \chi(x, \overline{\lambda}_0)^* J \chi(x, \overline{\lambda}_0) = 0.$$

This is a test for self-adjointness of the singular boundary condition $B_{\overline{\lambda}_0}(Y)$. It is the limiting form of $(B_1 B_2) J (B_1 B_2)^* = 0$, required in the regular case. $\square$

VII.6.5. Theorem. $(L - \lambda_0 I)^{-1}$ *is a bounded operator.* $\|(L - \lambda_0 I)^{-1}\| \leq 1/|Im\ \lambda_0|$.

Proof. Let $(L - \lambda_0 I)Y = F$. Then

$$\langle Y, F \rangle - \langle F, Y \rangle = \langle Y, (L - \lambda_0 I)Y \rangle - \langle (L - \lambda_0 I)Y, Y \rangle$$
$$= (\lambda_0 - \overline{\lambda}_0)\langle Y, Y \rangle \,.$$

Applying Schwarz's inequality to the left, we have

$$2|Im\ \lambda_0|\ \|Y\|^2 \leq 2\|Y\|\ \|F\| \,.$$

Cancelling $\|Y\|$ yields the result. (See Atkinson [1].) $\square$

VII.6.6. Theorem. *If* $JY' - BY = AF$, $AY = 0$ *implies* $Y = 0$, *then* D *is dense in* $L_A^2(a, b)$.

Proof. If D is not dense, then there is a G orthogonal to D. Let Y be in D, and let Z satisfy $Z \in D$, $JZ' - BZ = \overline{\lambda}_0 AZ + AG$ for $Im\ \lambda_0 \neq 0$. Then

$$0 = \langle Y, G \rangle = \int_a^b G^* AY\, dx$$
$$= \int_a^b [JZ' - BZ - \overline{\lambda}_0 AZ]^* Y\, dx$$
$$= \int_a^b Z^*[JY' - BY - \lambda_0 AY]\, dx \,.$$

Let $JY' - BY - \lambda_0 AY = AF$. Then

$$0 = \langle F, Z \rangle = \int_a^b Z^* AF\, dx \,.$$

Now F is arbitrary, so let $F = Z$. Thus $\int_a^b Z^* AZ\, dx = 0$ and $AZ = 0$. Thus $JZ' - BZ = AG$, and $Z = 0$. Since $Z = 0$, $AG = 0$ and $G = 0$ in $L_A^2(a, b)$. $\square$

VII.7 Extension of the Boundary Conditions

We have chosen $\overline{\lambda}_0$ to be fixed in generating the boundary condition at $x = b$. We would like to show now that properly extended, $\chi(x, \lambda) = \theta(x, \lambda) + \phi(x, \lambda)M(\lambda)$ remains in $L_A^2(a, b)$ and that if $\lim_{x \to b} \chi(x, \overline{\lambda}_0)^* JY(x) = 0$, then for all λ, $Im\ \lambda \neq 0$, $\lim_{x \to b} \chi(x, \overline{\lambda})^* JY(x) = 0$. In a sense, the boundary condition is independent of λ.

In the course of deriving the Green's function, we showed that $M(\lambda_0) = M(\overline{\lambda}_0)^*$. As a consequence we find the following.

VII.7.1. Theorem. *If $\chi(x, \lambda_0) = \theta(x, \lambda_0) + \phi(x, \lambda_0)M(\lambda_0)$, then*

$$\lim_{x \to b} \chi(x, \overline{\lambda}_0)^* J \chi(x, \lambda_0) = 0.$$

Proof.

$$\chi(x, \overline{\lambda}_0)^* J \chi(x, \lambda_0) = (I, M(\overline{\lambda}_0)^*)\mathcal{Y}(x, \overline{\lambda}_0)^* J \mathcal{Y}(x, \lambda_0) \begin{pmatrix} I \\ M(\lambda_0) \end{pmatrix}$$

$$= (I, M(\overline{\lambda}_0)^*) J \begin{pmatrix} I \\ M(\lambda_0) \end{pmatrix}$$

$$= 0,$$

so the limit is 0 also. $\qquad\qquad\square$

VII.7.2. Corollary. *Let the columns of $\chi(x, \lambda_0)$ be modified smoothly so that they vanish near $x = a$. Then, so modified, the columns of $\chi(x, \lambda_0)$ are in D.*

VII.7.3. Theorem. *If Y is in D, then*

$$\lim_{x \to b} \chi(x, \lambda_0)^* J Y(x) = 0.$$

Proof. *Since each column of $\chi(x, \lambda_0)$, appropriately modified, is in D, an application of Green's formula shows the limit is 0.*

VII.7.4. Definition. *We extend the definition of $\chi(x, \overline{\lambda}_0)$ to other values $\overline{\mu}$ in the same half plane as $\overline{\lambda}$, by*

$$\chi(x, \overline{\mu}) = \chi(x, \overline{\lambda}_0) + (\overline{\mu} - \overline{\lambda}_0) \int_a^b \mathbf{G}(\overline{\lambda}_0, x, \xi) A(\xi) \chi(\xi, \overline{\mu}) d\xi.$$

It is well known that if

$$|\overline{\mu} - \overline{\lambda}_0| < |1/\mathrm{Im}\,\lambda_0|,$$

then with initial estimate $\chi(x, \overline{\lambda}_0)$, a Neumann series may be used to show that the integral equation has a unique solution, analytic in $\overline{\mu}$ and in $L_A^2(a, b)$. Further, this solution can be extended analytically through the entire half plane containing $\overline{\lambda}_0$. It is also easy to show that $(L - \overline{\mu})\chi(x, \overline{\mu}) = 0$.

A simple calculation shows that $\chi(x, \overline{\mu}) = \theta(x, \overline{\mu}) + \phi(x, \overline{\mu})M(\overline{\mu})$, where

$$M(\overline{\mu}) = M(\overline{\lambda}_0) + (\overline{\mu} - \overline{\lambda}_0) \int_a^b \chi(x, \overline{\lambda}_0)^* A(\xi) \chi(\xi, \overline{\mu}) d\xi,$$

thus extending $M(\overline{\mu})$ as well.

VII.7.5. Theorem. $\displaystyle\lim_{x \to b} \chi(x, \overline{\lambda}_0)^* J \chi(x, \overline{\mu}) = 0.$

Proof. If $\lim\limits_{x\to b}\chi(x,\overline{\lambda}_0)^*J$ is applied to the integral equation, both terms on the right have limit zero. Extended analytically in $\overline{\mu}$, this remains 0.   □

VII.7.6. Corollary. *Let the columns of $\chi(x,\overline{\mu})$ be modified smoothly so that they vanish near $x = a$. Then, so modified, the columns of $\chi(x,\overline{\mu})$ are in D.*

VII.7.7. Theorem. *If Y is in D, then*

$$\lim_{x\to b}\chi(x,\overline{\mu})^*JY(x) = 0.$$

VII.7.8. Definition. *We extend the definition of $\chi(x,\mu)$ to other values μ in the same half plane as λ_0 by*

$$\chi(x,\mu) = \chi(x,\lambda) + (\mu - \lambda)\int_a^b \mathbf{G}(\lambda,x,\xi)A(\xi)\chi(\xi,\mu)d\xi.$$

The comments made after Definition VII.7.4 apply here as well, with the conjugate signs removed.

VII.7.9. Theorem. $\lim\limits_{x\to b}\chi(x,\lambda_0)^*J\chi(x,\mu) = 0.$

VII.7.10. Corollary. *Let the columns of $\chi(x,\mu)$ be modified smoothly so that they vanish near $x = a$. Then, so modified, $\chi(x,\mu)$ is in D.*

Proof. Clearly the columns are in D, associated with λ_0. But D, defined by $(B_{\lambda_0}(Y))$, and D, defined by $B_{\overline{\lambda}_0}(Y)$, are the same.   □

VII.7.11. Theorem. *If Y is in D, then*

$$\lim_{x\to b}\chi(x,\mu)JY(x) = 0.$$

Proof. As in Corollary VII.7.10, let the columns of $\chi(x,\mu)$ be modified smoothly so that $\chi(x,\mu)$ is in D. Let $L\chi = G$, and let $LY = F$. So $J\chi' - B\chi = AG$, and $JY' - BY = AF$. We compute the Green's formula

$$\begin{aligned}
0 &= \langle LY, \chi\rangle - \langle Y, L\chi\rangle \\
&= \int_a^b [\chi^*AF - G^*AY]dx \\
&= \int_a^b [\chi^*(JY' - BY) - (J\chi' - B\chi)^*Y]dx \\
&= \int_a^b [\chi^*JY' + \chi^{*'}JY]dx \\
&= \lim_{x\to b}\chi^*(x,\mu)JY(x).
\end{aligned}$$

                    □

In summary, we state the following.

VII.7.12. Theorem. *Let Y be in D. Then for all λ, Im $\lambda \neq 0$, $B_\lambda(Y) = 0$.*

VII.7.13. Theorem. *For all λ, Im $\lambda \neq 0$, $(L - \lambda I)^{-1}$ exists and is given by the formula in Theorem VII.6.3.*

$$\|(L - \lambda I)^{-1}\| \leq 1/|Im\ \lambda|\,.$$

Proof. The previous calculation holds, no matter what λ. $\qquad\square$

Finally we comment that for all μ, Im $\mu \neq 0$, $\chi(x,\mu) = \theta(x,\mu) + \phi(x,\mu)M(\mu)$ satisfies

$$\lim_{x \to b} \chi(x,\mu)^* J \chi(x,\mu) = 0\,.$$

This implies that $M(\mu)$ is on the μ limit-circle. Further the domain D, so long as Im $(\lambda) \neq 0$, is independent of the choice of the parameter λ.

VII.8 The Extended Green's Formula with one Singular Point

We are now in a position to extend Green's formula in the general singular case. What is necessary is to properly express the Lagrange bilinear form Z^*JY in terms which have limits as x approaches b.

As we have seen earlier, if Y and Z are in D_M, then a preliminary form of Green's formula is

$$\int_a^b [Z^*(JY' - BY) - (JZ' - BZ)^*Y]dx = Z^*JY\Big|_a^b\,.$$

Since Y and Z are in D_M, they satisfy

$$JY' - (\lambda A + B)Y = AF\,,$$
$$JZ' - (\overline{\lambda} A + B)Z = AG\,,$$

where Im $\lambda > 0$ and F and G are in $L_A^2(a,b)$. Using the Green's function for the problem

$$Jy' - (\lambda A + B)y = Ay\,,$$
$$(\alpha_1 \alpha_2)y(a) = 0\,,$$
$$\lim_{x \to b} \chi_b(x,\overline{\lambda})^* Jy(x) = 0\,,$$

we can write

$$Y(x) = \int_a^b \mathbf{G}(\lambda, x, \xi)A(\xi)F(\xi)d\xi + \chi_b(x,\lambda)C_1 + \phi(x,\lambda)C_2\,,$$

$$Z(x) = \int_a^b \mathbf{G}(\overline{\lambda}, x, \xi)A(\xi)G(\xi)d\xi + \chi_b(x,\overline{\lambda})D_1 + \phi(x,\overline{\lambda})D_2\,,$$

where C_2 and D_2 are chosen so that $\phi(x,\lambda)C_2$ and $\phi(x,\overline{\lambda})D_2$ consists only of $L_A^2(a,b)$ solutions.

We can solve for C_1, C_2, D_1, D_2. Since the integral in the formula for Y, as well as the term $\chi_b(x,\lambda)C_1$, satisfy the limit boundary condition, and

$$\lim_{x\to b}\chi_b(x,\overline{\lambda})J\phi(x,\lambda) = -I,$$

$$\lim_{x\to b}\chi_b(x,\overline{\lambda})^*JY(x) = -C_2\,.$$

Further

$$\lim_{x\to b}D_2^*\phi(x,\overline{\lambda})^*JY = \lim_{x\to b}D_2^*\phi(x,\overline{\lambda})^*JR_Y(x) + D_2^*C_1\,,$$

where

$$R_Y(x) = \int_a^b \mathbf{G}(\lambda,x\xi)A(\xi)F(\xi)d\xi\,.$$

Likewise

$$\lim_{x\to b}Z(x)^*J\chi_b(x,\lambda) = D_2^*\,,$$

and

$$\lim_{x\to b}Z(x)^*J\phi(x,\lambda)C_2 = \lim_{x\to b}R_Z(x)^*J\phi(x,\lambda)C_2 - D_1^*C_2\,,$$

where

$$R_Z(x) = \int_a^b \mathbf{G}(\overline{\lambda},x,\xi)A(\xi)G(\xi)d\xi\,.$$

We then compute

$$\lim_{x\to b}Z^*(x)JY(x) = \lim_{x\to b}R_Z(x)^*J\phi(x,\lambda)C_2$$
$$+ \lim_{x\to b}D_2^*\phi(x,\overline{\lambda})^*JR_Y(x) - D_1^*C_2 + D_2^*C_1\,,$$

the other terms cancelling because they satisfy the χ boundary condition at b.

Eliminating the terms $D_1^*C_2$ and $D_2^*C_1$ by substitution, we get

$$\lim_{x\to b}Z(x)^*JY(x) = \lim_{x\to b}Z(x)^*J\phi(x,\lambda)C_2 + \lim_{x\to b}D_2^*\phi(x,\overline{\lambda})^*JY(x)\,.$$

Now the term

$$\phi(x,\lambda)C_2 = -\phi(x,\lambda)\left[\lim_{x\to b}\chi_b(x,\overline{\lambda})^*JY(x)\right]\,.$$

Let $\phi(x,\lambda) = (\phi_1\phi_2)(x)E_b$, where $\phi_1(x,\lambda)$ consists of those $m-n$ solutions of $JY' = (\lambda A + B)Y$ in $L_A^2(a,b)$, and $\phi_2(x,\lambda)$ are not in $L_A^2(a,b)$. Let

$$\chi_b(x,\overline{\lambda})^*E_b^* = (\chi_1\chi_2)(x,\overline{\lambda})\,,$$

where χ_1 contains $m - n$ components, as well. Then

$$\phi(x, \lambda)C_2 = -\phi_1(x, \lambda)\left[\lim_{x \to b} \chi_1(x, \overline{\lambda})^* JY(x)\right].$$

Likewise, the term

$$D_2^* \phi(x, \overline{\lambda})^* = \left[\lim_{x \to b} Z(x)^* J\chi_b(x, \lambda)\right]\phi(x, \overline{\lambda})^*$$

$$= \left[\lim_{x \to b} Z(x)^* J\chi_1(x, \lambda)\right]\phi_1(x, \overline{\lambda})^*,$$

where $\phi_1(x, \overline{\lambda})$ contains $m - n$ components.

The terms

$$-\phi_2(x, \lambda)\left[\lim_{x \to b}\chi_2(x, \overline{\lambda})^* JY(x)\right] \quad \text{and} \quad \left[\lim_{x \to b} Z(x)^* J\chi_2(x, \lambda)\right]\phi_2(x, \overline{\lambda})^*$$

are not present because $\phi(x, \lambda)C_2$ and $\phi(x, \overline{\lambda})D_2$ contain only solutions in $L_A^2(a, b)$. Therefore we have the following theorem.

VII.8.1. Theorem. *Let $\phi(x, \lambda) = (\phi_1\phi_2)(x, \lambda)E_b$ where ϕ_1 consists of all of the $\phi - L_A^2(a, b)$ solutions of $JY' = (\lambda A + B)Y$. Let $\phi(x, \overline{\lambda}) = (\phi_1\phi_2)(x, \overline{\lambda})\widetilde{E}_b$, where ϕ_1 consists of all of the $\phi - L_A^2(a, b)$ solutions of $JY' = (\lambda A + B)Y$ with λ replaced by $\overline{\lambda}$. Let $(\chi_1\chi_2)(x, \overline{\lambda}) = \chi_b(x, \overline{\lambda})E_b^*$ and let $(\chi_1\chi_2)(x, \lambda) = \chi_b(x, \lambda)\widetilde{E}_b^*$. Then for all Y and Z in D_M*

$$\lim_{x \to b} \chi_2(x, \overline{\lambda})^* JY(x) = 0,$$

$$\lim_{x \to b} \chi_2(x, \lambda)^* JZ(x) = 0.$$

That these expressions are 0 answers a question raised by deficiency indices. When the χ boundary conditions were imposed, it appeared that we were imposing n boundary conditions when only $n - m$ were needed. As these limits show, however, the correct number are being imposed. These which are always 0 are sometimes called *annihilators*. They are the automatic "limit point" components of the boundary conditions.

Proof. This is the only way to ensure that $\phi_2(x, \lambda)$ and $\phi_2(x, \overline{\lambda})$ are not present. Returning to $\lim_{x \to b} Z(x)^* JY(x)$, we find

$$\lim_{x \to b} Z(x)^* JY(x) = \left[\lim_{x \to b} \phi_1(x, \lambda)^* JZ(x)\right]^* \left[\lim_{x \to b} \chi_1(x, \overline{\lambda})^* JY(x)\right]$$

$$- \left[\lim_{x \to b} \chi_1(x, \lambda)^* JZ(x)\right]^* \left[\lim_{x \to b} \phi_1(x, \overline{\lambda})^* JY(x)\right]. \qquad \square$$

VII.8.2. Theorem. *Under the conditions of Theorem VII.8.1, let*

$$B_b(Y) = \begin{pmatrix} \lim_{x \to b} \chi_1(x, \overline{\lambda})^* JY(x) \\ \lim_{x \to b} \phi_1(x, \overline{\lambda})^* JY(x) \end{pmatrix} \,,$$

$$\widetilde{B}_b(Z) = \begin{pmatrix} \lim_{x \to b} \chi_1(x, \lambda)^* JZ(x) \\ \lim_{x \to b} \phi_1(x, \lambda)^* JZ(x) \end{pmatrix} \,,$$

$$J_b = \begin{pmatrix} 0 & -I_{m-n} \\ I_{m-n} & 0 \end{pmatrix} \,,$$

where (m, m) are the defect indices of $JY' = (\lambda A + B)Y$ at b. Then

$$\lim_{x \to b} Z(x)^* JY(x) = \overline{\mathbf{B}}_b(z)^* J_b B_b(Y) \,.$$

The symbol indicates λ is used instead of $\overline{\lambda}$.

Some additional comments are in order. The components that multiply each other ϕ_{1j} and χ_{1j}, $j = 1, \ldots, m - n$, are in the same Niessen subspace of solutions. The components χ_{2j}, $j = 1, 2, \ldots, 2n - m$ lie in one-dimensional subspaces, the other component ϕ_{2j} not being in $L_A^2(a, b)$.

The end $x = a$ is regular. The computation is a bit easier. Note that

$$[\chi_b(x, \lambda), \phi(x, \lambda)] \, J[\chi_b(x, \overline{\lambda}), \phi(x, \overline{\lambda})]^* = J \,.$$

Therefore

$$Z^*(a)JY(a) = -Z^*(a)J[\chi_b(a, \lambda), \phi(a, \lambda)]J[\chi_b(a, \overline{\lambda}), \phi(a, \overline{\lambda})]^* JY(a) \,.$$

VII.8.3. Theorem. *Let*

$$B_a(Y) = \begin{pmatrix} \chi_b(a, \overline{\lambda})^* JY(a) \\ \phi(a, \overline{\lambda})^* JY(a) \end{pmatrix} \,,$$

$$\widetilde{B}_a(Z) = \begin{pmatrix} \chi_b(a, \lambda)^* JZ(a) \\ \phi(a, \lambda)^* JZ(a) \end{pmatrix} \,,$$

$$J = \begin{pmatrix} 0 & -I_n \\ I_n & 0 \end{pmatrix} \,.$$

Then

$$Z^*(a)JY(a) = \widetilde{B}_a(Z)^* \, J \, B_a(Y) \,.$$

Again, some comments. The terms $\phi(a, \overline{\lambda})JY(a)$ and $\phi(a, \lambda)JZ(a)$ are simply $(\alpha_1 \alpha_2)Y(a)$ and $(\alpha_1 \alpha_2)Z(a)$. The terms $\chi_b(a, \overline{\lambda})^* JY(a)$ and $\chi_b(a, \lambda)^* JZ(a)$ may be replaced by $\theta(a, \overline{\lambda})^* JY(a)$ and $\theta(a, \lambda)^* JZ(a)$, and the formula for $Z(a)^* JY(a)$ still holds. $\theta(a, \overline{\lambda})^* JY(a) = (-\alpha_2 \alpha_1)Y(a)$; $\theta(a, \overline{\lambda})^* JZ(a) = (-\alpha_2 \alpha_1)Z(a)$ as well. These are complementary boundary forms.

If the pieces are assembled together, we have the following theorem.

VII.8.4. Theorem. *Let Y and Z be in D_M. Then*

$$\int_a^b \{Z^*[JY' - BY] - [JZ' - BZ]^*Y\}dx$$
$$= -\widetilde{B}_a(Z)^* J B_a(Y) + \widetilde{B}_b(Z)^* J_b B_b(Y)$$
$$= (\widetilde{B}_a(Z)^*, \ \widetilde{B}_b(Z)^*) \begin{pmatrix} -J_a & 0 \\ 0 & J_b \end{pmatrix} \begin{pmatrix} B_a(Y) \\ B_b(Y) \end{pmatrix} .$$

Let M and N be $r \times 2n$ and $r \times (2m - 2n)$ matrices, respectively, where (m, m) are the defect indices of $JY' = (\lambda A + B)Y$ at b, with rank $(MN) = r$, $0 \le r \le 2m$. Further let P and Q be $(2m - r) \times 2n$ and $(2m - r) \times (2m - 2n)$ matrices such that $\begin{pmatrix} M & N \\ P & Q \end{pmatrix}$ is nonsingular. If $\widetilde{M}, \widetilde{N}, \widetilde{P}, \widetilde{Q}$ are $r \times 2n$, $r \times (2m - 2n)$, $(2m - r) \times 2n$, $(2m - r) \times (2m - 2n)$ matrices such that

$$\begin{pmatrix} \widetilde{M}^* & \widetilde{P}^* \\ \widetilde{N}^* & \widetilde{Q}^* \end{pmatrix} \begin{pmatrix} M & N \\ P & Q \end{pmatrix} = \begin{pmatrix} -J_a & 0 \\ 0 & J_b \end{pmatrix};$$

then inserting this gives Green's formula in its final form.

VII.8.5. Theorem. *(Green's Formula). Let Y and Z be in D_M. Then*

$$\int_a^b \{Z^*[JY' - BY] - [JZ' - BZ]^*Y\}dx$$
$$= [\widetilde{M}\widetilde{B}_a(Z) + \widetilde{N}\widetilde{B}_b(Z)]^*[MB_a(Y) + NB_b(Y)]$$
$$+ \widetilde{P}\widetilde{B}_a(Z) + \widetilde{Q}\widetilde{B}_b(Z)]^*[PB_a(Y) + QB_a(Y)] .$$

VII.8.6. Definition. We denote by D_1 those elements Y in $L_A^2(a, b)$ satisfying

(1) Y is in D_M,

(2) $MB_a(Y) + NB_b(Y) = 0$.

We denote by L_1 the operator defined by setting $L_1 Y = F$ whenever $JY' - BY = AF$ and Y is in D_1.

VII.8.7. Definition. We denote by D_1^* those elements Z in $L_A^2(a, b)$ satisfying

(1) Z is in D_M,

(2) $\widetilde{P}\widetilde{B}_a(Z) + \widetilde{Q}\widetilde{B}_b(Z) = 0$.

We note by L^* the operator defined by setting $L_1 Z = G$ whenever $JZ' - BZ = AG$ and Z is in D_1^*.

VII.8.8. Theorem. *The abuse of notation above is correct. The adjoint of L_1 in $L_A^2(a, b)$ is L_1^*. The adjoint of L_1^* in $L_A^2(a, b)$ is L.*

Proof. The form of the adjoint of L_1 is well known to be the same as that of L_1. From Green's formula it is clear that D_1^* is included in the domain of the adjoint.

Conversely, again from Green's formula, since $PB_a(Y)+QB_b(Y)$ is arbitrary, any element in the adjoint's domain must be in D_1^*.

There are parametric boundary conditions as well. Set

$$MB_a(Y) + NB_b(Y) = 0,$$
$$PB_a(Y) + QB_v(Y) = \Gamma,$$

where Γ is arbitrary. Multiply

$$\begin{pmatrix} M & N \\ P & Q \end{pmatrix} \begin{pmatrix} B_a(Y) \\ B_b(Y) \end{pmatrix} = \begin{pmatrix} 0 \\ \Gamma \end{pmatrix}$$

by

$$\begin{pmatrix} J & 0 \\ 0 & -J_{m-n} \end{pmatrix} \begin{pmatrix} \widetilde{M}^* & \widetilde{P}^* \\ \widetilde{N}^* & \widetilde{Q}^* \end{pmatrix} .$$

The result is

$$B_a(Y) = J\widetilde{P}^*\Gamma,$$
$$B_b(Y) = -J_b\widetilde{Q}^*\Gamma.$$

Likewise, if

$$\widetilde{M}\widetilde{B}_a(Z) + \widetilde{N}\widetilde{B}_b(Z) = \Delta,$$
$$\widetilde{P}\widetilde{B}_a(Z) + \widetilde{Q}\widetilde{B}_b(Z) = 0,$$

then post-multiply

$$(\widetilde{B}_a(Z)^*, \widetilde{B}_b^*(Z)) \begin{pmatrix} \widetilde{M}^* & \widetilde{P}^* \\ \widetilde{N}^* & \widetilde{Q}^* \end{pmatrix} = (\Delta^*, 0)$$

by $\begin{pmatrix} M & N \\ P & Q \end{pmatrix} \begin{pmatrix} J & 0 \\ 0 & -J_b \end{pmatrix}$. The result is

$$\widetilde{B}_a(Z) = -JM^*\Delta,$$
$$\widetilde{B}_b(Z) = J_b N^*\Delta. \qquad\qquad \square$$

VII.8.9. Theorem. *The parametric boundary conditions are fully equivalent to the original boundary conditions.*

VII.9 Self-Adjoint Boundary Value Problems With Mixed Boundary Conditions

VII.9.1. Theorem. *L is self adjoint if and only if $r = m$ and there exists an invertible transformation V such that $B_b(Y) = V\widetilde{B}_b(Y)$ and*

$$MJM^* = NVJ_bN^* .$$

Proof. Suppose L is self-adjoint. Then Y in D satisfies

$$MB_a(Y) + NB_b(Y) = 0$$

and

$$\widetilde{P}\widetilde{B}_a(Y) + \widetilde{Q}\widetilde{B}_b(Y) = 0 .$$

Since the boundary evaluations at each end must be the same, and $B_a(Y) = \widetilde{B}_a(Y)$, we find $B_b(Y) = V\widetilde{B}_b(Y)$ for all Y in D.

Then

$$\widetilde{B}_a(Y) = B_a(Y) = -JM^*\Delta ,$$
$$\widetilde{B}_b(Y) = V^{-1}B_b(Y) = J_bN^*\Delta .$$

We have

$$M(-JM^*\Delta) + N(VJ_bN^*\Delta) = 0 ,$$

or

$$[-MJM^* + NVJ_bN^*]\Delta = 0 .$$

Since Δ may be arbitrary,

$$MJM^* = NVJ_bN^* .$$

Conversely, if

$$MJM^* = NVJ_bN^* ,$$

then

$$(MN)\begin{pmatrix} -JM^* \\ VJ_bN^* \end{pmatrix} = 0 , \quad (MN)\begin{pmatrix} B_a(Y) \\ B_b(Y) \end{pmatrix} = 0 .$$

Thus there must exist a nonsingular Δ such that

$$\begin{pmatrix} -JM^* \\ VJ_bN^* \end{pmatrix} \Delta = \begin{pmatrix} B_a(Y) \\ B_b(Y) \end{pmatrix}$$

and $B_a(Y) = -JM^*\Delta$, $B_b(Y) = VJ_bN^*\Delta$. Hence $\widetilde{B}_a(Y) = -JM^*\Delta$, $\widetilde{B}_b(Y) = J_bN^*\Delta$. This implies

$$\widetilde{P}\widetilde{B}_a(Y) + \widetilde{Q}\widetilde{B}_b(Y) = 0 ,$$

and so Y is in D_1^*. A symmetrical argument then shows $D_1 = D_1^*$, and L is self-adjoint. $\qquad\square$

The existence of the transformation V can be guaranteed. We note in the evaluation of

$$\lim_{x\to b} Z^*(x)JY(x) = \left[\lim_{x\to b}\phi_1(x,\lambda)^*JZ(x)\right]^* \left[\lim_{x\to b}\chi_1(x,\overline{\lambda})JY(x)\right]$$
$$- \left[\lim_{x\to b}\chi_1(x,\lambda)^*JZ(x)\right]^* \left[\lim_{x\to b}\phi_1(x,\overline{\lambda})JY(x)\right]$$

that if this is thought of as using Z to define a boundary condition on Y, then this says that the boundary condition is linearly dependent on $B_b(Y)$.

VII.9.2. Theorem. *Let Z be in D_M,*

$$C = \left[\left(\lim_{x\to b}\phi_1(x,\lambda)^*JZ(x)\right)^*, -\left(\lim_{x\to b}\chi_1(x,\lambda)^*JZ(x)\right)\right]^*.$$

Then $B_z(Y) = \lim_{x\to b} Z(x)^*JY(x)$ satisfies

$$B_z(Y) = CB_b(Y).$$

Since all self-adjoint boundary conditions can be generated by using appropriate elements from D_M, this shows that solution generated boundary conditions also suffice.

As a special case, we let $Z = (\chi_1\phi_1)(x,\lambda)$.

VII.9.3. Theorem. $\widetilde{B}_b(Y) = VB_b(Y)$, *where*

$$V = \begin{pmatrix} \lim\limits_{x\to b}[\phi_1^*(X,\lambda)^*J\chi_1(x,\lambda)]^*, & -\lim\limits_{x\to b}[\chi_1(x,\lambda)^*J\chi_1(x,\lambda)]^* \\ \lim\limits_{x\to b}[\phi_1^*(X,\lambda)^*J\phi_1(x,\lambda)]^*, & -\lim\limits_{x\to b}[\chi_1(x,\lambda)^*J\phi_1(x,\lambda)]^* \end{pmatrix}.$$

The proof is tedious, but a mere computation. Note that the upper right V_{12} is 0. (See the definition of the limit circle.) This verifies that

$$\lim_{x\to b}[\chi_1(x,\lambda)^*JY] = V_{11}\lim_{x\to b}[\chi_1(x,\overline{\lambda})^*JY].$$

Hence if one is 0, so is the other.

VII.10   Examples

Most examples from the "real" world involve scalar problems, and those we present will be familiar, but with a slight twist. In order to have one regular end point, the intervals usually associated with these problems have been halved.

1. Consider the two-dimensional system

$$\begin{pmatrix} 0 & -1 \\ 1 & 0 \end{pmatrix}\begin{pmatrix} y_1 \\ y_2 \end{pmatrix}' = \left[\lambda\begin{pmatrix} 1 & 0 \\ 0 & 0 \end{pmatrix} + \begin{pmatrix} 0 & 0 \\ 0 & (x^2-1)^{-1} \end{pmatrix}\right]\begin{pmatrix} y_1 \\ y_2 \end{pmatrix}$$

on the interval $[0, 1)$, together with one of boundary conditions

$$O: \quad (1, 0) \begin{pmatrix} y_1(0) \\ y_2(0) \end{pmatrix} = 0,$$

or

$$E: \quad (0, 1) \begin{pmatrix} y_1(0) \\ y_2(0) \end{pmatrix} = 0$$

at $x = 0$, and

$$\lim_{x \to 1} (1, 0) \begin{pmatrix} 0 & -1 \\ 1 & 0 \end{pmatrix} \begin{pmatrix} y_1(x) \\ y_2(x) \end{pmatrix} = 0$$

at $x = 1$.

The differential system is equivalent to the Legendre differential equation. With the O boundary condition, the odd polynomial boundary value problem is generated. With the E boundary condition, the even polynomial boundary value problem arises. The boundary condition at $x = 1$ is the one satisfied by the Legendre polynomials; $x = 1$ is limit-circle.

2. Consider the two-dimensional system

$$\begin{pmatrix} 0 & -1 \\ 1 & 0 \end{pmatrix} \begin{pmatrix} y_1 \\ y_2 \end{pmatrix}' = \left[\lambda \begin{pmatrix} e^{-x^2} & 0 \\ 0 & 0 \end{pmatrix} + \begin{pmatrix} 0 & 0 \\ 0 & -e^{x^2} \end{pmatrix} \right] \begin{pmatrix} y_1 \\ y_2 \end{pmatrix}$$

on the interval $[0, \infty)$, together with one of the boundary conditions

$$O: \quad (1, 0) \begin{pmatrix} y_1(0) \\ y_2(0) \end{pmatrix} = 0,$$

or

$$E: \quad (0, 1) \begin{pmatrix} y_1(0) \\ y_2(0) \end{pmatrix} = 0$$

at $x = 0$. These give the odd (O) or even (E) Hermite polynomial problems. ∞ is the limit-point, and so no boundary condition is required.

3. Consider the four-dimensional system

$$\begin{pmatrix} 0 & 0 & -1 & 0 \\ 0 & 0 & 0 & -1 \\ 1 & 0 & 0 & 0 \\ 0 & 1 & 0 & 0 \end{pmatrix} \begin{pmatrix} y_1 \\ y_2 \\ y_3 \\ y_4 \end{pmatrix}'$$

$$= \left[\lambda \begin{pmatrix} 1 & 0 & 0 & 0 \\ 0 & 0 & 0 & 0 \\ 0 & 0 & 0 & 0 \\ 0 & 0 & 0 & 0 \end{pmatrix} + \begin{pmatrix} 0 & 0 & 0 & 0 \\ 0 & -2(1-x^2) & 1 & 0 \\ 0 & 1 & 0 & 0 \\ 0 & 0 & 0 & (1-x^2)^{-2} \end{pmatrix} \right] \begin{pmatrix} y_1 \\ y_2 \\ y_3 \\ y_4 \end{pmatrix}$$

on the interval $[0, 1)$, together with boundary conditions

$$O: \ (1,0,0,0)\begin{pmatrix} y_1(0) \\ y_2(0) \\ y_3(0) \\ y_4(0) \end{pmatrix} = 0\,, \quad (0,0,0,1)\begin{pmatrix} y_1(0) \\ y_2(0) \\ y_3(0) \\ y_4(0) \end{pmatrix} = 0$$

or

$$E: \ (0,1,0,0)\begin{pmatrix} y_1(0) \\ y_2(0) \\ y_3(0) \\ y_4(0) \end{pmatrix} = 0\,, \quad (0,0,1,0)\begin{pmatrix} y_1(0) \\ y_2(0) \\ y_3(0) \\ y_4(0) \end{pmatrix} = 0$$

at $x = 0$, and

$$\lim_{x\to 1} (\ln(1+x), (1+x)^{-1}, 0, -(1-x)^2)\begin{pmatrix} 0 & 0 & -1 & 0 \\ 0 & 0 & 0 & -1 \\ 1 & 0 & 0 & 0 \\ 0 & 1 & 0 & 0 \end{pmatrix}\begin{pmatrix} y_1 \\ y_2 \\ y_3 \\ y_4 \end{pmatrix} = 0\,,$$

$$\lim_{x\to 1} (1,0,0,0)\begin{pmatrix} 0 & 0 & -1 & 0 \\ 0 & 0 & 0 & -1 \\ 1 & 0 & 0 & 0 \\ 0 & 1 & 0 & 0 \end{pmatrix}\begin{pmatrix} y_1 \\ y_2 \\ y_3 \\ y_4 \end{pmatrix} = 0\,,$$

where at $x = 1$ the limit-4 case holds, and for $\lambda = 0$, $(1,0,0,0)^T$ and $(\ln(1+x), (1+x)^{-1}, 0, -(1-x)^2)^T$ are L^2 solutions. The problem with O boundary conditions is again the odd-degree Legendre polynomial boundary value problem. With E boundary conditions, the even-degree Legendre polynomial boundary value problem is the result.

We refer to Littlejohn and Krall [23] for additional examples. Modification to half the interval is required, but this is no problem.

References

[1] F. V. Atkinson, **Discrete and Continuous Boundary Value Problems**, Academic Press, New York, 1964.

[2] E. A. Coddington and N. Levinson, **Theory of Ordinary Differential Equations**, McGraw-Hill, N. Y., 1955.

[3] W. N. Everitt, *Integrable-square solutions of ordinary differential equations*, Quar. J. Math., Oxford (2) **10** (1959), pp. 145–155.

[4] ———, *Integrable-square solutions of ordinary differential equations II*, Quar. J. Math., Oxford (2) **13** (1962) pp. 217–220.

[5] ———, *A note on the self-adjoint domains of second-order differential equations*, Quar. J. Math., Oxford (2), **14**, (1963), pp. 41–45.

[6] ______, *Integrable-square solutions of ordinary differential equations III*, Quar. J. Math., Oxford (2) **14** (1963), pp. 170–180.

[7] ______, *Fourth order singular differential equations*, Math. Ann. **149** (1963), pp. 320–340.

[8] ______, *Singular differential equations I: The even order case*, Math. Ann. **156** (1964), pp. 9–24.

[9] ______, *Singular differential equations II: Some self-adjoint even order cases*, Quar. J. Math, Oxford **18** (1967), pp. 13–32.

[10] ______, *Legendre polynomials and singular differential operators, Lecture Notes in Mathematics*, Vol. 827, Springer-Verlag, New York, (1980), pp. 83–106.

[11] W. N. Everitt and V. K. Kumar, *On the Titchmarsh-Weyl theory of ordinary symmetric differential expressions, I and II*, Arch v. Wisk. **3** (1976), pp. 1–48, 109–145.

[12] D. B. Hinton and J. K. Shaw, *On Titchmarsh-Weyl $M(\lambda)$-functions for linear Hamiltonian systems*, J. Diff. Eq. **40** (1981), pp. 316–342.

[13] ______, *Titchmarsh-Weyl theory for Hamiltonian systems*, Spectral Theory Diff. Ops, North Holland, I. W. Knowles and R. E Lewis, eds., 1981, pp. 219–230 .

[14] ______, *On the spectrum of a singular Hamiltonian system*, Quaes. Math. **5** (1982), pp. 29–81.

[15] ______, *Titchmarsh's λ-dependent boundary conditions for Hamiltonian systems*, Lect. Notes in Math. (Springer-Verlag), 964 (1982), pp. 318–326.

[16] ______, *Well-posed boundary value problems for Hamiltonian systems of limit point or limit circle type*, Lect. Notes in Math. (Springer-Verlag), 964 (1982), pp. 614–631.

[17] ______, *Hamiltonian systems of limit point or limit circle type with both ends singular*, J. Diff. Eq. **50** (1983), pp. 444–464.

[18] ______, *Parameterization of the $M(\lambda)$ function for a Hamiltonian system of limit circle type*, Proc. Roy. Soc. Edin. **93** (1983), pp. 349–360.

[19] ______, *On boundary value problems for Hamiltonian systems with two singular points*, SIAM J. Math. Anal. **15** (1984), pp. 272–286.

[20] K. Kodaira *On ordinary differential equations of any even order and the corresponding eigenfunction expansions*, Amer. J. Math. **72** (1950), pp. 502–544.

[21] A. M. Krall, B. D. Hinton and J. K. Shaw, *Boundary conditions for differential systems in intermediate situations*, Proc. Conf. Diff. Eq. Alabama-Birmingham (1983), pp. 301–305.

[22] L. L. Littlejohn and A. M. Krall, *Orthogonal polynomials and higher order singular Sturm-Liouville systems*, Acta. Applicandae Math. **17** (1989), pp. 99–170.

[23] ______, *Orthogonal polynomials and singular Sturm-Liouville systems, II*.

[24] H. D. Niessen, *Singuläre S-hermitesche Rand-Eigenwert Probleme*, Manuscripta Math. **3** (1970), pp. 35–68.

[25] ______, *Zum verallgemeinerten zweiten Weylschen Satz*, Archiv der Math. **22** (1971), pp. 648–656.

[26] ———, *Greensche Matrix und die Formel von Titchmarch-Kodaira für singuläre S-hermitesche Eigenwert Probleme*, J. f. d. reine ang. Math. **261** (1972), pp. 164–193.

[27] E. C. Titchmarsh, **Eigenfunction Expansions**, Oxford Univ. Press, Oxford, 1962.

[28] H. Weyl, *Über gewöhnliche Differentialgleichungen mit Singularitäten und die zugehörigen Entwicklungen willkürlicher Funktionen*, Math. Ann. **68** (1910), pp. 220–269.

Chapter VIII

Hinton and Shaw's Extension
with Two Singular Points

This chapter extends the results of the previous one to cover the situation that occurs when both a and b are singular points. The technique is similar. We restrict our attention to an interval (a', b') within (a, b), develop two $M(\lambda)$ functions, one for generating L_A^2 solutions near a, one for generating L_A^2 solutions near b, by letting $a' \to a$, $b' \to b$.

We retain the assumption that if $JY' - BY = AF$ and $AY = 0$, then $Y = 0$.

As a preliminary step, consider $JY' = (\lambda A + B)Y$ over (a', b'), and impose at a' the self-adjoint boundary condition

$$(\alpha_1 \alpha_2)Y(a') = 0\,,$$

where α_1 and α_2 are $n \times n$ matrices, rank $(\alpha_1 \alpha_2) = n$, and

$$\alpha_1 \alpha_1^* + \alpha_2 \alpha_2^* = I_n\,,$$
$$\alpha_1 \alpha_2^* + \alpha_2 \alpha_1^* = 0\,.$$

Likewise at b' impose the self-adjoint boundary condition

$$(\beta_1 \beta_2)Y(b') = 0\,,$$

where $\beta_1 \beta_2$ are $n \times n$ matrices, rank $(\beta_1, \beta_2) = n$, and

$$\beta_1 \beta_1^* + \beta_2 \beta_2^* = I_n\,,$$
$$\beta_1 \beta_2^* - \beta_2 \beta_1^* = 0\,.$$

These are the most general regular, separated, self-adjoint boundary conditions. The differential equation, together with the two boundary conditions, defines a regular, self-adjoint boundary value (Sturm–Liouville) problem.

Let e be in (a', b'), and let $\mathcal{Y}(x, \lambda)$ be a fundamental matrix for $JY' = (\lambda A + B)Y$ satisfying $\mathcal{Y}(e, \lambda) = I_{2n}$. Then $\mathcal{Y}(x, \overline{\lambda})^* J\mathcal{Y}(x, \lambda) = J$ for all x. We decompose $\mathcal{Y}$ into $2n \times n$ matrices θ, ϕ, where θ and ϕ are further decomposed into $n \times n$ matrices $\theta_1, \theta_1, \phi_1, \phi_2$ as follows:

$$\mathcal{Y}(x, \lambda) = (\theta(x, \lambda), \phi(x, \lambda)) = \begin{pmatrix} \theta_1(x, \lambda) & \phi_1(x, \lambda) \\ \theta_2(x, \lambda) & \phi_2(x, \lambda) \end{pmatrix}.$$

We shall need $(J\mathcal{Y})^{-1}$ for future reference. It is easy to see that

$$(J\mathcal{Y}(x, \lambda))^{-1} = -J\mathcal{Y}(x, \overline{\lambda})^*.$$

VIII.1   M(λ) Functions, Limit-Circles, L^2 Solutions

If $\text{Im } \lambda \neq 0$, we attempt to satisfy the b' boundary condition by $\chi_b(x, \lambda) = \theta(x, \lambda) + \phi(x, \lambda)M_{b'}(\lambda)$. Insertion into $(\beta_1 \beta_2)Y(b') = 0$ shows that

$$M_{b'}(\lambda) = -[(\beta_1 \beta_2)\phi(b', \lambda))^{-1}[(\beta_1, \beta_2)\theta(b', \lambda)].$$

The inverse must exist, for otherwise the boundary value problem of $JY' = (\lambda A + B)Y$, the b' boundary condition and the e boundary condition $(I_n 0)Y(e) = 0$ would be self-adjoint, but would have a complex eigenvalue.

It is shown in the previous chapter that as b' approaches b, $M_{b'}$ can be made to approach $M_b = C_b + R_b U_b \overline{R}_b$, where

$$C_b = \lim_{b' \to b} -\left[2\text{Im } \lambda \int_e^{b'} \phi^* A\phi\, dt \right]^{-1} \left[2\text{Im } \lambda \int_e^{b'} \phi^* A\theta\, dt - iIn \right],$$

$$R_b = \lim_{b' \to b} \left[2|\text{Im } \lambda| \int_e^{b'} \phi^* A\phi\, dt \right]^{-1/2},$$

$$\overline{R}_b(\lambda) = R_b(\overline{\lambda}),$$

and U_b is any unitary matrix.

It is further shown that if $\chi_b(x, \lambda) = \theta(x, \lambda) + \phi(x, \lambda)M_b(\lambda)$, then

$$\int_e^b \chi_b^* A\chi_b\, dt \leq [M_b - M_b^*]/2i\text{Im } \lambda.$$

For $\chi_b(x, \lambda)$ and $\chi_b(x, \mu)$, $\text{Im } \lambda, \text{Im } \mu \neq 0$,

$$\lim_{x \to b} \chi_b(x, \mu)^* J\chi_b(x, \lambda) = 0,$$

provided the same unitary matrix μ_b is used in defining each.

Note that for all λ, Im $\lambda \neq 0$,

$$\frac{\text{Im } M_b}{\text{Im } \lambda} = \frac{M_b - M_b^*}{2i\text{Im } \lambda} > 0 \,.$$

The circle equation, satisfied by $M_{b'}$, is

$$\pm \chi_{b'}(b', \lambda)^*(J/i)\chi_{b'}\lambda) = 0 \,.$$

If at b', we let

$$\begin{pmatrix} \mathbf{A} & \mathbf{B}^* \\ \mathbf{B} & \mathbf{D} \end{pmatrix} = \begin{matrix} \mathcal{Y}^*(J/i)\mathcal{Y}, & \text{Im } \lambda > 0, \\ -\mathcal{Y}^*(J/i)\mathcal{Y}, & \text{Im } \lambda < 0, \end{matrix}$$

the circle equation can be rewritten as

$$(I_n, M_{b'}^*) \begin{pmatrix} \mathbf{A} & \mathbf{B}^* \\ \mathbf{B} & \mathbf{D} \end{pmatrix} \begin{pmatrix} I_n \\ M_{b'} \end{pmatrix} = 0 \,.$$

Expanded, this is

$$M_{b'}^* \mathbf{D} M_{b'} + M_{b'}^* \mathbf{B} + \mathbf{B}^* M_{b'} + \mathbf{A} = 0 \,.$$

It is possible to show that

$$\mathbf{A} = \pm\theta^*(J/i)\theta(b') \,,$$

$$\mathbf{B} = \pm\phi^*(J/i)\theta(b') = \pm[2\text{Im } \lambda \int_e^{b'} \phi^* A\theta dt - iI_n] \,,$$

$$\mathbf{D} = \pm\phi^*(J/i)\phi(b') = \pm[2\text{Im } \lambda \int_e^{b'} \phi^* A\phi dt] \,.$$

If $R_1 = \mathbf{D}^{-1/2}$, $R_2 = R_1(\overline{\lambda})$ and $C = -\mathbf{D}^{-1}\mathbf{B}$, then

$$M_{b'} = C + R_1 U_b R_2 \,,$$

where U_b is any unitary matrix. As b' approaches b, C, R_1 and R_2 have limits C_b, R_b, $\widetilde{R}_b$, giving M_b.

Similarly, we attempt to satisfy the a' boundary conditions by $\chi_{a'}(x, \lambda) = \theta(x, \lambda) + \phi(x, \lambda)M_{a'}(\lambda)$. Insertion into $(\alpha_1\alpha_2)Y(a') = 0$ shows that

$$M_{a'} = -[(\alpha_1\alpha_2)\phi(a', \lambda)]^{-1}[(\alpha_1\alpha_2)\phi(a', \lambda)] \,,$$

where, again, the inverse must exist.

As a' approaches a, $M_{a'}$ can be made to approach

$$M_a = C_a + R_a U_a \overline{R}_a \,,$$

where

$$C_a = \lim_{a' \to a} - \left[2\mathrm{Im}\, \lambda \int_{a'}^{e} \phi^* A\phi\, dt \right]^{-1} \left[2\mathrm{Im}\, \lambda \int_{a'}^{e} \phi^* A\theta\, dt + iI_n \right],$$

$$R_a = \lim_{a' \to a} \left[2|\mathrm{Im}\, \lambda| \int_{a'}^{e} \phi^* A\phi\, dt \right]^{-1/2},$$

$$\overline{R}_a = R_a(\overline{\lambda}),$$

and U_a is a (perhaps different) unitary matrix.

With $\chi_a(x, \lambda) = \theta(x, \lambda) + \phi(x, \lambda) M_a(\lambda)$ it is also true that

$$\int_a^e \chi_a^* A\chi_a\, dt \le [M_a^* - M_a]/2i\mathrm{Im}\, \lambda.$$

For $\chi_a(x, \lambda)$ and $\chi_a(x, \mu)$, $\mathrm{Im}\, \lambda$, $\mathrm{Im}\, \mu \ne 0$,

$$\lim_{x \to a} \chi(x, \mu)^* J\chi(x, \lambda) = 0,$$

provided the same unitary matrix U_a is used in defining each.

Note that

$$\frac{\mathrm{Im}\, M_a}{\mathrm{Im}\, \lambda} = \frac{M_a - M_a^*}{2i\mathrm{Im}\, \lambda} < 0.$$

The circle equation, satisfied by $M_{a'}$, is

$$\mp \chi_{a'}(a', \lambda)^* (J/i)\chi_{a'}(a', \lambda) = 0.$$

If, at a', we let

$$\begin{pmatrix} \mathbf{A} & \mathbf{B}^* \\ \mathbf{B} & \mathbf{D} \end{pmatrix} = \begin{array}{ll} -\mathcal{Y}^*(J/i)\mathcal{Y}, & \mathrm{Im}\, \lambda > 0, \\ \mathcal{Y}^*(J/i)\mathcal{Y}, & \mathrm{Im}\, \lambda < 0, \end{array}$$

the circle equation can be rewritten as

$$(I_n M_{a'}^*) \begin{pmatrix} \mathbf{A} & \mathbf{B}^* \\ \mathbf{B} & \mathbf{D} \end{pmatrix} \begin{pmatrix} I_n \\ M_{a'} \end{pmatrix} = 0.$$

Expanded, this is again

$$M_{a'}^* \mathbf{D} M_{a'} + M_{a'}^* \mathbf{B} + \mathbf{B}^* M_{a'} + \mathbf{A} = 0.$$

It is possible to show that

$$\mathbf{A} = \mp\theta^*(J/i)\theta(a'),$$

$$\mathbf{B} = \mp\phi^*(J/i)\theta(a') = \pm[2\mathrm{Im}\, \lambda \int_{a'}^{e} \phi^* A\theta\, dt + iI_n],$$

$$\mathbf{D} = \mp \phi^*(J/i)\phi(a') = \pm[2\mathrm{Im}\,\lambda \int_{a'}^{e} \phi^* A\phi dt]\,.$$

If $R_1 = \mathbf{D}^{-1/2}$ $R_2 = R_1(\overline{\lambda})$ and $C = -\mathbf{D}^{-1}\mathbf{B}$, then

$$M_{a'} = C + R_1 U_a R_2\,,$$

where U_a is any unitary matrix. As a' approaches a, C, R_1 and R_2 have limits C_a, R_a, $\widetilde{R}_a$, giving M_a.

There are some facts concerning $M_a(\lambda)$ and $M_b(\lambda)$ that we shall need.

VIII.1.1. Theorem. *For all λ, $\mathrm{Im}\,\lambda \neq 0$,*

(1) $M_a(\lambda) \neq M_b(\lambda)$, $[\mathrm{Im}\,\lambda][M_a(\lambda) - M_b(\lambda)] < 0$,

(2) $M_a(\lambda) = M_a(\overline{\lambda})^*$, $M_b(\lambda) = M_b(\overline{\lambda})^*$,

(3) $M_a(\lambda), M_b(\lambda), M_a(\lambda) - M_b(\lambda)$ *are all invertible,*

(4) $M_b(\lambda)[(M_a(\lambda) - M_b(\lambda)]^{-1} M_a(\lambda) = M_a(\lambda)[M_a(\lambda) - M_b(\lambda)]^{-1} M_b(\lambda)\,.$

Proof. The second part follows from the statements

$$\lim_{x \to a} \chi_a(x,\mu)^* J\chi_a(x,\lambda) = 0\,,$$

$$\lim_{x \to b} \chi_b(x,\mu)^* J\chi_b(x,\lambda) = 0\,,$$

letting $\mu - \overline{\lambda}$.

The third follows from noting that for any matrix $M = A + iB$, with $M^* = A - iB$, so $A = (M + M^*)/2$ and $B = (M - M^*)/2i$, if $B > 0$ or $B < 0$, then M is nonsingular. For suppose M is singular. Then there exists an eigenvector v such that $Mv = 0$. This implies

$$0 = \nu^* M\nu - \nu^* A\nu + i\nu^* B\nu\,.$$

Since $B > 0$ or $B < 0$, $i\nu^* B\nu$ is imaginary, while $\nu^* A\nu$ is real. This is impossible. $\square$

The fourth is an easy computation.

VIII.2 The Differential Operator

It is known that the number of solutions of $JY' = (\lambda A + B)Y$ in $L_A^2(a,b)$ is invariant provided $\mathrm{Im}\,\lambda > 0$ or $\mathrm{Im}\,\lambda < 0$. When A and B are real valued, the reflection principle shows that the number in each half plane is the same.

VIII.2.1. Definition. We denote by D_M those elements Y in $L_A^2(a,b)$ satisfying

$$(1) \quad lY = JY' - BY = AF$$

exists a.e. for some F in $L_A^2(a,b)$.

Although we shall not need it, the maximal operator L_M is defined by setting $L_M Y = F$ for all Y in D_M.

VIII.2.2. Theorem. *Let Y_j be a solution of*

$$JY_j' = (\overline{\lambda}A + B)Y_j, \qquad Im\ \lambda \neq 0.$$

Then for all Y in D_M,

$$B_{a_j}(Y) = \lim_{x \to a} Y_j^* JY$$

exists if and only if Y_j is in $L_A^2(a, e)$,

$$B_{b_j}(Y) = \lim_{x \to b} Y_j^* JY$$

exists if and only if Y_j is in $L_A^2(e, b)$.

The terms $B_{a_j}(Y)$ and $B_{b_j}(Y)$ are boundary conditions.

VIII.2.3. Definition. Let Im $\lambda \neq 0$. Let $M_a(\overline{\lambda}) = \widetilde{C}_a + \widetilde{R}_a U_a R_a$ be on the limit-circle at a. Let $\chi_a(x, \overline{\lambda}) = \theta(x, \overline{\lambda}) + \phi(x, \overline{\lambda}) M_a(\overline{\lambda})$ satisfy $JY' = (\lambda A + B)Y$ with λ replaced by $\overline{\lambda}$ and be in $L_A^2(e, b)$.

We define the boundary values $B_a(Y)$ and $B_b(Y)$ by setting

$$B_a(Y) = \lim_{x \to a} \chi_a(x, \overline{\lambda})^* JY(x),$$
$$B_b(Y) = \lim_{x \to b} \chi_b(x, \overline{\lambda})^* JY(x),$$

for all Y in D_M.

Note that $\overline{\lambda}$ is used in the definition. This is for convenience only and will be removed later.

VIII.2.4. Definition. We denote by D those elements Y in $L_A^2(a, b)$ satisfying

(1) $l(Y) = JY' - BY = AF$ exists a.e. for some F in $L_A^2(a, b)$.

(2) $B_a(Y) = 0$.

(3) $B_b(Y) = 0$ for some fixed λ, Im $\lambda \neq 0$.

We define the operator L by setting $LY = F$ for all Y in D.

VIII.3   The Resolvent. The Green's Function

The inverse of $(L - \lambda I)$ can be calculated with ease. We solve $JY' = (\lambda A + B)Y + AF$ together with the two boundary conditions. If we set $Y = \mathcal{Y}C$, variation of parameters shows $C' = -J\mathcal{Y}^* AF$, where $\mathcal{Y}^t(x, \lambda) = \mathcal{Y}(x, \overline{\lambda})^*$. Thus

$$Y(x) = -\mathcal{Y}(x, \lambda) \int_e^x J\mathcal{Y}^t(\xi, \lambda) A(\xi) F(\xi) d\xi + \mathcal{Y}(x, \lambda) K.$$

We multiply by $\begin{pmatrix} \chi_b^t \\ 0 \end{pmatrix} J$. So

$$\begin{pmatrix} \chi_b^t \\ 0 \end{pmatrix} JY(x) = - \begin{pmatrix} \chi_b^t \\ 0 \end{pmatrix} J\mathcal{Y} \int_e^x J\mathcal{Y}^t AF d\xi + \begin{pmatrix} \chi_b^t \\ 0 \end{pmatrix} J\mathcal{Y}K .$$

Now $\begin{pmatrix} \chi_b^t \\ 0 \end{pmatrix} J\mathcal{Y}$ is constant. At $x = e$

$$\begin{pmatrix} \chi_b^t \\ 0 \end{pmatrix} J\mathcal{Y} = \begin{pmatrix} M_b^t & -I_n \\ 0 & 0 \end{pmatrix} .$$

Thus

$$\begin{pmatrix} \chi_b^t \\ 0 \end{pmatrix} JY(x) = \int_e^x \begin{pmatrix} \chi_b^t \\ 0 \end{pmatrix} AF d\xi + \begin{pmatrix} M_b^t & -I_n \\ 0 & 0 \end{pmatrix} K .$$

Letting $x \to b$,

$$0 = \int_e^b \begin{pmatrix} \chi_b^t \\ 0 \end{pmatrix} AF d\xi + \begin{pmatrix} M_b & -I_n \\ 0 & 0 \end{pmatrix} K .$$

Note that the integral exists since χ_b is in $L_A^2(e,b)$.

Now return to Y and multiply by $\begin{pmatrix} 0 \\ \chi_a^t \end{pmatrix} J$. Then

$$\begin{pmatrix} 0 \\ \chi_a^t \end{pmatrix} JY(x) = \begin{pmatrix} 0 \\ \chi_a^t \end{pmatrix} J\mathcal{Y} \int_x^e J\mathcal{Y}^t AF d\xi + \begin{pmatrix} 0 \\ \chi_a^t \end{pmatrix} J\mathcal{Y}K .$$

Now $\begin{pmatrix} 0 \\ \chi_a^t \end{pmatrix} J\mathcal{Y}$ is constant. At $x = e$

$$\begin{pmatrix} 0 \\ \chi_a^t \end{pmatrix} J\mathcal{Y} = \begin{pmatrix} 0 & 0 \\ M_a^T & -I_n \end{pmatrix} .$$

Thus

$$\begin{pmatrix} 0 \\ \chi_a^t \end{pmatrix} JY(x) = - \int_x^e \begin{pmatrix} 0 \\ \chi_a^t \end{pmatrix} AF d\xi + \begin{pmatrix} 0 & 0 \\ M_a^t & -I_n \end{pmatrix} K .$$

Letting $x \to a$, we have

$$0 = - \int_a^e \begin{pmatrix} 0 \\ \chi_a^t \end{pmatrix} AF d\xi + \begin{pmatrix} 0 & 0 \\ M_a^t & -I_n \end{pmatrix} K .$$

Again the integral exists since χ_a is in $L_A^2(a,e)$.

Add the two limit equations together,

$$0 = \int_e^b \begin{pmatrix} \chi_b^t \\ 0 \end{pmatrix} AF d\xi - \int_a^e \begin{pmatrix} 0 \\ \chi_a^t \end{pmatrix} AF d\xi + \begin{pmatrix} M_b^t & -I_n \\ M_a^t & -I_n \end{pmatrix} K .$$

The coefficient of K is nonsingular, yielding

$$K = \begin{pmatrix} (M_a - M_b)^{-1} & -(M_a - M_b)^{-1} \\ M_a(M_a - M_b)^{-1} & -M_b(M_a - M_b)^{-1} \end{pmatrix} \int_e^{b'} \begin{pmatrix} I_n & M_b^t \\ 0 & 0 \end{pmatrix} \mathcal{Y}^t AF d\xi$$
$$- \begin{pmatrix} (M_a - M_b)^{-1} & -(M_a - M_b)^{-1} \\ M_a(M_a - M_b)^{-1} & -M_b(M_a - M_b)^{-1} \end{pmatrix} \int_a^e \begin{pmatrix} 0 & 0 \\ I_n & M_a^t \end{pmatrix} \mathcal{Y}^t AF d\xi .$$

Inserting this in Y, we get

$$Y(x) = \mathcal{Y} \begin{pmatrix} I_n & 0 \\ M_b & 0 \end{pmatrix} \begin{pmatrix} 0 & (M_a - M_b)^{-1} \\ (M_a - M_b)^{-1} & 0 \end{pmatrix} \int_a^x \begin{pmatrix} 0 & 0 \\ I_n & M_a^t \end{pmatrix} \mathcal{Y}^t AF d\xi$$
$$+ \mathcal{Y} \begin{pmatrix} 0 & I_n \\ 0 & M_a \end{pmatrix} \begin{pmatrix} 0 & (M_a - M_b)^{-1} \\ (M_a - M_b)^{-1} & 0 \end{pmatrix} \int_x^b \begin{pmatrix} I_n & M_b^t \\ 0 & 0 \end{pmatrix} \mathcal{Y}^t AF d\xi .$$

Further computation shows this can be written as

$$Y(x) = \chi_b(x, \lambda)(M_a(\lambda) - M_b(\lambda))^{-1} \int_a^x \chi_a(\xi, \overline{\lambda})^* A(\xi) F(\xi) d\xi$$
$$+ \chi_a(x, \lambda)(M_a(\lambda) - M_b(\lambda))^{-1} \int_x^b \chi_b(\xi, \overline{\lambda})^* A(\xi) F(\xi) d\xi .$$

VIII.3.1. Theorem. $(L - \lambda I)^{-1}$ *is given by*

$$(L - \lambda I)^{-1} F(x) = \int_a^b \mathbf{G}(\lambda, x, \xi) A(\xi) F(\xi) d\xi ,$$

where

$$\mathbf{G}(\lambda, x, \xi) = \chi_b(x, \lambda)(M_a(\lambda) - M_b(\lambda))^{-1} \chi_a(\xi, \overline{\lambda})^* , \quad a < \xi < x < b ,$$
$$= \chi_a(x, \lambda)(M_a(\lambda) - M_b(\lambda))^{-1} \chi_b(\xi, \overline{\lambda})^* , \quad a < x < \xi < b .$$

VIII.3.2. Theorem. $\mathbf{G}$ *is symmetric.*

$$\mathbf{G}(\lambda, x, \xi) = \mathbf{G}(\overline{\lambda}, \xi, x)^*$$

VIII.3.3. Theorem. $(L - \lambda I)^{-1}$ *is bounded.*

$$\|(L - \lambda I)^{-1}\| \leq 1/Im\,\lambda|$$

VIII.3.4. Theorem. L *is self-adjoint.*

VIII.4   Parameter Independence of the Domain

It appears that D is dependent on the parameter λ used in the boundary conditions. Indeed, the Green's function was calculated only for λ. We show now that λ may vary with impunity so long as Im $\lambda \neq 0$.

VIII.4.1. Theorem. *Let Y be in D. Then for all λ, Im $\lambda \neq 0$, $B_a(Y) = 0$, $B_b(Y) = 0$.*

Proof. Recall that if λ, μ have nonzero imaginary parts,

$$\lim_{x \to a} \chi_a(x, \mu)^* J \chi_a(x, \lambda) = 0,$$

$$\lim \chi_b(x, \mu)^* J \chi_b(x, \lambda) = 0.$$

If λ is replaced by $\overline{\lambda}$ and $\chi_a(x, \mu)$ and $\chi_b(x, \mu)$ are modified smoothly to be zero near b and a, respectively, then, so modified, $\chi_a(x, \mu)$ and $\chi_b(x, \mu)$ are in D. An application of Green's formula shows that

$$\begin{aligned}
0 &= \langle LY, \chi_a \rangle - \langle Y, L\chi_a \rangle a \\
&= \int_a^b \{\chi_a^*[JY' - BY] - [J\chi_a' - \chi_a]^* Y\} dx \\
&= -\lim_{x \to a} \chi_a^*(x, \mu) JY(x).
\end{aligned}$$

Likewise

$$\begin{aligned}
0 &= \langle LY, \chi_b \rangle - \langle Y, L\chi_b \rangle \\
&= \lim_{x \to b} \chi_b^*(x, \mu) JY(x).
\end{aligned}$$ $\square$

VIII.4.2. Theorem. *For all λ, Im $\lambda \neq 0$, $(L - \lambda I)^{-1}$ exists and is given by Theorem VIII.3.1. $\|(L - \lambda)^{-1}\| \leq 1/|Im \; \lambda|$.*

Proof. It is permissible to use the previous calculation, no matter what λ. $\square$

VIII.5 The Extended Green's Formula with Two Singular Points

We can use the results of the previous chapter with only minor modifications. Since a is also singular, in order to evaluate $Z^* JY$ at $x = a$, assume there are p solutions, $n \leq p \leq 2n$, of $JY' = (\lambda A + B)Y$ in $L_A^2(a, e) < 0$. Then there are $p - n$ combinations from $\phi(x, \lambda)$ in $L^2(a, e)$ when Im $\lambda < 0$. If we write

$$\phi(x, \lambda) = (\phi_1 \phi_2)(x, \lambda) E_a, \quad \chi_a(x, \overline{\lambda}) E_a^* = (\chi_1 \chi_2)(x, \overline{\lambda}),$$

$$\phi(x, \overline{\lambda}) = (\phi_1 \phi_2)(x, \overline{\lambda}) \widetilde{E}_a, \quad \chi_a(x, \lambda) \widetilde{E}_a^* = (\chi_1 \chi_2)(x, \lambda),$$

then we find that

$$B_a(Y) = \begin{pmatrix} \lim_{x \to a} \chi_1(x, \overline{\lambda})^* JY(x) \\ \lim_{x \to a} \phi_1(x, \overline{\lambda})^* JY(x) \end{pmatrix},$$

$$\widetilde{B}_a(Z) = \begin{pmatrix} \lim_{x \to a} \chi_1(x, \lambda)^* JZ(x) \\ \lim_{x \to a} \phi_1(x, \lambda)^* JZ(x) \end{pmatrix},$$

$$\lim_{x \to a} \chi_2(x, \overline{\lambda})^* JY(x) = 0 \,,$$

$$\lim_{x \to a} \chi_2(x, \lambda)^* JZ(x) = 0 \,,$$

and

$$\lim_{x \to a} Z^*(x) JY(x) = \widetilde{B}_a(Z)^* J_a B_a(Y) \,,$$

where

$$J_a = \begin{pmatrix} 0 & -I_{p-n} \\ I_{p-n} & 0 \end{pmatrix} \,.$$

M, N, P, Q are $r \times (2p - 2n)$, $r \times (2m - 2n)$, $(2m + 2p - 4n - r) \times (2p - 2n)$ and $(2m - 2p - 4n - r) \times (2m - 2n)$ matrices, $0 \le r \le (2m + 2p)$.

Finally, in the parametric boundary conditions, replace J by J_a. Self-adjointness requires $m + p - 2n = r$ With these modifications, the following results still hold.

VIII.5.1. Theorem. *(Green's Formula). Let Y and Z be in D_M. Then*

$$\int_a^b \{Z^*[JY' - BY] - [JZ' - BZ]^* Y\} dx$$

$$= [\widetilde{M}\widetilde{B}_a(Z) + \widetilde{N}B_b(Z)]^* [MB_a(Y) + NB_b(Y)]$$

$$+ [\widetilde{P}\widetilde{B}_a(Z) + \widetilde{Q}B_b(Z)]^* [PB_a(Y) + QB_b(Y)] \,.$$

Parametric boundary conditions look the same as in the regular and singly singular cases.

$$\begin{pmatrix} B_a(Y) \\ B_b(Y) \end{pmatrix} = \begin{pmatrix} J_a \widetilde{P}^* \Gamma \\ -J_b Q^* \Gamma \end{pmatrix} = -\mathbf{J} \begin{pmatrix} \widetilde{P}^* \\ \widetilde{Q}^* \end{pmatrix} \Gamma \,,$$

$$\begin{pmatrix} \widetilde{B}_a(Y) \\ \widetilde{B}_b(Y) \end{pmatrix} = \begin{pmatrix} -J_a M^* \Delta \\ J_b N^* \Delta \end{pmatrix} = \mathbf{J} \begin{pmatrix} M^* \\ N^* \end{pmatrix} \Delta \,,$$

where

$$\mathbf{J} = \begin{pmatrix} -J_a & 0 \\ 0 & J_b \end{pmatrix} \,.$$

VIII.5.2. Definition. We denote by D those elements Y in $L_A^2(a, b)$ satisfying:

(1) Y is in D_M.

(2) $MB_a(Y) + NB_b(Y) = 0$.

We denote by L the operator defined by setting $LY = F$ whenever $JY' = BY = AF$ and Y is in D.

VIII.5.3. Definition. We denote by D^* those elements Z in $L_A^2(a, b)$ satisfying:

(1) Z is in D_M.

(2) $\widetilde{P}\widetilde{B}_a(Z) + \widetilde{Q}B_b(Z) = 0$.

We denote by L^* the operator defined by setting $L^*Z = G$ whenever $JZ' - BZ = AG$ and Z is in D^*.

VIII.5.4. Theorem. *The abuse of notation is correct. The adjoint of L in $L_A^2(a,b)$ is L^*. The adjoint of L^* in $L_A^2(a,b)$ is L.*

VIII.5.5. Theorem. *L is self-adjoint if and only if $r = m + p - 2n$, there exists a transformation U such that*

$$\begin{pmatrix} B_a(Y) \\ B_b(Y) \end{pmatrix} = U \begin{pmatrix} \widetilde{B}_a(Y) \\ \widetilde{B}_b(Y) \end{pmatrix}$$

and

$$(MN)U\mathbf{J}\begin{pmatrix} M^* \\ N^* \end{pmatrix} = 0.$$

Proof. If L is self-adjoint, then

$$B_a(Y) = J_a P^* \Gamma, \quad \widetilde{B}_a(Y) = -J_a M^* \Delta,$$
$$B_b(Y) = -J_a Q^* \Gamma, \quad \widetilde{B}_b(Y) = -J_b N^* \Delta.$$

There must be a transformation U such that $\begin{pmatrix} B_a(Y) \\ B_b(Y) \end{pmatrix} = U \begin{pmatrix} \widetilde{B}_a(Y) \\ \widetilde{B}_b(Y) \end{pmatrix}$ so that no

additional constraints other than those on $\begin{pmatrix} B_a(Y) \\ B_b(Y) \end{pmatrix}$ are placed on Y. Then

$$\begin{pmatrix} B_a(Y) \\ B_b(Y) \end{pmatrix} = U \begin{pmatrix} \widetilde{B}_a(Y) \\ \widetilde{B}_b(Y) \end{pmatrix} = U\mathbf{J}\begin{pmatrix} M^* \\ N^* \end{pmatrix} \Delta$$

and

$$(MN) \begin{pmatrix} B_a(Y) \\ B_b(Y) \end{pmatrix} = (MN)U\mathbf{J}\begin{pmatrix} M^* \\ N^* \end{pmatrix} \Delta = 0.$$

Since Δ is arbitrary, $(MN)U\mathbf{J}\begin{pmatrix} M^* \\ N^* \end{pmatrix} = 0$.

Conversely, let $\begin{pmatrix} B_a(Y) \\ B_b(Y) \end{pmatrix} = U \begin{pmatrix} \widetilde{B}_a(Y) \\ \widetilde{B}_b(Y) \end{pmatrix}$, and $(MN)U\mathbf{J}\begin{pmatrix} M^* \\ N^* \end{pmatrix} = 0$. If Z is

in D^*, then $(MN)U \begin{pmatrix} \widetilde{B}_a(Z) \\ \widetilde{B}_b(Z) \end{pmatrix} = 0$, and $(MN) \begin{pmatrix} B_a(Y) \\ B_b(Y) \end{pmatrix} = 0$. So Z is in D. If

Y is in D, then $(MN) \begin{pmatrix} B_a(Y) \\ B_b(Y) \end{pmatrix} = 0$ and $(MN)U\mathbf{J}\begin{pmatrix} M^* \\ N^* \end{pmatrix} = 0$ imply there is a

constant Δ such that

$$\begin{pmatrix} B_a(Y) \\ B_b(Y) \end{pmatrix} = U\mathbf{J}\begin{pmatrix} M^* \\ N^* \end{pmatrix} \Delta,$$

and

$$\begin{pmatrix} \widetilde{B}_a(Y) \\ \widetilde{B}_b(Y) \end{pmatrix} = \mathbf{J} \begin{pmatrix} M^* \\ N^* \end{pmatrix} \Delta \,.$$

Hence Y is in D^*.

The matrix U has the form

$$U = \begin{bmatrix} V_a & 0 \\ 0 & V_b \end{bmatrix} \,.$$

Just as in the previous chapter, its existence is guaranteed. The self-adjointness criteria can be rewritten as

$$M V_a J_a M^* = N V_b J_b M^* \,.$$

If a is regular, then $V_a = I_n$, and

$$U = \begin{bmatrix} I_n & 0 \\ 0 & V_b \end{bmatrix} \,.$$

If both a and b are regular, then $V_a = V_b = I_n$ and

$$U = \begin{bmatrix} I_n & 0 \\ 0 & I_n \end{bmatrix} = I_{2n} \,. \qquad \qquad \Box$$

VIII.6 Examples

A vast number of examples await our discussion. Scalar second order problems abound. Further, we now have available a number of fourth order and sixth order scalar examples. We present a variety.

The classic second order problems include:

(1) The Jacobi polynomials, with their special cases the Legendre and Tchebycheff polynomials.

(2) The ordinary and generalized Laguerre polynomials.

(3) The Hermite polynomials.

(4) Bessel functions.

These satisfy a second order differential equation of the form

$$\ell y = [-(py')' + qy]/w = \lambda y \,,$$

which, as we have noted earlier, can be written in system format as

$$\begin{pmatrix} 0 & -1 \\ 1 & 0 \end{pmatrix} \begin{pmatrix} y_1 \\ y_2 \end{pmatrix}' = \left[\lambda \begin{pmatrix} w & 0 \\ 0 & 0 \end{pmatrix} + \begin{pmatrix} -q & 0 \\ 0 & 1/p \end{pmatrix} \right] \begin{pmatrix} y_1 \\ y_2 \end{pmatrix} ,$$

$\ell y = [-(py')' + qy]/w = f$, in system format, becomes

$$\begin{pmatrix} 0 & -1 \\ 1 & 0 \end{pmatrix} \begin{pmatrix} y_1 \\ y_2 \end{pmatrix}' - \begin{pmatrix} -q & 0 \\ 0 & 1/p \end{pmatrix} \begin{pmatrix} y_1 \\ y_2 \end{pmatrix} = \begin{pmatrix} w & 0 \\ 0 & 0 \end{pmatrix} \begin{pmatrix} f \\ 0 \end{pmatrix}.$$

The system operators L were defined by setting $LY = F$ when $JY' - BY = AF$, i.e., $LY = F$ when

$$-y_2' + qy_1 = wf, \quad y_1' - (1/p)y_2 = 0,$$

or when

$$-(py_1')' + qy_1 = wf.$$

This is the same as

$$\ell y = [-(py_1')' + qy_1]/w = f.$$

Thus we see that $LY = F$ if and only if $\ell y_1 = f$. The two operators are "isomorphic."

Since L_A^2 is generated by

$$\langle Y, Z \rangle_A = \int_I (\bar{z}_1 \bar{z}_2) \begin{pmatrix} w & 0 \\ 0 & 0 \end{pmatrix} \begin{pmatrix} y_1 \\ y_2 \end{pmatrix} dx$$

$$= \int_I y_1 \bar{z}_1 w \, dx$$

$$= \langle y_1, z_1 \rangle_w.$$

The system and scalar Hilbert spaces are also "isomorphic." The point of all this is that we may freely use the scalar representation without loss of generality.

The same is true of the examples involving higher order differential equations.

VIII.6.1 The Jacobi Boundary Value Problem

The Jacobi differential operator is

$$Ly = -(1 - x)^{-\alpha}(1 + x)^{-\beta}((1 - x)^{1+\alpha}(1 + x)^{1+\beta}y')',$$

set in $L^2(-1, 1; (1 - x)^\alpha(1 + x^\beta))$. 1 is in the limit circle case when $-1 < \alpha < 1$. It is limit point when $1 \le \alpha$. -1 is in the limit circle case when $-1 < \beta < 1$. It is limit circle case when $1 \le \beta$. Two solutions to $Ly = 0$ are $u = \tilde{u} = 1$ and $v = \tilde{v} = Q_0^{(\alpha,\beta)}(x)$. Boundary conditions, therefore, are given by

$$B_u(y) = -\lim_{x \to 1}(1 - x)^{1+\alpha}(1 + x)^{1+\beta}y'(x),$$

$$B_v(y) = \lim_{x \to 1}(1 - x)^{1+\alpha}(1 + x)^{1+\beta}[Q_0^{(\alpha,\beta)'}(x)y(x) - Q_0^{(\alpha,\beta)}(x)y'(x)],$$

$$B_{\tilde{u}}(y) = -\lim_{x \to 1}(1 - x)^{1+\alpha}(1 + x)^{1+\beta}y'(x),$$

$$B_{\tilde{v}}(y) = \lim_{x \to 1}(1 - x)^{1+\alpha}(1 + x)^{1+\beta}[Q_0^{(\alpha,\beta)'}(x)y(x) - Q_0^{(\alpha,\beta)}(x)y'(x)],$$

when they are needed.

The boundary value problem with the Jacobi polynomials as eigenfunctions is

$$Ly = -(1-x)^{-\alpha}(1+x)^{-\beta}((1-x)^{1+\alpha}(1+x)^{1+\beta}y')' = -n(n+\alpha+\beta+1)y,$$

$$B_u(y) = -\lim_{x\to 1}(1-x)^{1+\alpha}(1+x)^{1+\beta}y'(x) = 0, \qquad -1 < \alpha < 1,$$

$$B_{\widetilde{u}}(y) = -\lim_{x\to 1}(1-x)^{1+\alpha}(1+x)^{1+\beta}y'(x) = 0, \qquad -1 < \beta < 1.$$

VIII.6.2   The Legendre Boundary Value Problem

In the Jacobi problem, if $\alpha = \beta = 0$, we find immediately the Legendre boundary value problem. The Legendre operator is

$$Ly = -((1-x^2)y')'$$

in $L^2(-1,1;1)$. Both -1 and 1 are in the limit circle case. Two independent solutions to $((1-x^2)y')' = 0$ are $u = \widetilde{u} = 1$ and $v = \widetilde{v} = \frac{1}{2}\ln([(1+x)/(1-x)]$. Both are in $L^2(-1,1;1)$.

Boundary conditions are

$$B_u(y) = -\lim_{x\to 1}(1-x^2)y'(x),$$

$$B_v(y) = -\lim_{x\to 1}(1-x^2)\left[(1-x^2)^{-1}y(x) - \frac{1}{2}\ln\left(\frac{1+x}{1-x}\right)y'(x)\right],$$

$$B_{\widetilde{u}}(y) = -\lim_{x\to -1}(1-x^2)y'(x),$$

$$B_{\widetilde{v}}(y) = -\lim_{x\to -1}(1-x^2)\left[(1-x^2)^{-1}y(x) - \frac{1}{2}\ln\left(\frac{1+x}{1-x}\right)y'(x)\right].$$

The boundary value problem which has the Legendre polynomials as eigenfunctions is

$$Ly = -((1-x^2)y')' = n(n+1)y,$$

$$B_u(y) = -\lim_{x\to 1}(1-x^2)y'(x) = 0,$$

$$B_{\widetilde{u}}(y) = -\lim_{x\to -1}(1-x^2)y'(x) = 0.$$

There are other self-adjoint problems involving L with mixed boundary conditions.

VIII.6.3   The Tchebycheff Problem of the First Kind

In the Jacobi problem, if $\alpha = \beta = -\frac{1}{2}$, we find the first Tchebycheff boundary value problem. Set in $L^2(-1,1;(1-x^2)^{-\frac{1}{2}})$ the Tchebycheff differential operator is

$$Ly = -(1-x^2)^{\frac{1}{2}}((1-x^2)^{\frac{1}{2}}y')'.$$

Both -1 and 1 are limit circles. Two independent solutions to $((1-x^2)^{-\frac{1}{2}}y')' = 0$ are $u = \widetilde{u} = 1$ and $v = \widetilde{v} = \sin^{-1}x$. Both are in $L^2(-1,1,(1-x^2)^{-\frac{1}{2}})$.

Boundary conditions are

$$B_u(y) = -\lim_{x \to 1}(1-x^2)^{\frac{1}{2}}y'(x),$$

$$B_v(y) = \lim_{x \to 1}(1-x^2)^{\frac{1}{2}}\left[(1-x^2)^{-\frac{1}{2}}y(x) - (\sin^{-1}x)y'(x)\right],$$

$$B_{\widetilde{u}}(y) = \lim_{x \to -1}(1-x^2)^{\frac{1}{2}}y'(x),$$

$$B_{\widetilde{v}}(y) = \lim_{x \to -1}(1-x^2)^{\frac{1}{2}}\left[(1-x^2)^{-\frac{1}{2}}y(x) - (\sin^{-1}x)y'(x)\right].$$

The boundary value problem which has the Tchebycheff polynomials of the first kind as eigenfunctions is

$$Ly = -(1-x^2)^{\frac{1}{2}}((1-x^2)^{\frac{1}{2}}y')' = n^2 y,$$

$$B_u(y) = -\lim_{x \to 1}(1-x^2)^{\frac{1}{2}}y'(x) = 0,$$

$$B_{\widetilde{u}}(y) = -\lim_{x \to -1}(1-x^2)^{\frac{1}{2}}y'(x) = 0.$$

Again there are other self-adjoint problems involving the first Tchebycheff operator.

VIII.6.4 The Tchebycheff Problem of the Second Kind

In the Jacobi problem, if $\alpha = \beta = 1/2$, we find the second Tchebycheff boundary value problem. The Tchebycheff differential operator is

$$Ly = -(1-x^2)^{-\frac{1}{2}}((1-x^2)^{\frac{3}{2}}y')'.$$

Both -1 and 1 are limit circles. Two independent solutions to $((1-x^2)^{\frac{3}{2}}y')' = 0$ are $u = \widetilde{u} = 1$ and $v = \widetilde{v} = (x/(1-x^2)^{\frac{1}{2}})$. Both are in $L^2(-1,1;(1-x^2)^{\frac{1}{2}})$.

Boundary conditions are

$$B_u(y) = \lim_{x \to 1}(1-x^2)^{\frac{3}{2}}y'(x),$$

$$B_v(y) = \lim_{x \to 1}(1-x^2)^{\frac{3}{2}}((1-x^2)^{-\frac{3}{2}}y(x) - (x/(1-x^2)^{\frac{1}{2}})y'(x)),$$

$$B_{\widetilde{u}}(y) = -\lim_{x \to -1}(1-x^2)^{\frac{3}{2}}y'(x),$$

$$B_{\widetilde{v}}(y) = \lim_{x \to -1}(1-x^2)^{\frac{3}{2}}((1-x^2)^{-\frac{3}{2}}y(x) - (x/(1-x^2)^{\frac{1}{2}})y'(x)).$$

The boundary value problem which has the Tchebycheff polynomials of the second kind as eigenfunctions is

$$Ly = -(1-x^2)^{-\frac{1}{2}}((1-x^2)^{\frac{3}{2}}y')' = n(n+2)y,$$

$$B_u(y) = -\lim_{x \to 1}(1-x^2)^{\frac{3}{2}}y'(x) = 0,$$

$$B_{\widetilde{u}}(y) - -\lim_{x \to -1}(1-x^2)^{\frac{3}{2}}y'(x) = 0.$$

VIII.6.5   The Generalized Laguerre Boundary Value Problem

The generalized Laguerre differential operator is

$$Ly = -x^{-\alpha}e^{x}(x^{\alpha+1}e^{-x}y')',$$

set in $L^2(0,\infty; x^{\alpha}e^{-x})$. The point 0 is in the limit circle case if $-1 < \alpha < 1$, and it is a limit point if $1 \le \alpha$, where the point ∞ is always limit point.

The boundary conditions at 0, $-1 < \alpha < 1$, are given by choosing $\tilde{u} = 1$ and

$$\tilde{v} = \int_a^x \frac{e^{\xi}}{\xi^{\alpha+1}}\, d\xi$$

as independent solutions to $Ly = 0$. Then

$$B_{\tilde{u}}(y) = -\lim_{x\to 0} x^{\alpha+1}e^{-x}y'(x),$$

$$B_{\tilde{v}}(y) = \lim_{x\to 0} x^{\alpha+1}e^{-x}\left[\frac{e^{x}}{x^{\alpha+1}}\, y(x) - \int_a^x \frac{e^{\xi}}{\xi^{\alpha+1}}\, d\xi y'(x)\right].$$

The boundary condition satisfied by the Laguerre polynomials is $B_{\tilde{u}}(y) = 0$. Hence the boundary value problem with the generalized Laguerre polynomials as eigenfunctions is

$$Ly = -x^{-\alpha}e^{x}(x^{\alpha+1}e^{-x}y')' = ny,$$
$$B_{\tilde{u}}(y) = -\lim_{x\to 0} x^{\alpha+1}e^{-x}y'(x) = 0, \qquad (\text{required when } -1 < \alpha < 1).$$

VIII.6.6   The Ordinary Laguerre Boundary Value Problem

The ordinary Laguerre differential operator is

$$Ly = -e^{x}(xe^{-x}y')',$$
$$B_{\tilde{u}}(y) = -\lim_{x\to 0} xe^{-x}y'(x) = 0,$$

set in $L^2(0,\infty; e^{-x})$. 0 is the limit circle case: ∞ is in the limit point case, and so no boundary conditions are required.

VIII.6.7   The Hermite Boundary Value Problem

The Hermite differential operator is

$$Ly = -e^{x^2}(e^{-x^2}y')',$$

set in $L^2(-\infty,\infty; e^{-x^2})$. Both $\pm\infty$ are in the limit point case, and so no boundary conditions are required.

The conditions $\lim_{x\to\pm\infty} e^{-x^2}y'(x) = 0$ are automatically satisfied. Consequently the boundary value problem with the Hermite polynomials as eigenfunctions is

$$Ly = e^{x^2}(e^{-x^2}y')' = ny, \qquad -\infty < x < \infty.$$

VIII.6.8 Bessel Functions

Although they are not polynomials, we include a brief discussion of the Bessel boundary value problems because of their importance. There are four different problems encountered. They differ over the intervals involved.

(1) If the interval is $[a, b]$, $0 < a < b < \infty$, the problem is regular.

(2) If the interval is $[0, b]$, $0 < b < \infty$, the problem is singular at 0, regular at b.

(3) If the interval is $[a, \infty)$, $0 < a < \infty$, the problem is regular at a, limit point at ∞.

(4) If the interval is $[0, \infty)$, the problem is singular at both ends.

The Bessel differential operator is

$$Ly = -x^{-1}((xy')' + (n^2/x)y),$$

set in $L^2(a, b; x)$. Two convenient solutions of $Ly = 0$ for finite a and/or b are $u = \tilde{u} = x^n$ and $v = \tilde{v} = x^{-n}$, $n \neq 0$, and $u = \tilde{u} = 1$, $v = \tilde{v} = \log x$ if $n = 0$.

If $a = 0$, it is easy to see that the endpoint 0 is in the limit circle case only if $|n| < 1$, while the limit point case holds when $|n| \geq 1$. Assuming $|n| < 1$ the boundary conditions at 0 are

$$B_{\tilde{u}}(y) = \lim_{x \to 0} x \left[nx^{n-1}y(x) - x^n y'(x) \right],$$
$$B_{\tilde{v}}(y) = \lim_{x \to 0} x \left[-nx^{-n-1}y(x) - x^{-n}y'(x) \right],$$

if $n \neq 0$.

    If $n = 0$,

$$B_{\tilde{u}}(y) = \lim_{x \to 0} xy'(x),$$
$$B_{\tilde{v}}(y) = \lim_{x \to 0} x \left[x^{-1}y(x) - \log x y'(x) \right].$$

The boundary condition traditionally associated with Bessel functions of the first kind is

$$B_{\tilde{u}}(y) = \lim_{x \to 0} x \left[nx^{n-1}y(x) - x^n y'(x) \right],$$

when $n \neq 0$, or

$$B_{\tilde{u}}(y) = \lim_{x \to 0} xy'(x) = 0,$$

when $n = 0$.

At a regular point, the conditions involving $y(a)$ and $y'(a)$ or $y(b)$ and $y'(b)$ are preferred.

At ∞ the limit point case always holds, and so no boundary condition is required.

VIII.6.9   The Legendre Squared Problem

The square of the Legendre operator is

$$Ly = [(1 - x^2)^2 y'']'' - 2[(1 - x^2)y']'.$$

We put this in system format by setting $y_1 = y$, $y_2 - y'$, $y_3 = -((1 - x^2)^2 y'')' - 2(1 - x^2)y'$, $y_4 = (1 - x^2)y''$. Then $(L - \lambda)y = 0$ becomes

$$\begin{pmatrix} 0 & 0 & -1 & 0 \\ 0 & 0 & 0 & -1 \\ 1 & 0 & 0 & 0 \\ 0 & 1 & 0 & 0 \end{pmatrix} \begin{pmatrix} y_1 \\ y_2 \\ y_3 \\ y_4 \end{pmatrix}'$$

$$= \left[\lambda \begin{pmatrix} 1 & 0 & 0 & 0 \\ 0 & 0 & 0 & 0 \\ 0 & 0 & 0 & 0 \\ 0 & 0 & 0 & 0 \end{pmatrix} + \begin{pmatrix} 0 & 0 & 0 & 0 \\ 0 & 2(1 - x^2) & 1 & 0 \\ 0 & 1 & 0 & 0 \\ 0 & 0 & 0 & 1/(1 - x^2)^2 \end{pmatrix} \right] \begin{pmatrix} y_1 \\ y_2 \\ y_3 \\ y_4 \end{pmatrix}.$$

With $\lambda = 0$ there are four solutions, all of which are square integrable over $(-1, 1)$. They are

$$Y_1 = \begin{pmatrix} 1 \\ 0 \\ 0 \\ 0 \end{pmatrix}, \quad Y_2 = \begin{pmatrix} \ln(1 + x) \\ (1 + x)^{-1} \\ 0 \\ -(1 - x)^2 \end{pmatrix}, \quad Y_3 = \begin{pmatrix} \ln(1 - x) \\ -(1 - x)^{-1} \\ 0 \\ -(1 + x)^2 \end{pmatrix} \quad \text{and} \quad Y_4,$$

whose initial component is

$$y_1 = \int^x (1 - t)^{-1} \int^t [\ln(1 - s)/(1 + s)^2]\, ds\, dt$$

but nonetheless is square integrable. Each generates boundary values at ± 1. We content ourselves with displaying those satisfied by the Legendre polynomials. They are in terms of scalar y,

$$\lim_{x \to -1} [(1 - x^2)^2 y''' - 4x(1 - x^2)y'' - 2(1 - x^2)y'] = 0,$$

$$\lim_{x \to -1} [(1 + x)^2(-y' + (1 - x)y'')] = 0,$$

$$\lim_{x \to 1} [-(1 - x^2)^2 y''' + 4x(1 - x^2)y'' + 2(1 - x^2)y'] = 0,$$

$$\lim_{x \to 1} [(1 - x)^2(y' + (1 + x)y'')] = 0.$$

For system Y, they are, of course, $\lim Y_j^* JY$, where $j = 1, 2$ as $x \to 1$, $j = 1, 3$ as $x \to -1$.

VIII.6.10 The Laguerre-Type Problem

In addition to the classic orthogonal polynomials, that satisfy a differential equation of any even order, there are three new collections of orthogonal polynomials that satisfy a differential equation of fourth order. We present those named Laguerre-type (see Krall [5]).

The scalar differential equation for the Laguerre-type polynomials is

$$(x^2 e^{-x} y'')'' - (([2R+2]x + 2)e^{-x} y')' = \lambda e^{-x} y$$

on $[0, \infty)$. The weight function for the polynomials has a $1/R$ jump at 0 and is

$$w(x) = \frac{1}{R}\, \delta(x) + e^{-x} H(x),$$

where $H(x)$ is the Heaviside function and $\delta(x)$ is its derivative, the Dirac delta function. If

$$y_1 = y \qquad\qquad , \quad y_2 = y',$$
$$y_3 = -(x^2 e^{-x} y'')' + ([2R+2]x + 2)e^{x} y' \quad , \quad y_4 = x^2 e^{-x} y''.$$

The problem can be written in system format

$$\begin{pmatrix} 0 & 0 & -1 & 0 \\ 0 & 0 & 0 & -1 \\ 1 & 0 & 0 & 0 \\ 0 & 1 & 0 & 0 \end{pmatrix} \begin{pmatrix} y_1 \\ y_2 \\ y_3 \\ y_4 \end{pmatrix}'$$
$$= \left[\lambda \begin{pmatrix} e^{-x} & 0 & 0 & 0 \\ 0 & 0 & 0 & 0 \\ 0 & 0 & 0 & 0 \\ 0 & 0 & 0 & 0 \end{pmatrix} + \begin{pmatrix} 0 & 0 & 0 & 0 \\ 0 & -([2R+2]x+2)e^{-x} & 1 & 0 \\ 0 & 1 & 0 & 0 \\ 0 & 0 & 0 & e^{-x}/x^2 \end{pmatrix} \right] \begin{pmatrix} y_1 \\ y_2 \\ y_3 \\ y_4 \end{pmatrix}.$$

If $x = 0$, the indicial roots to the scalar equation are always $-1, 0, 1, 2$. Thus 0 is in a limit 3 case; ∞ is limit 2. The boundary conditions generated by the solutions associated with the indicial roots $0, 1, 2$ are equivalent to boundary conditions associated with vectors whose first component is 1 or x or x^2. These vectors are

$$Y_1 = \begin{pmatrix} 1 \\ 0 \\ 0 \\ 0 \end{pmatrix}, \quad Y_2 = \begin{pmatrix} x \\ 1 \\ ([2R+2]x+2)e^{-x} \\ x^2 e^{-x} \end{pmatrix}, \quad Y_3 = \begin{pmatrix} x^2 \\ 2x \\ [4R+6]x^2 e^{-x} \\ 2x^2 e^{-x} \end{pmatrix}.$$

The boundary condition associated with Y_3, $\lim\limits_{x \to 0} Y_1^* JY$, has a limit which for "nice" functions may be identified with $-2y'(0)$. Likewise the boundary condition associated with Y_2, $\lim\limits_{x \to 0} Y_2^* JY$ may be identified with $2y(0)$. Finally we note that for "nice" functions the scalar differential equation at $x = 0$ becomes $-2Ry'(0) = \lambda y(0)$. Thus we may re-express the differential equation at $x = 0$ in terms of boundary conditions by setting $2R \lim_{x \to 0} Y_1^* JY = \lambda \lim_{x \to 0} Y_2^* JY$.

The boundary value problem, therefore, consists of the differential system, $0 < x < \infty$, together with the λ-dependent boundary condition above.

The polynomial problems continue to generate examples. We list some known to satisfy differential equations through the eighth order. Jacobi, Laguerre and Hermite satisfy second order equations. The Legendre, Laguerre and Jacobi type satisfy fourth order equations. (So do the second order examples.) They also satisfy sixth order equations, but there are two known sixth order examples, the H. L. Krall and Littlejohn I polynomials. The Littlejohn II polynomials satisfy an eighth order equation. All may be written as linear Hamiltonian systems. There are others (to follow).

Polynomials	Weight	Interval
Jacobi	$(1-x)^A(1+x)^B; A, B > -1$	$(-1,1)$
Laguerre	$x^\alpha e^{-x}; \alpha > -1$	$(0,\infty)$
Hermite	e^{-x^2}	$(-\infty,\infty)$
Legendre type	$1/2\left[\delta(x+1) + \delta(x-1)\right] + \alpha/2$	$(-1,1)$
Laguerre type	$1/R\,\delta(x) + e^{-x}$	$(0,\infty)$
Jacobi type	$1/M\,\delta(x) + (1-x)^\alpha; \alpha > -1$	$(0,1)$
H. L. Krall	$1/A\,\delta(x+1) + 1/B\,\delta(x-1) + C$	$(-1,1)$
Littlejohn I	$1/R\,\delta(x) + xe^{-x}$	$(0,\infty)$
Littlejohn II	$1/R\,\delta(x) + x^2e^{-x}$	$(0,\infty)$

References

[1] W. N. Everitt, *Integrable-square solutions of ordinary differential equations*, Quar. J. Math., Oxford (2) **10** (1959), 145–155.

[2] D. B. Hinton and J. K. Shaw, *Titchmarsh's λ-dependent boundary conditions for Hamiltonian systems*, Lect. Notes in Math. (Springer-Verlag), 964 (1982), 318–326.

[3] ———, *Hamiltonian systems of limit point or limit circle type with both ends singular*, J. Diff. Eq. **50** (1983), 444–464.

[4] ———, *On boundary value problems for Hamiltonian systems with two singular points*, SIAM J. Math. Anal. **15** (1984), 272–286.

[5] A. M. Krall, *Orthogonal polynomials satisfying fourth order differential equations*, Proc. Roy. Soc. Edin. **87** (1981), 271–288.

[6] A. M. Krall, B. D. Hinton and J. K. Shaw, *Boundary conditions for differential systems in intermediate limit situations*, Proc. Conf. on Ordinary and Partial Differential Equations, ed. I. W. Knowles and R. E. Lewis, North Holland, 1984, 301–305.

[7] L. L. Littlejohn and A. M. Krall, *Orthogonal polynomials and singular Sturm–Liouville systems, I* Rocky Mt. J. Math., 1987.

[8] ______, *Orthogonal polynomials and higher order singular Sturm–Liouville systems,* Acta Applicandae Mathematicae, (1989), 99–170.

[9] E. C. Titchmarsh, **Eigenfunction Expansions** Oxford Univ. Press, Oxford, 1962.

Chapter IX

The M(λ) Surface

It is well known that if $n = 1$, that is if the dimension of the system is 2, then the function $M(\lambda)$ lies on a circle or is merely a point. A cursory check of the $M(\lambda)$ equation, or its representation as

$$M = C + R_1 U R_2 = C + \rho e^{i\theta},$$

easily shows this to be true. In higher derivations, however, the surface is so complicated that it is impossible to visualize in any real sense. For example, if $n = 4$, the $M(\lambda)$ function is a 2×2 complex matrix. It involves four complex components, and is too much for us in our three-dimensional world.

It is possible to describe mathematically, however. Its description is rather intriguing and worth a look.

We first use the Niessen theory in conjunction with the Hinton–Shaw development to establish the number of variable components in $M(\lambda)$. This gives a so-so picture of how it behaves. Then we take a detailed direct look, knowing what we are looking for.

Because of Niessen's work, we consider $M_a(\lambda)$, the $M(\lambda)$ function needed as x approaches a.

IX.1   The Connection Between the Hinton–Shaw and Niessen Approaches

It is apparent that the spaces generated by the columns of $\chi = \mathcal{Y} \begin{pmatrix} I_n \\ M \end{pmatrix}$ of Hinton–Shaw and the eigenvector solutions $y_j = \mathcal{Y} v_j$, $j = 1, \ldots, m$, of Niessen should be related. We shall show that indeed they are, but the situation is more complicated than merely choosing appropriate solutions $\{y_j\}_{j=1}^{n}$, one from each of the Niessen subspaces $N_j \cap L_A^2(a,b)$, $j = 1, \ldots, n$. Some χ solutions are to be found in this manner, but others are not.

We decompose χ into two components χ_1 and χ_2, as was done in Chapters VII and VIII by multiplying by an appropriate matrix E_a.

$$(\chi_1\chi_2)(x,\lambda) = \chi(x,\lambda)E_a \, .$$

The elements in χ_2 generate automatic boundary conditions. These are generated only by Niessen solutions from the one-dimensional subspaces $N_j \cap L_A^2(a,b)$, $j = m-n+1,\ldots,n$, which are associated with the eigenvalues $\mu_j = 0$, $j = m+n-1,\ldots,n$. We shall return to these shortly.

Our current attention, however, must be focused on the elements represented by χ_1. These depend upon *all* the solutions $\{y_j\}_{j=1}^m$. If $V = (v_1,\ldots,v_n)$ represents the matrix of all limiting eigenvectors, then $\chi_1 = \mathcal{Y}VC$, where $C = (C_1C_2C_3C_4)^T$. C_1 is $2n\times(m-n)$, C_2 is $2n\times(2n-m)$, C_3 is $2n\times(m-n)$ and C_4 is $2n\times(2n-m)$ (all zeros). C is, of course, fully determined by the first $m-n$ columns of $V^* \begin{pmatrix} I_n \\ M \end{pmatrix} E_a$. C_1 multiplies $y_1,\ldots,y_{m-n}$, those L_A^2 solutions corresponding to positive eigenvalues $\mu_1,\ldots,\mu_{m-n}$. C_2 multiplies $y_{m-n+1},\ldots,y_n$, those L_A^2 solutions corresponding to zero eigenvalues $\mu_{m-n+1},\ldots,\mu_n$. C_3 multiplies $y_{n+1},\ldots,y_m$, those L_A^2 solutions corresponding to negative eigenvalues $\mu_{n+1},\ldots,\mu_m$. $C_4 = 0$. Recall

$$\mu_{n+j} = -1/[4(\operatorname{Im}\lambda)^2\mu_j], \ (-\mu_j/\mu_{n+j}) = \rho_j^2, \quad j = 1,\ldots,n \, .$$

Now we have

$$= C_1^* \begin{pmatrix} \mu_1 & & \\ & \ddots & \\ & & \mu_{m-n} \end{pmatrix} C_1 + C_3^* \begin{pmatrix} \mu_{n+1} & & \\ & \ddots & \\ & & \mu_m \end{pmatrix} C_3$$

$$= C_1^* \begin{pmatrix} \mu_1 & & \\ & \ddots & \\ & & \mu_{m-n} \end{pmatrix} C_1 - C_3^* \begin{pmatrix} \mu_1/\rho_1 & & \\ & \ddots & \\ & & \mu_{m-n}/\rho_{m-n} \end{pmatrix} C_3 \, ,$$

where $[Y,Z] = \lim_{x\to a} Z^*(J(i)Y$. Further, since χ_1 is $m-n$-dimensional and $[\chi_1,\chi_1] = 0$, both C_1 and C_3 are nonsingular. Hence, letting D_1 represent

$$\begin{pmatrix} \mu_1 & & \\ & \ddots & \\ & & \mu_{m-n} \end{pmatrix}^{1/2} ,$$

D_2 represent

$$\begin{pmatrix} \mu_1/\rho_1 & & \\ & \ddots & \\ & & \mu_{m-n}/\rho_{m-n} \end{pmatrix}^{1/2} ,$$

we have

$$(D_1^{*-1}C_1^{*-1}C_3^*D_2^*)\,(D_2C_2C_1^{-1}D_1^{-1}) = I\,,$$

and $D_2C_2C_1^{-1}D_1^{-1} = U$, a unitary matrix. Hence $C_3 = D_2^{-1}UD_1C_1$, or

$$C_3 = \begin{pmatrix} \rho_1 u_{11} & \cdots & \rho_1(\mu_{m-n}/\mu_1)^{1/2}u_{1,m-n} \\ \vdots & & \vdots \\ \rho_{m-n}(\mu_1/\mu_{m-n})^{1/2}u_{m-n,1} & \cdots & \rho_{m-n}u_{m-n,m-n} \end{pmatrix} C_1\,.$$

The general term in the first matrix is $(\rho_i(\mu_j/\mu_i)^{1/2}u_{ij})$. We set

$$P = (\rho_i(\mu_j/\mu_i)^{1/2}u_{ij})\,.$$

χ_2 consists only of elements from one-dimensional spaces $N_j \cap L_A^2(a,b)$, which correspond to solutions $y_{m-n+1},\ldots,y_n$. So $\chi_2 = \mathcal{Y}VD$, where

$$D = (D_1, D_2, D_3, D_4)^T \quad \text{and} \quad D_1 = 0,\ D_3 = 0,\ D_4 = 0.$$

We therefore have

IX.1.1. Theorem. *Let $\mu_1,\ldots,\mu_m$ be the finite Niessen eigenvalues, let $\rho_J = \sqrt{-\mu_j/\mu_{n+j}}$, $j = 1,\ldots,m-n$, and let $V = (v_1, v_2,\ldots,v_n)$ be the limiting Niessen eigenvectors. Then for some unitary $(m-n) \times (m-n)$ matrix $U = (u_{ij})$,*

$$\chi(x,\lambda) = \mathcal{Y}(x,\lambda)V \begin{pmatrix} C_1 & 0 \\ C_2 & D_2 \\ PC_1 & 0 \\ 0 & 0 \end{pmatrix} E_a^{-1}\,,$$

where $P = (\rho_i(\mu_j/\mu_i)^{1/2}u_{ij})$. Further

$$\begin{pmatrix} I \\ M \end{pmatrix} = V \begin{pmatrix} C_1 & 0 \\ C_2 & D_2 \\ PC_1 & 0 \\ 0 & 0 \end{pmatrix} E_a^{-1}\,.$$

If we let $\mathcal{Y}_+ = \mathcal{Y}(V_1,\ldots,V_{m-n})$, $\mathcal{Y}_0 = \mathcal{Y}(V_{m-n+1},\ldots,V_n)$, $\mathcal{Y}_- = \mathcal{Y}(V_{n=1},\ldots,V_j)$, then the variable component (as U varies), χ_1, can be expressed by

$$\chi_1(x,\lambda) = (\mathcal{Y}_+ + \mathcal{Y}_-P)C_1 + \mathcal{Y}_0C_2\,.$$

The component

$$\mathcal{Y}_+ + \mathcal{Y}_-P = \mathcal{Y}V \begin{pmatrix} I_{m-n} \\ 0 \\ P \\ 0 \end{pmatrix} C_1$$

is isomorphic to P. As U varies, P describes the intersection of $(m - n)$ mutually orthogonal complex ellipsoids. Each point on the surface of intersection gives a representation of χ_1.

When U is diagonal $(U = \delta_{jk} \exp(i\theta_j))$, the components of $\mathcal{Y}_+ + \mathcal{Y}_- P$ are indeed Niessen circle solutions. In this case P is diagonal,

$$P = (\rho_j \delta_{jk} \exp(i\theta_j)),$$

and

$$\mathcal{Y}_+ + \mathcal{Y}_- P = ([y_1 + \rho_1 e^{i\theta_1} y_{n+1}] \cdots [y_{m-n} + \rho_{m-n} e i\theta_{m-n} y_m])C_1,$$

where the bracketed terms are in $N_1 \cap L_A^2(a, b), \ldots, N_{m-n} \cap L_A^2(a, b)$. The diagonal choice of P is in a sense like choosing points on the major and minor axes of an ellipse.

IX.2   A Direct Approach to the $M(\lambda)$ Surface

Having determined that the size of the unitary matrix U, which allows variability over the $M(\lambda)$ surface, is $m - n \times m - n$, we are now in a position to examine $M(\lambda)$ directly using the theory of Hinton and Shaw.

We know that $M(\lambda)$ is nonsingular. Therefore if there are additional square integrable solutions of $(JY' = (\lambda A + B)Y)$ when Im $\lambda \neq 0$, they must be combinations of columns of ϕ.

IX.2.1. Theorem. *Let $\mu_1(b') \leq \cdots \leq \mu_n(b')$ be the eigenvalues of $\mathcal{D}(b', \lambda)$. Let there be $m, n \leq m \leq 2n$, solutions of $(JY' = (\lambda A + B)Y$, Im $\lambda \neq 0$, in $L_A^2(a, b)$. Then as b' approaches b, $\mu_1(b'), \ldots, \mu_{m-n}(b')$ remain finite and $\mu_{m-n+1}(b'), \ldots, \mu_n(b')$ approach ∞.*

Proof. Linearly indepenent solutions generated by ϕ are of the form $y_j = \phi v_j$, where v_j is a unit eigenvector of $\mathcal{D}$ associated with μ_j. Since

$$2i \, \text{Im} \, \lambda \int_a^{b'} y_j^* A y_j \, dt = i(\, \text{sgn} \, \lambda) \, \mu_j,$$

we see that as $b' \to b$ (through subsequences if necessary) that y_j is square integrable whenever μ_j remains bounded.

We summarize: For Im $\lambda \neq 0$, $(JY' = (\lambda A + B)Y)$ has n solutions in $L_A^2(a, b)$ generated by χ, $m - n$ solutions in $L_A^2(a, b)$ generated by ϕ.

Since $\mathcal{D}$ is positive definite at b', it can be written as

$$\mathcal{D} = V \begin{pmatrix} \mu_1 & & \\ & \ddots & \\ & & \mu_n \end{pmatrix} V^*,$$

where $V = (V_1, \ldots, V_n)$ are normalized mutually orthogonal eigenvectors for $\mu_1, \ldots, \mu_n$. Note that V is a unitary matrix.

Since $R = \mathcal{D}^{-1/2}$, it has the representation

$$R = V \begin{pmatrix} r_1 & & \\ & \ddots & \\ & & r_n \end{pmatrix} V^*,$$

where $r_j^2 = 1/\mu_j$, $j = 1, \ldots, n$. As b' approaches b, $r_{m-n+1}, \ldots, r_n$ approach O.

We now insert the decomposition of R into the expression for $M_{b'} = C + RUR$

$$M_{b'} = C + V \begin{pmatrix} r_1 & & \\ & \ddots & \\ & & r_n \end{pmatrix} V^*.$$

We let $X = V^*(M_{b'} - C)V$ and $\widetilde{U} = V^*UV$. Then, noting that $\widetilde{U}$ is also unitary,

$$X = \begin{pmatrix} r_1 & & \\ & \ddots & \\ & & r_n \end{pmatrix} \widetilde{U} \begin{pmatrix} r_1 & & \\ & \ddots & \\ & & r_n \end{pmatrix}. \qquad \square$$

IX.2.2. Theorem. *For finite b', $M_{b'}(\lambda) = C_{b'}$ is unitary equivalent to*

$$X = \begin{pmatrix} r_1 & & \\ & \ddots & \\ & & r_n \end{pmatrix} \widetilde{U} \begin{pmatrix} r_1 & & \\ & \ddots & \\ & & r_n \end{pmatrix}.$$

X represents the intersection of n mutually orthogonal complex ellipsoidal cylinders in n^2-dimensional complex space as $\widetilde{U}$ varies.

Proof. If we solve for $\widetilde{U}$, we find

$$\begin{pmatrix} x_{11}/r_1^2 & \cdots & x_{1n}\ldots x_{1n}/r_1 r_n \\ \vdots & & \vdots \\ x_{n1}/r_n r_1 & & x_{nn}/r_n^2 \end{pmatrix} = \begin{pmatrix} \widetilde{u}_{11} & \cdots & \widetilde{u}_{in} \\ \vdots & & \vdots \\ \widetilde{u}_{n1} & & \widetilde{u}_{nn} \end{pmatrix}.$$

Since unitary matrices have columns that are orthonormal, we find

$$\sum_{j=1}^n \frac{|x_{jk}|^2}{r_j^2 \, r_k^2} = 1, \quad k = 1, \ldots, n,$$

and

$$\sum_{j=1}^n \frac{x_{jk}\overline{x}_{jl}}{r_j^2 \, r_k \, r_l} = 0, \quad k, l = 1, \ldots, n, \quad k \neq l.$$

(We have used columns here. Rows would do just as well. To arrive at the equivalent row equations, merely replace x_{jk} by x_{kj} throughout.)

We are also in a position to see what the limiting $M(\lambda)$ surface (as $b' \to b$) is.

$$\square$$

IX.2.3. Theorem. *Let $(JY' = (\lambda A + B)Y)$ have m solutions in $L^2_A(a,b)$ when* $Im\ \lambda \neq 0$. *The limiting $M(\lambda)$ surface is unitarily equivalent to*

$$
X = \begin{pmatrix} r_1 & & & & \\ & \ddots & & & \\ & & r_{m-n} & & \\ & & & 0 & \\ & & & & \ddots \\ & & & & & 0 \end{pmatrix} U \begin{pmatrix} r_1 & & & & \\ & \ddots & & & \\ & & r_{m-n} & & \\ & & & 0 & \\ & & & & \ddots \\ & & & & & 0 \end{pmatrix}.
$$

X represents the intersection of $m - n$ mutually orthogonal complex ellipsoidal cylinders in $(m - n)$-dimensional complex space as $\widetilde{U}$ varies. The components of X, x_{jk}, are 0 if j or k exceeds $m - n$.

Proof. The equations previously derived still hold provided the limit on the sums (for j) is $m-n$, and k and ℓ are not greater than $m-n$. The limiting representation for X clearly shows that $x_{jk} = 0$ if j or k exceeds $m - n$.

In some sense, therefore the limiting surface is a "flattened football (American style)".

Note in the limiting case $\widetilde{U}$ is not completely determined. Since some of the components of X vanish as $b' \to b$, those components of $\widetilde{U}$ may disappear, making the limiting $\widetilde{U}$ possibly contracting. $\qquad\square$

IX.3　Examples

1. We note that if $\widetilde{U}$ is diagonal ($\widetilde{U} = (\delta_{jk} \exp i\theta_j)$), then

$$
X = \begin{pmatrix} r_1 & & & & \\ & \ddots & & & \\ & & r_{m-n} & & \\ & & & 0 & \\ & & & & \ddots \\ & & & & & 0 \end{pmatrix} \begin{pmatrix} \exp i\theta_1 & & \\ & \ddots & \\ & & \exp i\theta_n \end{pmatrix}
$$

$$
\times \begin{pmatrix} r_1 & & & & \\ & \ddots & & & \\ & & r_{m-n} & & \\ & & & 0 & \\ & & & & \ddots \\ & & & & & 0 \end{pmatrix}
$$

$$
= \begin{pmatrix}
r_1^2 \exp i\theta_1 & & & & & & \\
& \ddots & & & & & \\
& & r_{m-n}^2 \exp i\theta_1 & & & & \\
& & & \ddots & & & \\
& & & & r_{m-n}^2 \exp i\theta_{m-n} & & \\
& & & & & 0 & \\
& & & & & & \ddots \\
& & & & & & & 0
\end{pmatrix}.
$$

The equations $X_{jj} = r_j^2 \exp i\theta_j$, $j = 1,\ldots, m-n$, represent $m-n$ mutually orthogonal "circles". (In particular, when $m = 2$, $n = 1$, $x_{11} = r_1^2 \exp i\theta_1$ is the original Weyl circle.) Note that these circles are really "extremals" on the ellipsoids, that is, extreme "points" taken along the axes of ellipsoids. In fact, we are looking at $m-n$ one-dimensional ellipsoids.

2. The previous example can be clarified if we examine the situation when $n = 2$. Before the limit is taken,

$$
\begin{pmatrix} x_{11} & x_{12} \\ x_{21} & x_{22} \end{pmatrix} = \begin{pmatrix} r_1 & 0 \\ 0 & r_2 \end{pmatrix} \begin{pmatrix} \widetilde{u}_{11} & \widetilde{u}_{12} \\ \widetilde{u}_{21} & \widetilde{u}_{22} \end{pmatrix} \begin{pmatrix} r_1 & 0 \\ 0 & r_2 \end{pmatrix}.
$$

The elliptical and orthogonal equations are

$$
\frac{|x_{11}|^2}{r_1^4} + \frac{|x_{21}|^2}{r_1^2 r_2^2} = 1, \qquad \frac{|x_{12}|^2}{r_1^2 r_2^2} + \frac{|x_{22}|^2}{r_2^4} = 1, \qquad \frac{x_{11}\,\overline{x}_{12}}{r_1^3 r_2} + \frac{x_{21}\,\overline{x}_{22}}{r_1 r_2^3} = 0.
$$

If $m = 4$, this is also the limiting situation. If $\widetilde{U}$ is diagonal,

$$
U = \begin{pmatrix} \exp i\theta_1 & 0 \\ 0 & \exp i\theta_2 \end{pmatrix}.
$$

Then $x_{12} = x_{21} = 0$, $x_{11} = r_1^2 \exp i\theta_1$, $x_{22} = r_2^2 \exp i\theta_2$. The equations for x_{11} and x_{22} represent two Weyl circles. That these are extremals of the ellipses is easily seen. When $x_{12} = x_{21} = 0$, we have $|x_{11}|^2/r_1^4 = 1$, $|x_{22}|^2/r_2^4 = 1$. We are along the major or minor axis in each case.

If $m = 3$, then $r_2 \to 0$ as $b' \to b$. This forces $x_{12} = x_{21} = x_{22} = 0$. Only the equation $x_{11} = r_1^2 u_{11}$ remains. Since $|u_{11}|^2 \leq 1$, we may be at any point inside the limiting "circle" $x_{11} = r_1^2 \exp i\theta_1$. A contracting $\widetilde{U}$ may be used.

If $m = 2$, both r_1 and $r_2 = 0$. $X = 0$. This is the limit point case $M = C_b$.

References

[1]   A. M. Krall, *The decomposition of $M(\lambda)$ surfaces using Niessen' limit circles*, J. Math. Anal. Appl. **154** (1991), 292–300.

[2]   ______, *Direct decomposition of $M(\lambda)$ matrices for Hamiltonian systems*, Appl. Anal. **41** (1991), 237–245.

Chapter X

The Spectral Resolution for Linear Hamiltonian Systems with One Singular Point

We have reached an interesting point. The next goal is to find out exactly what the abstract spectral resolution, derived for arbitrary self-adjoint operators in a Hilbert space, looks like when applied to the self-adjoint linear Hamiltonian systems of Hinton and Shaw. Remarkably we can find detailed formulas for the spectral measure and the Hilbert space it generates, far more than is possible for the setting employed by Niessen.

X.1 The Specific Problem

Consider the boundary value problem

$$JY' = (\lambda A + B)Y\,, \qquad x \in [a,b)\,,$$
$$(\alpha_1 \alpha_2)Y(a) = 0\,, \qquad (\beta_1 \beta_2)Y(b') = 0\,, \qquad b' < b\,,$$

where Y is $2n$-dimensional, $J = \begin{pmatrix} 0 & -I_n \\ I_n & 0 \end{pmatrix}$, $A = A^* \geq 0$ and $B = B^*$ are real and locally integrable, and $\alpha_1, \alpha_2, \beta_1, \beta_2$ satisfy

$$\alpha_1 \alpha_1^* + \alpha_2 \alpha_2^* = I_n\,, \qquad\qquad \beta_1 \beta_1^* + \beta_2 \beta_2^* = I_n\,,$$
$$\alpha_1 \alpha_2^* - \alpha_2 \alpha_1^* = 0\,, \qquad\qquad \beta_1 \beta_2^* - \beta_2 \beta_1^* = 0\,.$$

In addition, for technical reasons, assume that if Y satisfies $JY' - BY = AF$, $AY = 0$, then $Y = 0$. In Chapter VII it was shown that the boundary value problem is self-adjoint in $L_A^2(a,b')$ and that, as b' approaches a singularity b, solutions in $L_A^2(a,b)$ can be found (Im $\lambda \neq 0$), boundary conditions at b can be defined, and a suitable self-adjoint boundary value problem on $[a,b)$ can be defined.

Our purpose here is to use the $[a,b')$ eigenfunction expansion to derive a spectral resolution for the singular problem on $[a,b)$. Much of what follows is similar to the problems discussed in Brauer [2], Coddington and Levinson [3], and Krall [8]. But at a number of critical places new reasons to sidestep difficulties had to be invented.

X.2   The Spectral Expansion

Recall that we denote by $\mathcal{Y}(x, \lambda)$ the $2n \times 2n$ matrix solution of $JY' = (\lambda A + B)Y$ that satisfies $\mathcal{Y}(a, \lambda) = \begin{pmatrix} \alpha_1^* & -\alpha_2^* \\ \alpha_2^* & \alpha_1^* \end{pmatrix}$. $\mathcal{Y}(x, \lambda)$ is then partitioned into two $2n \times n$ components θ, ϕ where $\theta = \begin{pmatrix} \theta_1 \\ \theta_2 \end{pmatrix}$, $\phi = \begin{pmatrix} \phi_1 \\ \phi_2 \end{pmatrix}$, with each piece θ_1, θ_2, ϕ_1, ϕ_2 being $n \times n$. Hence

$$\mathcal{Y}(x, \lambda) = (\theta(x, \lambda)\phi(x, \lambda)) = \begin{pmatrix} \theta_1(x, \lambda) & \phi_1(x, \lambda) \\ \theta_2(x, \lambda) & \phi_2(x, \lambda) \end{pmatrix}.$$

Note that ϕ satisfies the boundary condition at $x = a$.

For Im $\lambda \neq 0$, the solution satisfying the boundary condition at $x = b'$ is

$$\chi_{b'} = \theta + \phi M_{b'},$$

where

$$M_{b'} = -((\beta_1\beta_2)\phi(b', \lambda))^{-1}((\beta_1\beta_2)\theta(b', \lambda)).$$

As b' approaches a singularity b, it can be shown that $\chi_{b'}$ converges to a solution χ which is in $L_A^2(a, b)$. Further the singular boundary condition we assume at $x = b$ is

$$\lim_{x \to b} \chi^* JY(x) = 0.$$

For all λ, $M_{b'}$ is analytic in λ except for real poles, which are eigenvalues for the regular problem on $[a, b']$. We denote these eigenvalues by $\{\lambda_k\}_{k=1}^{\infty}$. Their corresponding eigenfunctions are $\{V_k\}_{k=1}^{\infty}$. For each λ_k,

$$V_k = \phi(x, \lambda_k)K_k,$$

where K_k is an $n \times 1$ matrix. We assume that at multiple eigenvalues, the corresponding eigenfunctions have been made orthogonal.

X.2.1. Theorem.   *Let $F(x)$ be an arbitrary element in $L_A^2(a, b')$. Then*

$$F(x) = \sum_{k=1}^{\infty} \phi_k(x)C_k,$$

where $\phi_k(x) = \phi(x, \lambda_k)$ and

$$C_k = \left[\int_a^{b'} \phi_k^* A\phi_k d\xi \right]^{-1} \int_a^{b'} \phi_k AF d\xi.$$

This is, of course, just the standard eigenfunction expansion. Note C_k is an $n \times 1$ matrix.

X.2.2. Theorem. *Let $F(x)$ be an arbitrary element in $L_A^2(a, b')$. Then*

$$\int_a^{b'} F^* A F d\xi = \sum_{k=1}^{\infty} \left[\int_a^{b'} F^* A \phi_k d\xi \right] \left[\int_a^{b'} \phi_k^* A \phi_k d\xi \right]^{-1} \left[\int_a^{b} \phi_k^* A F d\xi \right] .$$

This is Parseval's equality.

X.2.3. Definition. Let R_k^2 denote the $n \times n$ matrix

$$\left[\int_a^{b'} \phi_k^* A \phi_k d\xi \right]^{-1} .$$

X.2.4. Definition. Let

$$G(\lambda) = \int_a^{b'} \phi(\xi, \lambda)^* A(\xi) F(\xi) d\xi .$$

X.2.5. Definition. Let $P_{b'}(\lambda)$ be an $n \times n$ matrix-valued function satisfying

(1) $P_{b'}(0+) = 0$,

(2) $P_{b'}(\lambda)$ is increasing, jumping R_k^2 at $\lambda = \lambda_k$, but otherwise constant, continuous from above.

Thus

$$P_{b'}(\lambda) = \sum_{0 < \lambda_k \leq \lambda} R_k^2, \quad \lambda \geq 0,$$
$$= - \sum_{\lambda < \lambda_k \leq 0} R_k^2, \quad \lambda < 0.$$

X.2.6. Theorem (Parseval's Equality). *Let F be an arbitrary element of $L_A^2(a, b')$. Then*

$$\int_a^{b'} F^* A F d\xi = \int_{-\infty}^{\infty} G^*(\lambda) dP_{b'}(\lambda) G(\lambda) .$$

This can be extended by the polarization identities to inner products.

X.2.7. Corollary. *Let F_1, F_2 be arbitrary elements in $L_A^2(a, b')$. Let $G_1(\lambda)$ and $G_2(\lambda)$ correspond to them according to Definition X.2.4. Then*

$$\int_a^{b'} F_2^* A F_1 d\xi = \int_{-\infty}^{\infty} G_2^*(\lambda) dP_{b'}(\lambda) G_1(\lambda) .$$

We shall need this in the theorem to follow.

X.2.8. Theorem. *There exists a nondecreasing $n \times n$ matrix-valued function $P(\lambda)$, defined on $(-\infty, \infty)$, such that*

(1) $P(0+) = 0,$

(2) $P(\lambda) - P(\mu) = \lim\limits_{b' \to b}[P_{b'}(\lambda) - P_{b'}(\mu)], \qquad \lambda > \mu.$

Proof. Let $M_{b'}$ be on the circle $E(M)$, defined by setting $\chi_{b'}^*(b', \lambda)(J/i)\chi_{b'}(b', \lambda) = 0$. Set λ equal to λ_0, and $\chi_{b'}(x, \lambda_0) = \theta(x, \lambda_0) + \phi(x, \lambda_0)M_{b'}(\lambda_0)$. Then

$$J\chi_{b'}' = (\lambda_0 A + B)\chi_{b'} \,.$$

Apply Parseval's equality to $\chi_{b'}$:

$$\int_a^{b'} \chi_{b'}^* A\chi_{b'}\,d\xi = \int_{-\infty}^{\infty} \mathcal{G}(\lambda)^* dP_{b'}(\lambda)\mathcal{G}(\lambda) \,,$$

where

$$\mathcal{G}(\lambda) = \int_a^{b'} \phi^*(\xi, \lambda)A(\xi)\chi_{b'}(\xi, \lambda_0)d\xi \,.$$

Now

$$J\chi_b' = (\lambda_0 A + B)\chi_{b'},$$
$$J\phi_k' = (\lambda_k A + B)\phi_k$$

imply

$$\phi_k^* J\chi_{b'}\Big|_a^{b'} = (\lambda_0 - \lambda_k) \int_a^{b'} \phi_k^* A\chi_{b'}\,d\xi \,.$$

At b' both ϕ_k and $\chi_{b'}$ satisfy $(\beta_1\beta_2)Y(b') = 0$, so the upper limit on the left is 0. At $x = a$,

$$\phi_k^* J\chi_{b'}(a) = (-\alpha_2\alpha_1)\begin{pmatrix} 0 & -I_n \\ I_n & 0 \end{pmatrix}\begin{pmatrix} \alpha_1^* & -\alpha_2^* \\ \alpha_2^* & \alpha_1^* \end{pmatrix}\begin{pmatrix} I_n \\ M_{b'} \end{pmatrix} = I_n \,.$$

So

$$\mathcal{G}(\lambda_k) = \int_a^{b'} \phi_k^* A\chi_{b'}\,d\xi = (\lambda_k - \lambda_0)^{-1}I_n \,.$$

Parseval's equality becomes

$$\int_a^{b'} \chi_{b'}^* A\chi_{b'}\,d\xi = \int_{-\infty}^{\infty} |\lambda - \lambda_0|^{-2}dP_{b'}(\lambda) \,.$$

An easy calculation using the differential equation for $\chi_{b'}$ shows

$$\int_a^{b'} \chi_{b'}^* A\chi_{b'}\,d\xi = [M_{b'} - M_{b'}^*]/(2i\mathrm{Im}\,\lambda_0) \,.$$

If $\lambda_0 = i = \sqrt{-1}$, then

$$\int_{-\infty}^{\infty} |i - \lambda|^{-2} dP_{b'}(\lambda) = [M_{b'}(i) - M_{b'}^*(i)]/(2i) .$$

Since $|i - \lambda|^2 = \lambda^2 + 1$, we see there is a $K > 0$ such that

$$\int_{-\infty}^{\infty} (\lambda^2 + 1)^{-1} dP_{b'}(\lambda) < K .$$

This implies for $\mu > 0$ that

$$\int_{-\mu}^{\mu} dP_{b'}(\lambda) < \mathrm{K}[1 + \mu^2] .$$

This implies since $P_{b'}$ is increasing,

$$0 \le P_{b'}(\mu) < \mathrm{K}[1 + \mu^2],$$
$$0 \le -P_{b'}(-\mu) < \mathrm{K}[1 + \mu^2]$$

for all $\mu \ge 0$. Therefore $P_{b'}(\lambda)$ is uniformly bounded on compact subintervals of
the real line. Helly's first theorem shows there is a subsequence that converges
"weakly" to $P(\lambda)$ with the properties stated. $\square$

X.2.9. Theorem. *If F is in $L_A^2(a, b)$, there is a function $G(\lambda)$ in $L_p^2(-\infty, \infty)$, with
inner product and norm square*

$$\langle G, H \rangle_p = \int_{-\infty}^{\infty} H^* dPG , \quad \|G\|_p^2 = \left[\int_{-\infty}^{\infty} G^* dPG \right]^{1/2}$$

such that if

$$E(\lambda) = G(\lambda) - \int_a^{b'} \phi^*(\xi, \lambda) A(\xi) F(\xi) d\xi ,$$

then

$$\lim_{b' \to b} \int_{-\infty}^{\infty} E(\lambda)^* dP(\lambda) E(\lambda) = 0 ,$$

and

$$\int_a^b F(\xi)^* A(\xi) F(\xi) d\xi = \int_{-\infty}^{\infty} G(\lambda)^* dP(\lambda) G(\lambda) .$$

Proof. 1. Let F be in $C_0^1(a, b)$. Then F is in D, the domain of functions satisfying
$JF' - BF = AH$ for some H in $L_A^2(a, b)$, together with the boundary condition
at $x = a$ and the singular boundary condition at $x = b$. The operator L is defined
by setting $LF = H$ if and only if $JF' - BF = AH$ for all F in D.

If b' is large enough,

$$\int_a^{b'} (LF)^* A(LF) = \int_a^{b'} H^* A H \, d\xi$$

$$= \sum_{k=1}^{\infty} \left(\int_a^b \phi_k^* A H \, d\xi \right)^* R_k^2 \left(\int_a^b \phi_k^* A H \, d\xi \right) .$$

But

$$\int_a^b \phi_k^* A H \, d\xi = \int_a^{b'} \phi_k^* [JF' - BF] \, d\xi$$

$$= \int_a^{b'} [J\phi_k' - B\phi_k]^* F \, d\xi$$

$$= \lambda_k \int_a^{b'} \phi_k^* A F \, d\xi$$

$$= \lambda_k G(\lambda_k) .$$

Thus

$$\int_a^b (LF)^* A(LF) d\xi = \int_{-\infty}^{\infty} \lambda^2 G(\lambda)^* dP_{b'}(\lambda) G(\lambda) .$$

Let $N > 0$. Then

$$\left(\int_{-\infty}^{-N} + \int_N^{\infty} \right) G^* dP_{b'} G \le \frac{1}{N^2} \left(\int_{-\infty}^{-N} + \int_N^{\infty} \right) \lambda^2 G^* dP_{b'} G$$

$$\le \frac{1}{N^2} \int_{-\infty}^{\infty} \lambda^2 G^* dP_{b'} G$$

$$\le \frac{1}{N^2} \int_a^b (LF)^* A(LF) d\xi .$$

So, since

$$\int_a^b F^* A F \, d\xi = \left(\int_{-\infty}^{-N} + \int_{-N}^N + \int_N^{\infty} \right) G^* dP_{b'} G ,$$

$$\left| \int_a^b F^* A F \, d\xi - \int_{-N}^N G^* dP_{b'} G \right| = \left(\int_{-\infty}^{-N} + \int_N^{\infty} \right) G^* dP_{b'} G$$

$$\le \frac{1}{N^2} \int_a^b (LF)^* A(LF) d\xi .$$

Let b' approach b. Helly's second theorem (Hinton and Shaw [7]) implies

$$\left| \int_a^b F^* A F \, d\xi - \int_{-N}^N G^* dP G \right| \le \frac{1}{N^2} \int_a^b (LF)^* A(LF) d\xi .$$

Let N approach ∞.

$$\int_a^b F^* A F d\xi = \int_{-\infty}^\infty G^* dPG\,,$$

provided F is in $C_0^1(a,b)$.

2. Let F vanish near b but otherwise be arbitrary in $L_A^2(a,b)$. Choose $\{F_j\}_{j=1}^\infty$ in $C_0^1(a,b)$ such that

$$\lim_{j\to\infty}\int_a^b (F_j - F)^* A(F_j - F)d\xi = 0\,.$$

Apply Parseval's equality to $F_j - F_k$:

$$\int_a^b (F_j - F_k)^* A(F_j - F_k)d\xi = \int_{-\infty}^\infty (G_j - G_k)^* dP(G_j - G_k)\,,$$

where

$$G_j = \int_a^b \phi^* A F_j d\xi\,, \quad G_k = \int_a^b \phi^* A F_k d\xi\,.$$

Since $\lim_{j\to\infty} F_j = F$, $\{G_j\}_{j=1}^\infty$ is also a Cauchy sequence in $L_p^2(-\infty,\infty)$. Thus there is a G in $L_p^2(-\infty,\infty)$ such that $\lim_{j\to\infty} G_j = G$. Since

$$\left| G_j - \int_a^{b'} \phi^* A F d\xi \right| = \left| \int_a^{b'} \phi^* A(F_j - F)d\xi \right|$$

$$\leq \left(\int_a^{b'} \phi^* A\phi d\xi \right)^{1/2} \left(\int_a^{b'} [F_j - F]^* A[F_j - F]d\xi \right)^{1/2}$$

implies

$$\lim_{j\to\infty} G_j = \int_a^{b'} \phi^* A F d\xi\,,$$

which is continuous, we find

$$G = \int_a^b \phi^* A F d\xi\,, \quad \text{a.e.}$$

Thus if F vanishes near $x = b$,

$$\int_a^b F^* A F d\xi = \lim_{j\to\infty} \int_a^b F_j^* A F_j d\xi$$

$$= \lim_{j\to\infty} \int_{-\infty}^\infty G_j^* dPG_j$$

$$= \int_{-\infty}^\infty G^* dPG\,.$$

3. Finally, if F is arbitrary in $L_A^2(a, b)$, let

$$F_{b'} = F, \quad x \le b'$$
$$= 0, \quad x > b'.$$

Let

$$G_{b'} = \int_a^b \phi^* A F_{b'}\, d\xi = \int_a^{b'} \phi^* A F d\xi.$$

Since

$$\int_{-\infty}^{\infty} (G_c - G_d)^* dP(G_c - G_d) = \int_c^d F^* A F d\xi,$$

$\{G_{b'}\}$ is a Cauchy sequence as b' approaches b. Let $\lim_{b' \to b} G_{b'} = G$ in $L_p^2(-\infty, \infty)$. Letting b' approach b in the previous result,

$$\int_a^b F^* A F d\xi = \int_{-\infty}^{\infty} G^* dPG.$$

4. Since $G_{b'}$ approaches G in $L_p^2(-\infty, \infty)$,

$$\lim_{b' \to b} \int_{-\infty}^{\infty} [G(\lambda) - \int_a^{b'} \phi^* A F d\xi]^* dP(\lambda)[G(\lambda) - \int_a^{b'} \phi^* A F d\xi] = 0.$$

We remark that if the condition $JY' - BY = AF$, $AY = 0$ implies $Y = 0$ fails to hold, then the approximation used in the proof of this theorem may hold only on a subspace of $L_A^2(a, b)$. The subspace may be as small as having only one dimension. We shall give examples later. $\qquad\square$

X.2.10. Theorem. *If $G(\lambda)$ is the limit of $\displaystyle\int_a^{b'} \phi(\xi, \lambda)^* A(\xi) F(\xi) d\xi$ in $L_p^2(-\infty, \infty)$, then*

$$\int_{-\infty}^{\infty} \phi(x, \lambda) dP(\lambda) G(\lambda) = F(x)$$

in $L_A^2(a, b)$. That is,

$$\lim_{I \to (-\infty, \infty)} \int_a^b \left[F - \int_I \phi dPG\right]^* A \left[F - \int_I \phi dPG\right] d\xi = 0.$$

Proof. Let $I = (\mu, \nu)$, and

$$F_I(x) = \int_I \phi(x, \lambda) dP(\lambda) G(\lambda).$$

If b' is in $[a, b)$, then

$$\int_a^{b'} [F - F_I]^* A F_I d\xi = \int_a^{b'} [F - F_i]^* A \left[\int_I \phi dPG \right] d\xi$$

$$= \int_I \left[\int_a^{b'} [F - F_i]^* A \phi d\xi \right] dPG.$$

Likewise

$$\int_a^{b'} [F - F_I]^* A F d\xi = \int_{-\infty}^{\infty} \left[\int_a^{b'} [F - F_I]^* A \phi d\xi \right] dPG.$$

Subtracting, we get

$$\int_a^{b'} [F - F_i]^* A [F - F_i] d\xi = \int_{(-\infty,\infty)-I} \left[\int_a^{b'} [F - F_I]^* A \phi d\xi \right] dPG.$$

Now $\int_a^{b'} \phi^* A [F - F_I] d\xi$ is the transform of a function in $L_A^2(a, b)$ that vanishes
on (b', b). Consequently the integral from a to b' in brackets is in $L_p^2(-\infty, \infty)$.
Applying Schwarz's inequality,

$$\left(\int_a^{b'} [F - F_I]^* A [F - F_I] d\xi \right)^2$$

$$\leq \left(\int_{(-\infty,\infty)-I} \left[\int_a^{b'} \phi^* A [F - F_I] d\xi \right]^* dP \left[\int_a^{b'} \phi^* A [F - F_I] d\xi \right] \right)$$

$$\left(\int_{(-\infty,\infty)-I} G^* dPG \right).$$

The first integral on the right is less than or equal to

$$\int_a^{b'} [F - F_I]^* A [F - F_I] d\xi.$$

If this is inserted and cancelled,

$$\int_a^{b'} [F - F_I] A [F - F_I] d\xi \leq \int_{(-\infty,\infty)-I} G^* dPG.$$

Let b' approach b. Then let I approach $(-\infty, \infty)$. The result is that

$$F = \lim_{I \to (-\infty,\infty)} F_I, \quad \text{or} \quad F(x) = \lim_{(\mu,\nu) \to (-\infty,\infty)} \int_\mu^\nu \phi(x,\lambda) dP(\lambda) G(\lambda)$$

in $L_A^2(a,b)$. $\qquad\qquad\qquad\qquad\qquad\qquad\qquad\qquad\qquad\qquad\qquad\qquad\qquad\quad\square$

Theorem X.2.9 may be extended to involve inner products by use of the polarization identity. The inner product form of Parseval's equality is

$$\int_a^b F_2(\xi)^* A(\xi) F_1(\xi) d\xi = \int_{-\infty}^\infty G_2(\lambda)^* dP(\lambda) G_1(\lambda) \,,$$

where

$$G_j(\lambda) = \int_a^b \phi(\xi,\lambda)^* A(\xi) F_j(\xi) d\xi \,, \qquad j = 1,2 \,.$$

Theorems X.2.9 and X.2.10 may be extended to represent the resolvent operator $(L - \lambda_0 I)^{-1}$ when λ_0 is not in the support of $dP(\lambda)$. Parseval's equality is

$$\int_a^b [(L - \lambda_0 I)^{-1} F(\xi)]^* A(\xi) [(L - \lambda_0 I)^{-1} F(\xi)] d\xi = \int_{-\infty}^\infty \frac{G(\lambda)^* dP(\lambda) G(\lambda)}{|\lambda - \lambda_0|^2} \,.$$

The resolvent expansion is

$$(L - \lambda_0 I)^{-1} F(x) = \int_{-\infty}^\infty \phi(x,\lambda) dP(\lambda) \frac{G(\lambda)}{\lambda - \lambda_0} \,.$$

X.3   The Converse Problem

The preceding section began with choosing an F in $L_A^2(a,b)$, producing G in $L_p^2(-\infty, \infty)$, and then showing that F could be recovered from G. In this section we begin with G, produce F, and then recover G.

Without the assumption that $JY' - BY = 0$, $AY = 0$ implies $Y = 0$, $L_p^2(-\infty, \infty)$ may be too large in the sense that $G \to F \to \widetilde{G}$, but $\widetilde{G}$ may not equal G. $\widetilde{G}$ may be only in a subspace of $L_p^2(-\infty, \infty)$.

With the assumption that we made in the introduction (let $F = 0$), there is no difficulty.

X.3.1. Lemma. *Let $G(\lambda)$ be in $L_p^2(-\infty, \infty)$. Let*

$$F_I(x) = \int_I \phi(x,\lambda) dP(\lambda) G(\lambda) \,.$$

Then $\displaystyle\lim_{I \to (-\infty,\infty)} F_I(x)$ *exists in* $L_A^2(a,b)$.

Proof. Let $I_1 \subset I_2$. Then

$$
\begin{aligned}
F_{I_2} - F_{I_1} &= \int_{I_2 - I_1} \phi(x, \lambda) dP(\lambda) G(\lambda) \\
&= \int_{-\infty}^{\infty} \phi(x, \lambda) dP(\lambda) K_{I_2 - I_1}(\lambda) G(\lambda) ,
\end{aligned}
$$

where

$$
\begin{aligned}
K_{I_2 - I_1}(\lambda) &= 1 , \qquad \lambda \in I_2 - I_1 \\
&= 0 , \qquad \lambda \notin I_2 - I_1 .
\end{aligned}
$$

Let R be an arbitrary element of $L_A^2(a, b)$ that vanishes near b. Then

$$
\begin{aligned}
\int_a^b R^* A[F_{I_2} - F_{I_1}] d\xi &= \int_a^b R^* A \left[\int_{I_2 - I_1} \phi dPG \right] d\xi \\
&= \int_{I_2 - I_1} \left[\int_a^b \phi^* A R d\xi \right]^* dPG \\
&= \int_{I_2 - I_1} S^* dPG ,
\end{aligned}
$$

where S is the transform of R:

$$
S = \int_a^b \phi^* A R d\xi .
$$

We now let $R = F_{I_2} - F_{I_1}$ on $[a, b')$, but set $R = 0$ near b. Then

$$
\begin{aligned}
\int_a^b [F_{I_2} - F_{I_1}]^* A[F_{I_2} - F_{I_1}] d\xi &= \int_{I_2 - I_1} S^* dPG \\
&\leq \left(\int_{I_2 - I_1} S^* dPS \right)^{1/2} \left(\int_{I_2 - I_1} G^* dPG \right)^{1/2} \\
&\leq \left(\int_{-\infty}^{\infty} S^* dPS \right)^{1/2} \left(\int_{I_2 - I_1} G^* dPG \right)^{1/2} .
\end{aligned}
$$

But by Parseval's equality

$$
\int_a^b [F_{I_2} - F_{I_1}^* A[F_{I_2} - F_{I_1}] d\xi = \int_{-\infty}^{\infty} S^* dPS .
$$

Take the square root, cancel with the inequality above, then square:

$$
\int_a^b [F_{I_2} - F_{I_1}]^* A[F_{I_2} - F_{I_1}] d\xi \leq \int_{I_2 - I_1} G^* dPG .
$$

Since the right side is independent of b', let b' approach b. Then $F_{I_2} - F_{I_1}$ is in $L_A^2(a, b)$. As I approaches $(-\infty, \infty)$, the inequality also shows that $\{F_I\}$ forms a Cauchy sequence in $L_A^2(a, b)$, and therefore

$$\lim_{I \to (-\infty, \infty)} F_I = F$$

in $L_A^2(a, b)$. $\square$

X.3.2. Lemma. *Let $G(\lambda)$ be in $L_p^2(-\infty, \infty)$. Let*

$$F_I(x) = \int_I \phi(x, \lambda) dP(\lambda) G(\lambda) \,,$$

$$F(x) = \lim_{I \to (-\infty, \infty)} \int_I \phi(x, \lambda) dP(\lambda) G(\lambda) \,.$$

Let

$$\widetilde{G}(\lambda) = \int_a^b \phi^*(\xi, \lambda) A(\xi) F(\xi) d\xi \,,$$

and let

$$\widetilde{F}_I(x) = \int_I \phi(x, \lambda) dP(\lambda) \widetilde{G}(\lambda) \,.$$

Then

$$\lim_{I \to (-\infty, \infty)} \int_a^b [F_I - \widetilde{F}_I]^* A [F_I - \widetilde{F}_I] d\xi = 0 \,.$$

Proof. From Theorem X.2.10, $F(x) = \lim_{I \to (-\infty, \infty)} \widetilde{F}_I(x)$ in $L_A^2(a, b)$. But by definition, $F(x) = \lim_{I \to (-\infty, \infty)} F_I(x)$. The triangle inequality

$$\|F_I - \widetilde{F}_I\|_A \leq \|F_I - F\|_A + \|F - \widetilde{F}_I\|_A$$

shows that as I approaches $(-\infty, \infty)$, $\|F_I - \widetilde{F}_I\|_A$ approaches 0.

At this point we have, given G, there is an F. F yields $\widetilde{G}$, which again yields F. And so the process stops. We continue to show that G and $\widetilde{G}$ coincide. $\square$

X.3.3. Lemma. *Let λ_0 be a complex number with positive imaginary part, and let*

$$H_I(x, \lambda_0) = \int_I \phi(x, \lambda) dP(\lambda) \left[\frac{G(\lambda) - \widetilde{G}(\lambda)}{\lambda - \lambda_0} \right] \,.$$

Then for all fixed λ_0, $\lim_{I \to (-\infty, \infty)} H_I(x, \lambda_0) = 0$.

Proof. H_I satisfies

$$JH_I' - [\lambda_0 A + B]H_I = A(x)\int_I \phi(x,\lambda)dP(\lambda)[G(\lambda) - \widetilde{G}(\lambda)]\,.$$

Further, H_I satisfies the boundary condition

$$(\alpha_1\alpha_2)H_I(a,\lambda_0) = 0\,.$$

Thus

$$H_I(x) = \int_a^b G(\lambda_0,x,\xi)A(\xi)[F_I(\xi) - \widetilde{F}_I(\xi)]d\xi + \phi(x,\lambda_0)C\,,$$

where here $G(\lambda_0,x,\xi)$ is the Green's function for the singular boundary value problem. As I approaches $(-\infty,\infty)$, the integral approaches 0. Thus

$$H(x,\lambda_0) = \lim_{I\to(-\infty,\infty)} H_I(x,\lambda_0) = \phi(x,\lambda_0)C\,.$$

But

$$\lim_{x\to b}\chi(x,\overline{\lambda}_0)^* JH(x,\lambda_0) = 0\,,$$

for, since H_I satisfies the boundary condition as x approaches b, H must satisfy the singular boundary condition as x approaches b as well. Since

$$\lim\chi(x,\overline{\lambda}_0)^* J\phi(x,\lambda_0) = M(\lambda)\,,$$

which is nonsingular, $C = 0$. $\qquad\square$

X.3.4. Lemma. $G(\lambda) = \widetilde{G}(\lambda)$ *in* $L_p^2(-\infty,\infty)$.

Proof. Let K be a constant $2n \times 1$ matrix; let

$$Y_s(\lambda) = \int_a^s \phi(\xi,\lambda)^* A(\xi)KD\xi\,.$$

Then $Y_s(\lambda)$ is in $L_p^2(-\infty,\infty)$, since it is the transform of a function that vanishes near b. Then

$$\int_a^s K^* A(\xi)H_I(\xi,\lambda_0)d\xi = \int_I Y_s(\lambda)^* dP(\lambda)\left[\frac{G(\lambda) - \widetilde{G}(\lambda)}{\lambda - \lambda_0}\right]\,.$$

Let I approach $(-\infty,\infty)$. Then

$$0 = \int_{-\infty}^{\infty} Y_s(\lambda)^* dP(\lambda)\left[\frac{G(\lambda) - \widetilde{G}(\lambda)}{\lambda - \lambda_0}\right]\,.$$

We may assume without loss of generality that $G(\lambda) - \widetilde{G}(\lambda)$ is real. Consequently, if we take the imaginary part, we see that

$$0 = \int_{-\infty}^{\infty} Y_s(\lambda)^* dP(\lambda)[G(\lambda) - \widetilde{G}(\lambda)] \left[\frac{\nu_0}{(\lambda - \mu_0)^2 + \nu_0^2} \right].$$

If we integrate with respect to μ_0 from α to β, and reverse the order of integration, we have

$$0 = \int_{-\infty}^{\infty} Y_s(\lambda)^* dP(\lambda)[G(\lambda) - \widetilde{G}(\lambda)] \left[\tan^{-1}\left(\frac{\beta - \lambda}{\nu_0} \right) - \tan^{-1}\left(\frac{\alpha - \lambda}{\nu_0} \right) \right].$$

Letting ν_0 approach 0, we get

$$0 = \int_{\alpha}^{\beta} Y_s(\lambda)^* dP(\lambda)[G(\lambda) - \widetilde{G}(\lambda)],$$

or

$$0 = \int_{a}^{s} \left[\int_{\alpha}^{\beta} K^* A(\xi)\phi(\xi, \lambda) dP(\lambda)[G(\lambda) - \widetilde{G}(\lambda)] \right] d\xi.$$

Differentiate with respect to s,

$$0 = \int_{\alpha}^{\beta} K^* A(x)\phi(x, \lambda) dP(\lambda)[G(\lambda) - \widetilde{G}(\lambda)].$$

Now apply an extension of the mean value theorem, remembering that the expression is 1×1 and ϕ is analytic in λ. We find for some λ_0 in $[\alpha, \beta]$

$$K^* A(x)\phi(x, \lambda_0) \int_{\alpha}^{\beta} dP(\lambda)[G(\lambda) - \widetilde{G}(\lambda)] = 0.$$

Since $A(x)\phi(x, \lambda_0)v$ is only zero for all x when $v = 0$, we can choose K appropriately to conclude that

$$\int_{\alpha}^{\beta} dP(\lambda[G(\lambda) - \widetilde{G}(\lambda)] = 0$$

for all α, β. We use this to build up integrals involving step functions, dense in $L_p^2(-\infty, \infty)$, that have as their limit

$$\int_{-\infty}^{\infty} [G(\lambda) - \widetilde{G}(\lambda)]^* dP(\lambda)[G(\lambda) - \widetilde{G}(\lambda)] = 0.$$

Hence $G = \widetilde{G}$ in $L_p^2(-\infty, \infty)$. $\qquad\square$

We summarize.

X.3.5. Theorem. *If $G(\lambda)$ is in $L_p^2(-\infty, \infty)$, there is a unique $F(x)$ in $L_A^2(a, b)$ such that*

$$F(x) = \int_{-\infty}^{\infty} \phi(x, \lambda) dP(\lambda) G(\lambda)$$

and

$$G(\lambda) = \int_a^b \phi(\xi, \lambda)^* A(\xi) F(\xi) d\xi.$$

X.4   The Relation Between $\mathbf{M(\lambda)}$ and $\mathbf{P(\lambda)}$

The matrix $M(\lambda)$ can frequently be determined by careful inspection of the solutions of $JY' = (\lambda A + B)Y$ to determine appropriate solutions in $L_A^2(a, b)$. More difficult is the determination of the spectral measure $P(\lambda)$, since its existence follows from Helly's selection theorems. Fortunately, they are intimately connected. The following is inserted as a reminder.

X.4.1. Theorem. *Let $M(\lambda)$ be on the limit-circle and let $\chi(x, \lambda) = \theta(x, \lambda) + \phi(x, \lambda)M(\lambda)$ for $\lambda = \mu + i\nu$, $\nu \neq 0$. Then*

$$\int_a^b \chi(\xi, \lambda)^* A(\xi) \chi(\xi, \lambda) d\xi = [M(\lambda) - M(\lambda)^*]/(2i\text{Im }\lambda).$$

Proof. This follows from Green's formula, initial conditions at $x = a$ and Weyl's second theorem in its extended form. $\qquad\square$

X.4.2. Theorem. *Let $\lambda_0 = \mu + i\nu$, $\nu \neq 0$. Then*

$$\int_a^b \chi(\xi, \lambda_0)^* A(\xi) \chi(\xi, \lambda_0) d\xi = \int_{-\infty}^{\infty} |\lambda - \lambda_0|^{-2} dP(\lambda).$$

Proof. The theorem is true if b is replaced by b'. Let b' approach b. $\qquad\square$

X.4.3. Theorem. *If λ_1 and λ_2 are real, then*

$$P(\lambda_2) - P(\lambda_1) = \lim_{\nu \to 0+} \frac{1}{\pi} \int_{\lambda_1}^{\lambda_2} Im\ M(\mu + i\nu) d\mu.$$

Further, if λ_1 and λ_2 have nonzero imaginary parts, then

$$M(\lambda_2) - M(\lambda_1) = \int_{-\infty}^{\infty} [(\lambda - \lambda_1)^{-1} - (\lambda - \lambda_1)^{-1}] dP(\lambda).$$

Proof. We have

$$\int_{-\infty}^{\infty} |\lambda - \lambda_0|^{-2} dP(\lambda) = [M(\lambda_0) - M(\lambda_0)^*]/(2i\text{Im }\lambda_0).$$

So Im $M(\mu + i\nu) = [M(\lambda_0) - M(\lambda_0)^*]/2i$ satisfies

$$\text{Im } M(\mu + i\nu) = \int_{-\infty}^{\infty} \frac{\nu dP(\lambda)}{(\lambda - \mu)^2 + \nu^2} \, .$$

Integrate both sides from λ_1 to λ_2 with respect to μ.

$$\int_{\lambda_1}^{\lambda_2} \text{Im } M(\mu + i\nu)d\mu = \int_{\lambda_1}^{\lambda_2} \int_{-\infty}^{\infty} \frac{\nu dP(\lambda)}{(\lambda - \mu)^2 1 + \nu^2} \, d\mu$$
$$= \int_{-\infty}^{\infty} \left[\tan^{-1}\left(\frac{\lambda_2 - \lambda}{\nu}\right) - \tan^{-1}\left(\frac{\lambda_1 - \lambda}{\nu}\right) \right] dP(\lambda) \, .$$

Letting ν approach 0 from above, we get

$$\lim_{\nu \to 0+} \int_{\lambda_1}^{\lambda_2} \text{Im } M(\mu + i\nu)d\mu = \pi \int_{\lambda_1}^{\lambda_2} dP(\lambda) \, ,$$

$$= \pi[P(\lambda_2) - P(\lambda_1)] \, .$$

To validate the second part, note

$$\text{Im } [M(\lambda_0) - M(\lambda_1)] = \int_{-\infty}^{\infty} \left[\frac{\nu_2}{|\lambda - \lambda_2|^2} - \frac{\nu}{|\lambda - \lambda_1|^2} \right] dP(\lambda) \, ,$$
$$= \text{Im } \int_{-\infty}^{\infty} \left[\frac{(\lambda - \lambda_2)^*}{|\lambda - \lambda_2|^2} - \frac{(\lambda - \lambda_1)^*}{|\lambda - \lambda_1|^2} \right] dP(\lambda) \, ,$$
$$= \text{Im } \int_{-\infty}^{\infty} [(\lambda - \lambda_2)^{-1} - (\lambda - \lambda_1)^{-1}]dP(\lambda) \, .$$

We therefore have two functions, analytic in λ_1 and λ_2, whose imaginary parts are equal. The real parts can differ only by a constant. Letting $\lambda_2 = \lambda_1$ shows this constant to be 0. $\qquad\square$

X.5   The Spectral Resolution

We connect the result of Sections X.2, X.3 and X.4 to the resolution of the identity as an integral generated by a projection valued measure E_λ. Given

$$F(x) = \int_{-\infty}^{\infty} \phi(x, \lambda)dP(\lambda)G(\lambda) \, ,$$

where

$$G(\lambda) = \int_{a}^{b} \phi^*(\xi, \lambda)A(\xi)F(\xi)d\xi \, ,$$

we define

$$E_\lambda F(x) = \int_{-\infty}^{\lambda+} \phi(x, \lambda) dP(\lambda) G(\lambda) .$$

If we consider E_λ as the limit of eigenfunction expansions, we can show that it is a projection, is continuous from above, and satisfies $E_{\lambda_1} E_{\lambda_2} = E_{\lambda_1}$ when $\lambda_1 \leq \lambda_2$, as well as $E_{-\infty} = O$, $E_\infty = I$. If we let $\{\lambda_j\}_{j=-\infty}^\infty$ be a partition of $(-\infty, \infty)$, $\lambda_i < \lambda_j$ if $i < j$, and

$$\Delta_j EF(x) = \int_{\lambda_j}^{\lambda_{j+1}} \phi((x, \lambda) dP(\lambda) G(\lambda) ,$$

then $F(x) = \sum_{j=-\infty}^{\infty} \Delta_j EF(x)$. As $\{\lambda_j\}_{j=-\infty}^\infty$ becomes finer, we may write

$$F(x) = \int_{-\infty}^{\infty} dE_\lambda F(x)$$

as the limit of the decomposition above.

If Y is in D, it has the representation

$$Y(x) = \int_{-\infty}^{\infty} \phi(x, \lambda) dP(\lambda) G(\lambda) ,$$

where

$$G(\lambda) = \int_{a}^{b} \phi^*(\xi, \lambda) A(\xi) Y(\xi) d\xi .$$

Then

$$LY(x) = \int_{-\infty}^{\infty} \lambda \phi(x, \lambda) dP(\lambda) G(\lambda) .$$

This translates into

$$LY(x) = \int_{-\infty}^{\infty} \lambda dE_\lambda Y(x) .$$

The resolvent operator also has the standard representation. If λ_0 is complex,

$$(L - \lambda_0 I)^{-1} F(x) = \int_{-\infty}^{\infty} \frac{1}{\lambda - \lambda_0} dE_\lambda F(x) .$$

It is apparent that λ_0 is in the spectrum of L if and only if it is in the support of dE_λ or of $dP(\lambda)$.

X.6   An Example

Consider the fourth-order scalar problem $(y'')'' = \lambda y$ on $[0, \infty)$, together with boundary conditions $y(0) = 0$, $y'''(0) = 0$. ∞ is a limit-point, so no boundary condition at ∞ is required.

If we let $y_1 = y$, $y_2 = y'$, $y_3 = -y'''$, $y_4 = y''$, the problem is equivalent to

$$\begin{pmatrix} 0 & 0 & -1 & 0 \\ 0 & 0 & 0 & -1 \\ 1 & 0 & 0 & 0 \\ 0 & 1 & 0 & 0 \end{pmatrix} \begin{pmatrix} y_1 \\ y_2 \\ y_3 \\ y_4 \end{pmatrix} = \left[\begin{pmatrix} 1 & 0 & 0 & 0 \\ 0 & 0 & 0 & 0 \\ 0 & 0 & 0 & 0 \\ 0 & 0 & 0 & 0 \end{pmatrix} + \begin{pmatrix} 0 & 0 & 0 \\ 0 & 0 & 0 \\ 0 & 1 & 0 \\ 0 & 0 & 1 \end{pmatrix} \right] \begin{pmatrix} y_1 \\ y_2 \\ y_3 \\ y_4 \end{pmatrix},$$

with boundary condition

$$\begin{pmatrix} 1 & 0 & 0 & 0 \\ 0 & 0 & 0 & 1 \end{pmatrix} \begin{pmatrix} y_1 \\ y_2 \\ y_3 \\ y_4 \end{pmatrix} (0) = \begin{pmatrix} 0 \\ 0 \\ 0 \\ 0 \end{pmatrix}.$$

The initial condition for the fundamental matrix is

$$\begin{pmatrix} 1 & 0 & 0 & 0 \\ 0 & 0 & 0 & -1 \\ 0 & 0 & 1 & 0 \\ 0 & 1 & 0 & 0 \end{pmatrix}.$$

The components of the four 4×1 matrices that make its columns are combinations of $\sinh z_0 x$, $\cosh z_0 x$, $\sin z_0 x$, and $\cos z_0 x$, where $z_0 = \lambda^{1/4}$. Multiplying the fundamental matrix by $\begin{pmatrix} I_2 \\ M \end{pmatrix}$ and requiring the result to be in $L_A^2(0, \infty)$ show that

$$M = \begin{pmatrix} z_0^3(1+i)/2 & z_0(1-i)/2 \\ z_0(-i+1)/2 & (1+i)/2z_0 \end{pmatrix}.$$

As $\lambda = z_0^4$ approaches the positive real axis,

$$\operatorname{Im} M \to \begin{pmatrix} \lambda^{3/4}/2 & -\lambda^{1/4}/2 \\ -\lambda^{1/4}/2 & 1/2\lambda^{1/4} \end{pmatrix}.$$

As λ approaches the negative real axis, $\operatorname{Im} M \to 0$. Consequently,

$$dP(\lambda) = \frac{1}{\pi} \begin{pmatrix} \lambda^{3/4}/2 & -\lambda^{1/4}/2 \\ -\lambda^{1/4}/2 & 1/2\lambda^{1/4} \end{pmatrix} d\lambda, \quad \lambda \geq 0.$$

Elements in $L_A^2(0, \infty)$ are dependent only on their first component, so

$$F = (f(x), 0, 0, 0)^T.$$

Then

$$G(\lambda) = \frac{1}{2\lambda^{3/4}} \int_0^\infty \begin{pmatrix} \sin \lambda^{1/4}\xi \\ -\lambda^{1/2}\sin \lambda^{1/4}\xi \end{pmatrix} f(\xi)d\xi,$$

$$F(x) = \begin{pmatrix} \dfrac{2}{\pi}\displaystyle\int_0^\infty \sin \mu x \left[\int_0^\infty \sin \mu\xi f(\xi)d\xi\right] d\mu \\ 0 \\ 0 \\ 0 \end{pmatrix}$$

where we have replaced λ by μ^4.

While this example is obviously the Fourier sine transform, the calculations are tedious. We invite the reader to calculate additional examples.

X.7 Subspace Expansions

At the beginning the assumption that $JY' = BY = AF$, $AY = 0$ implies $Y = 0$ was made. We would like to comment briefly on what occurs if the assumption fails to hold.

First A must be singular. Thus if F is in the maximal domain $JF' - BF = AH$, the dimension of AH is less than $2n$. If functions $\gamma_1,\ldots,\gamma_{2n}$ are chosen so that

$$(\gamma_1,\ldots,\gamma_{2n})A = 0,$$

then

$$(\gamma_1,\ldots,\gamma_{2n})(JF' - BF) = 0.$$

Letting $(\gamma_1,\ldots,\gamma_{2n})J = K$, $(\gamma_1,\ldots,\gamma_{2n})B = L$, we have $K_j F_j' - L_j F_j = 0$, $j = 1,\ldots,2n$.

Three possibilities occur:

(1) If $K_j \neq 0$, $F_j = c_j \exp \displaystyle\int_a^x (L_j/K_j)d\xi$.

(2) If $K_j = 0$, $L_j = 0$, F_j is arbitrary.

(3) If $K_j = 0$, $L_j \neq 0$, $F_j = 0$.

Since not all $K_j = 0$, F_j is restricted and may not be dense in $L_A^2(a,b)$. We provide two examples to illustrate. First consider

$$\begin{pmatrix} 0 & -1 \\ 1 & 0 \end{pmatrix}\begin{pmatrix} y_1 \\ y_2 \end{pmatrix}' = \lambda \begin{pmatrix} 4 & -1 \\ -1 & 1/4 \end{pmatrix}\begin{pmatrix} y_1 \\ y_2 \end{pmatrix} + \begin{pmatrix} 4 & -1 \\ -1 & 1/4 \end{pmatrix}\begin{pmatrix} f_1 \\ f_2 \end{pmatrix}.$$

The fundamental matrix is

$$\mathcal{Y}(x,\lambda) = \begin{pmatrix} 1 - \lambda x & 1/4\lambda x \\ -4\lambda x & 1 + \lambda x \end{pmatrix}.$$

Elements in $L_A^2(0,1)$ have an inner product

$$\left\langle \begin{pmatrix} f_1 \\ f_2 \end{pmatrix}, \begin{pmatrix} g_1 \\ g_2 \end{pmatrix} \right\rangle = \int_0^1 (\bar{g}_1, \bar{g}_2) \begin{pmatrix} 4 & -1 \\ -1 & 1/4 \end{pmatrix} \begin{pmatrix} f_1 \\ f_2 \end{pmatrix} d\xi,$$

an element being 0 if $2f_1 = \dfrac{1}{2} f_2$.

    If boundary conditions

$$(1,0) \begin{pmatrix} y_1 \\ y_2 \end{pmatrix}(0) = 0, \quad (0,1) \begin{pmatrix} y_1 \\ y_2 \end{pmatrix}(1) = 0,$$

are imposed, there is only one eigenvalue $\lambda = -1$ and one eigenfunction $\begin{pmatrix} y_1 \\ y_2 \end{pmatrix} =$
$\begin{pmatrix} -1/4x \\ 1-x \end{pmatrix}$. The solution ϕ is $\begin{pmatrix} 1/4\lambda x \\ 1+\lambda x \end{pmatrix}$, so for $F = \begin{pmatrix} f_1 \\ f_2 \end{pmatrix}$,

$$G(\lambda) = \int_0^1 \left(-f_1 + \frac{1}{4} f_2 \right) d\xi.$$

dP is 0 unless $\lambda = -1$. $dP(-1) = 4$. So

$$F(x) = \begin{pmatrix} -1/4x \\ 1-x \end{pmatrix} \cdot 4 \cdot + \int_0^1 \left(-f_1 + \frac{1}{4} f_2 \right) d\xi.$$

Clearly this is one-dimensional. In fact, elements in the maximal domain are of the
form $\begin{pmatrix} c \\ 0 \end{pmatrix}$, where c is constant. Such elements are equivalent to the eigenfunction
$\begin{pmatrix} -1/4x \\ 1-x \end{pmatrix}$. Only these can be expanded.

    Secondly, consider

$$\begin{pmatrix} 0 & -1 \\ 1 & 0 \end{pmatrix} \begin{pmatrix} y_1 \\ y_2 \end{pmatrix}' = \left[\lambda \begin{pmatrix} e^{-x} & -1 \\ -1 & e^x \end{pmatrix} + \begin{pmatrix} -e^{-x} & 1 \\ 1 & 0 \end{pmatrix} \right] \begin{pmatrix} y_1 \\ y_2 \end{pmatrix} + \begin{pmatrix} e^{-x} & -1 \\ -1 & e^x \end{pmatrix} \begin{pmatrix} f_1 \\ f_2 \end{pmatrix},$$
$$(1,0) \begin{pmatrix} y_1 \\ y_2 \end{pmatrix}(0) = 0, \quad (0,1) \begin{pmatrix} y_1 \\ y_2 \end{pmatrix}(1) = 0.$$

The only eigenvalue is at $\lambda = (1-e)^{-1}$. Its eigenvector is $\begin{pmatrix} [e^x - 1]/[1-e] \\ [1 - ee^{-x}]/[1-e] \end{pmatrix}$. If
$\begin{pmatrix} y_1 \\ y_2 \end{pmatrix}$ is an element of $L_A^2(0,1)$, then

$$G(\lambda) = \int_0^1 (-e^{-\xi} f_1 + f_2) d\xi.$$

dP is 0 except at $\lambda = (1 - e)^{-1}$, where it is $e/(1 - e)$. Thus if $F(x) = \begin{pmatrix} y_1 \\ y_2 \end{pmatrix}$ is in the maximal domain, D_M,

$$F(x) = \begin{pmatrix} [e^x - 1]/[1 - e] \\ [1 - ee^x]/[1 - e] \end{pmatrix} \left(\frac{e}{1 - e} \right) \int_0^1 (-e^{-\xi} f_1 + f_2) d\xi \,.$$

A quick calculation shows that elements in D_M have the form $\begin{pmatrix} c \\ 0 \end{pmatrix}$, where c is constant. Arbitrary elements in $L_A^2(0,1)$ have the form $\begin{pmatrix} f_1 - e^x f_2 \\ 0 \end{pmatrix}$, so D_M is not dense.

X.8 Remarks

More can be said concerning the expansion in the limit-circle case (Hinton and Shaw [4, 5, 6, 7]) since $M(\lambda)$ is meromorphic with real poles.

Recently the problem

$$-y^{iv} + (xy')' = \lambda y$$

on $L^2[1, \infty)$ was discovered to have essential spectrum at $\lambda = 0$. The problem is regular at $x = 1$, while it is limit 3 at $x = \infty$. We invite the reader to contact B. Schultze in Essen for details of this work.

References

[1] F. V. Atkinson, **Discrete and Continuous Boundary Value Problems**, Academic Press, New York, 1964.

[2] F. Brauer, *Spectral theory for linear systems of differential equations*, Pacific J. Math. **10** (1960), 17–34.

[3] E. A. Coddington and N. Levinson, **Theory of Ordinary Differential Equations**, McGraw-Hill, New York, 1955.

[4] D. B. Hinton and J. K. Shaw, *Titchmarsh-Weyl theory for Hamiltonian Systems*, Spectral Theory of Differential Operations, ed. I. W. Knowles and R. T. Lewis, North Holland.

[5] ——, *On the spectrum of a singular Hamiltonian system*, Quaes. Math. **5** (1982), 29–81.

[6] ——, *Well-posed boundary value problems for Hamiltonian systems of limit point or limit circle type*, Lecture Notes on Mathematics, Vol. 964, Springer-Verlag, Berlin, 1982, 614–631.

[7] ——, *Parameterization of the $M(\lambda)$ function for a Hamiltonian system of limit circle type*, Proc. Roy. Soc. Edinburgh **93** (1983), 349–360.

[8] A. M. Krall, **Applied Analysis**, D. Reidel, Dordrecht, Netherlands, 1987.

Chapter XI

The Spectral Resolution for
Linear Hamiltonian Systems with
Two Singular Points

This chapter in large measure repeats the ideas and techniques of the previous chapter with but minor modifications due to the second singular point. Again we employ the setting due to Hinton and Shaw.

The main difference is that the spectral measure is now $2n \times 2n$, rather than $n \times n$.

XI.1 The Specific Problem

Consider the problem

$$JY' = (\lambda A + B)Y, \quad x \in (a',b'),$$
$$(\alpha_1 \alpha_2)Y(a') = 0, \quad (\beta_1 \beta_2)Y(b'), \qquad a < a' < b' < b,$$

where Y is $2n$-dimensional, $J = \begin{pmatrix} 0 & -I_n \\ I_n & 0 \end{pmatrix}$, $A = A^* \geq 0$ and $B = B^*$ are real and locally integrable, and $\alpha_1, \alpha_2, \beta_1, \beta_2$ satisfy

$$\alpha_1 \alpha_1^* + \alpha_2 \alpha_2^* = I_n, \quad \beta_1 \beta_1^* + \beta_2 \beta_2^* = I_n,$$
$$\alpha_1 \alpha_2^* - \alpha_2 \alpha_1^* = 0, \quad \beta_1 \beta_2^* - \beta_2 \beta_1^* = 0.$$

In addition, for technical reasons assume that if Y satisfies $JY' - BY = AF$, $AY = 0$, then $Y = 0$. This ensures the density of the domain of the maximal operator in $L_A^2(a,b)$.

We know that the boundary value problem is self-adjoint on (a',b') and that, as (a',b') approaches (a,b), where a and b are singular points, solutions in $L_A^2(a,e)$

and $L_A^2(e, b)$, $a < e < b$, can be found, boundary conditions at a and b can be defined, and a suitable self-adjoint boundary value problem on (a, b) can be defined.

Our purpose here is to use the (a', b') eigenfunction expansion to derive the spectral resolution for the singular problem on (a, b).

XI.2   The Spectral Expansion

Recall that we denote by $\mathcal{Y}(x, \lambda)$ the $2n \times 2n$ matrix solution of $JY' = (\lambda A + B)Y$ that satisfies $\mathcal{Y}(e, \lambda) = I_{2n}$, where $e \in (a', b')$. $\mathcal{Y}(x, \lambda)$ is then partitioned into two $2n \times n$ matrices θ, ϕ where $\theta = \begin{pmatrix} \theta_1 \\ \theta_2 \end{pmatrix}$, $\phi = \begin{pmatrix} \phi_1 \\ \phi_2 \end{pmatrix}$, with each piece $\theta_1, \theta_2, \phi_1, \phi_2$ being $n \times n$. Hence

$$\mathcal{Y}(x, \lambda) = (\theta(x, \lambda), \phi(x, \lambda)) = \begin{pmatrix} \theta_1(x, \lambda) & \phi_1(x, \lambda) \\ \theta_2(x, \lambda) & \phi_2(x, \lambda) \end{pmatrix} .$$

For $\operatorname{Im} \lambda \neq 0$, the solution satisfying the boundary condition at $x = b'$ is $\chi_{b'} = \theta + \phi M_{b'}$, where

$$M_{b'} = -\left((\beta_1, \beta_2)\phi(b', \lambda)\right)^{-1} \left((\beta_1, \beta_2)\theta(b', \lambda)\right) .$$

The inverse exists; otherwise there is a self-adjoint boundary value problem with λ (complex) as an eigenvalue.

As b' approaches the singularity b, $M_{b'}$ can be made to approach M_b, a limiting value, with $\chi_b = \theta + \phi M_b$ in $L_A^2(e, b)$. Further, χ_b generates a singular boundary condition

$$\lim_{x \to b} \chi_b(x, \lambda)^* JY(x) = 0 ,$$

which we require. Finally, χ_b satisfies

$$\int_e^b \chi_b^* A \chi_b d\xi = [M_b - M_b^*]/2i\operatorname{Im} \lambda .$$

Likewise, for $\operatorname{Im} \lambda \neq 0$, the solution satisfying the boundary condition at $x = a'$ is $\chi_{a'} = \theta + \phi M_{a'}$, where

$$M_{a'} = -\left((\alpha_1 \alpha_2)\phi(a', \lambda)\right)^{-1} \left((\alpha_1 \alpha_2)\theta(a', \lambda)\right) .$$

As a' approaches the singularity a, $M_{a'}$ can be made to approach M_a, a limiting value, with $\chi_a = \theta + \phi M_a$ in $L_A^2(a, e)$. Further, χ_a generates a singular boundary condition

$$\lim_{x \to a} \chi_{a'}(x, \lambda) JY(x) = 0 ,$$

which we require. Finally, χ_a satisfies

$$\int_a^e \chi_a^* A \chi_a d\xi = [M_a^* - M_a]/2i\operatorname{Im} \lambda .$$

M_a and M_b have additional properties:

(1) $M_a \neq M_b$, $[\text{Im } \lambda][M_a - M_b] < 0$.

(2) $M_a(\lambda) = M_a(\overline{\lambda})^*$, $M_b(\lambda) = M_b(\overline{\lambda})^*$.

(3) M_a, M_b, $M_a - M_b$ are all invertible.

(4) $M_b[M_a - M_b]^{-1}M_a = M_a[M_a - M_b]^{-1}M_b$.

We can now state the boundary value problem we wish to consider. It is

$$JY' = (\lambda A + B)Y, \qquad a < x < b.$$

$$\lim_{x \to a} \chi_a^* JY = 0,$$

$$\lim_{x \to b} \chi_b^* JY = 0.$$

The problem is self-adjoint. It is its "eigenfunction expansion" or, more properly, its spectral resolution we wish to derive.

We are considering the operator L in $L_A^2(a, b)$ having domain D: functions satisfying

(1) $JY' - BY = AF$ exists a.e. for some F in $L_A^2(a, b)$.

(2) $\lim_{x \to a} \chi_a^* JY = 0$.

(3) $\lim_{x \to b} \chi_b^* JY = 0$.

For such Y, LY is defined by setting $LY = F$ when $JY' - BY = AF$. L is self-adjoint in $L_A^2(a, b)$.

The boundary value problem on (a', b')

$$JY' = (\lambda A + B)Y, \qquad x \in (a', b'),$$
$$(\alpha_1 \alpha_2)Y(a') = 0, \qquad (\beta_1 \beta_2)Y(b') = 0,$$

is self-adjoint. It has a countable number of eigenvalues $\{\lambda_k\}_{k=1}^{\infty}$ with normalized orthogonal eigenfunctions

$$\chi_k = \theta(x, \lambda_k)S_k + \phi(x, \lambda_k)T_k, \qquad k = 1, \ldots.$$

Of course, λ_k, S_k, T_k depend upon $(a', b') = I$, but we suppress this dependence. Note that

$$\chi_k = \mathcal{Y}(x, \lambda_k)\begin{pmatrix} S_k \\ T_k \end{pmatrix},$$

and that

$$\int_{a'}^{b'} (S_k^*, T_k^*)\mathcal{Y}(\xi, \lambda_k)^* A \mathcal{Y}(\xi, \lambda_k)\begin{pmatrix} S_k \\ T_k \end{pmatrix} d\xi = 1.$$

Arbitrary elements F in $L_A^2(a', b')$ can be expanded in terms of these eigenfunctions.

XI.2.1. Theorem. *For all F in $L_A^2(a',b')$,*

$$F(x) = \sum_{k=1}^{\infty} \mathcal{Y}(x,\lambda_k) \begin{pmatrix} S_k \\ T_k \end{pmatrix} (S_k^*, T_k^*) \int_{a'}^{b'} \mathcal{Y}(\xi,\lambda_k)^* A(\xi)F(\xi)d\xi .$$

Parseval's equality also holds.

XI.2.2. Theorem. *For all F in $L_A^2(a',b')$,*

$$\int_{a'}^{b'} F^* AFd\xi = \sum_{k=1}^{\infty} \left[\int_{a'}^{b'} \mathcal{Y}(\xi,\lambda_k)^* AFd\xi \right]^* \begin{pmatrix} S_k S_k^* & S_k T_k^* \\ T_k S_k^* & T_k T_k^* \end{pmatrix} \left[\int_{a'}^{b'} \mathcal{Y}(\xi,\lambda_k)^* AFd\xi \right].$$

XI.2.3. Definition. Let R_k^2 denote the $2n \times 2n$ matrix

$$\begin{pmatrix} S_k S_k^* & S_k T_k^* \\ T_k S_k^* & T_k T_k^* \end{pmatrix} .$$

XI.2.4. Definition. Let

$$G_I(\lambda) = \int_{a'}^{b'} \mathcal{Y}(\xi,\lambda)^* AFd\xi , = \int_{a'}^{b'} \begin{pmatrix} \theta(\xi,\lambda)^* \\ \phi(\xi,\lambda)^* \end{pmatrix} AFd\xi .$$

We can rewrite these statements using a kind of counting measure P_I.

XI.2.5. Definition. Let $P_I(\lambda)$ be a $2n \times 2n$ matrix-valued function satisfying

(1) $P_I(0+) = 0$,

(2) $P_I(\lambda)$ is increasing, jumping R_k^2 at $\lambda = \lambda_k$, but otherwise constant, continuous from above.

Thus

$$P_I(\lambda) = \sum_{0 < \lambda_k \leq \lambda} R_k^2 , \qquad \lambda \geq 0$$

$$= - \sum_{\lambda < \lambda_k \leq 0} R_k^2 , \qquad \lambda < 0 .$$

XI.2.6. Theorem (Parseval's Equality). *Let F be an arbitrary element in $L_A^2(a',b')$. Then*

$$\int_{a'}^{b'} F^* AFd\xi = \int_{-\infty}^{\infty} G_I^*(\lambda) dP_I(\lambda) G_I(\lambda) .$$

This can be extended by the polarization identities.

XI.2.7. Corollary. *Let F_1, F_2 be arbitrary elements in $L_A^2(a',b')$. Let G_{I1}, G_{I2} correspond to them according to Definition XI.2.4. Then*

$$\int_{a'}^{b'} F_2^* AF_1 d\xi = \int_{-\infty}^{\infty} G_{I2}(\lambda)^* dP_I(\lambda) G_{I1}(\lambda) .$$

XI.2.8. Theorem. *There exists a nondecreasing $2n \times 2n$ matrix-valued function $P(\lambda)$, defined on $(-\infty, \infty)$ satisfying*

(1) $P(0+) = 0$,

(2) $P(\lambda) - P(\mu) = \lim\limits_{I \to (a,b)} [P_I(\lambda) - P_I(\mu)]$, $\lambda > \mu$.

Proof. Recall that the Green's function $\mathbf{G}_I(\lambda, x, \xi)$ for the interval $I = (a', b')$ is given by

$$\mathbf{G}_I(\lambda, x, \xi) = \chi_{b'}(x, \lambda)[M_{a'} - M_{b'}]^{-1}\chi_{a'}(\xi, \overline{\lambda})^*, \quad a' < \xi < x < b'$$
$$= \chi_{a'}(x, \lambda)[M_{a'} - M_{b'}]^{-1}\chi_{b'}(\xi, \overline{\lambda})^*, \quad a' < x < \xi < b'.$$

Fix $\xi = e$, $\lambda = \lambda_0$, Im $\lambda_0 \neq 0$ and let $H(x)$ be the result.

$$H(x) = \chi_{b'}(x, \lambda_0)[M_{a'} - M_{b'}]^{-1}[I_n, M_{a'}], \quad e < x < b'$$
$$= \chi_{a'}(x, \lambda_0)[M_{a'} - M_{b'}]^{-1}[I_n, M_{b'}], \quad a' < x < e.$$

Note that H satisfies the boundary conditions

$$\chi'_a(a', \lambda)^* J H(a') = 0 \quad \text{or} \quad (\alpha_1 \alpha_2) H(a') = 0,$$
$$\chi'_b(b', \lambda)^* J H(b') = 0 \quad \text{or} \quad (\beta_1 \beta_2) H(b') = 0.$$

Further, except for $x = e$, $JH' = (\lambda_0 A + B)H$. Manipulation of this equation shows that

$$(H^* J H)' = 2i\text{Im } \lambda_0 H^* A H.$$

Consequently,

$$\int_{a'}^{e} H^* A H \, d\xi$$
$$= (2i\text{Im } \lambda_0)^{-1} \begin{pmatrix} I_n \\ M^*_{b'} \end{pmatrix} (M^*_{a'} - M^*_{b'})^{-1}(M^*_{a'} - M^*_{a'})(M^*_{a'} - M^*_{b'})^{-1}(I_n M_{b'}),$$

since the limit of $H^* J H$ at $x = a'$ is 0. Also,

$$\int_{e}^{b'} H^* A H \, d\xi$$
$$= (2i\text{Im } \lambda_0)^{-1} \begin{pmatrix} I_n \\ M^*_{a'} \end{pmatrix} (M^*_{a'} - M^*_{b'})^{-1}(-M^*_{b'} + M_{b'})(M_{a'} - M_{b'})^{-1}(I_n M_{a'}).$$

Adding, we find

$$\int_{a'}^{b'} H^* A H \, d\xi < K,$$

where K is greater than the sum of the two right sides. We next compute G_I for H. Manipulation of the differential equations for H and $\mathcal{Y}$ over (a', e) and (e, b') shows that

$$G_I = (\lambda_0 - \lambda)^{-1} \left[\begin{pmatrix} -M_{a'} \\ I_n \end{pmatrix} (M_{a'} - M_{b'})^{-1}(I_n M_{b'}) \right.$$

$$\left. + \begin{pmatrix} M_{b'} \\ -I_n \end{pmatrix} (M_{a'} - M_{b'})^{-1}(I M_{a'}) \right]$$

$$= -(\lambda_0 - \lambda)^{-1} I_{2n}.$$

Inserting these into Parseval's equality, we get

$$\int_{-\infty}^{\infty} \frac{dP_I(\lambda)}{|\lambda_0 - \lambda|^2} < K.$$

This implies that for $\mu > 0$,

$$\int_{-\mu}^{\mu} dP_I(\lambda) < K[1 + \mu^2],$$

which shows that $P_I(\lambda)$ is uniformly bounded on compact subintervals of the real line. Helly's first convergence theorem [5] shows there is a subsequence that converges weakly to $P(\lambda)$ with the properties stated. $\qquad\square$

XI.2.9. Theorem. *If F is in $L_A^2(a, b)$, there is a function $G(\lambda)$ in $L_p^2(-\infty, \infty)$, with inner product*

$$\langle G, H \rangle_P = \int_{-\infty}^{\infty} H^* dPG, \quad \|G\|_P^2 = \int_{-\infty}^{\infty} G^* dPG,$$

such that if

$$E(\lambda) = G(\lambda) - \int_{a'}^{b'} \mathcal{Y}(\xi, \lambda)^* A(\xi) F(\xi) d\xi,$$

then

$$\lim_{(a', b') \to (a, b)} \int_{-\infty}^{\infty} E(\lambda)^* dP(\lambda) E(\lambda) = 0,$$

and

$$\int_a^b F(\xi)^* A(\xi) F(\xi) d\xi = \int_{-\infty}^{\infty} G(\lambda)^* dP(\lambda) G(\lambda).$$

Proof. 1. Let F be in $C_0^1(a, b)$. Then F is in D. If (a', b') is sufficiently near (a, b), then if $LF = H$,

$$\int_a^b (LF)^* A(LF) d\xi = \int_{a'}^{b'} H^* A H d\xi$$

$$= \sum_{k=1}^{\infty} \left[\int_{a'}^{b'} \mathcal{Y}_k^* A H d\xi \right]^* R_k^2 \left[\int_{a'}^{b'} \mathcal{Y}_k^* A H d\xi \right],$$

where $\mathcal{Y}_k = \mathcal{Y}(x, \lambda_k)$. But

$$\int_{a'}^{b'} \mathcal{Y}_k^* A H d\xi = \int_{a'}^{b'} \mathcal{Y}_K^* [JF' - BF] d\xi$$

$$= \int_{a'}^{b'} [J\mathcal{Y}_k' - B\mathcal{Y}_k]^* F d\xi$$

$$= \lambda_k \int_{a'}^{b'} \mathcal{Y}_k^* A F d\xi$$

$$= \lambda_k G_I(\lambda_k) .$$

Thus

$$\int_a^b (LF)^* A(LF) d\xi = \int_{-\infty}^{\infty} \lambda^2 G_I(\lambda)^* dP_I(\lambda) G_I(\lambda) .$$

Letting

$$G(\lambda) = \int_a^b \mathcal{Y}(\xi, \lambda) A(\xi) F(\xi) d\xi ,$$

we can drop the subscript in G_I. If $N > 0$, then

$$\left(\int_{-\infty}^{-N} + \int_N^{\infty} \right) G^* dP_I G \le \frac{1}{N^2} \left(\int_{-\infty}^{-N} + \int_N^{\infty} \right) \lambda^2 G^* dP_I G$$

$$\le \frac{1}{N^2} \int_{-\infty}^{\infty} \lambda^2 G^* dP_I G$$

$$\le \frac{1}{N^2} \int_a^b (LF)^* A(LF) d\xi .$$

Since

$$\int_a^b F^* A F d\xi = \left(\int_{-\infty}^{-N} + \int_{-N}^{N} + \int_N^{\infty} \right) G^* DP_I G ,$$

we have

$$\left| \int_a^b F^* A F d\xi - \int_{-N}^{N} G^* dP_I G \right| \le \frac{1}{N^2} \int_a^b (LF)^* A(LF) d\xi .$$

Let $I \to (a, b)$. Helly's second theorem (Hinton and Shaw [5]) implies

$$\left| \int_a^b F^* A F d\xi - \int_{-N}^{N} G^* dPG \right| < \frac{1}{N^2} \int_a^b (LF)^* A(LF) d\xi .$$

Let N approach ∞. Then

$$\int_a^b F^* A F d\xi = \int_{-\infty}^{\infty} G^* dPG ,$$

provided F is in $C_0^1(a, b)$.

2. Let F vanish near a and b, but otherwise be arbitrary in $L_A^2(a,b)$. Choose $\{F_j\}_{j=1}^\infty$ in $C_0^1(a,b)$ such that

$$\lim_{j\to\infty} \int_a^b (F_j - F)^* A(F_j - F)d\xi = 0\,.$$

Apply Parseval's equality to $F_j - F_k$.

$$\int_a^b (F_j - F_k)^* A(F_j - F_k)d\xi = \int_{-\infty}^\infty (G_j - G_k)^* dP(G_j - G_k)\,,$$

where

$$G_j = \int_a^b \mathcal{Y}^* AF_j d\xi\,, \quad G_k = \int_a^b \mathcal{Y}^* AF_k d\xi\,.$$

Since $\lim_{j\to\infty} F_j = F$, $\{G_j\}_{j=1}^\infty$ form a Cauchy sequence in $L_p^2(-\infty,\infty)$. Thus there is a G in $L_p^2(-\infty,\infty)$ such that $\lim_{j\to\infty} G_j = G$. Since

$$\left| G_j - \int_a^b \mathcal{Y}^* AF d\xi \right| = \left| \int_a^b \mathcal{Y}^* A(F_j - F)d\xi \right|$$

$$\leq \left(\int_{a'}^{b'} \mathcal{Y}^* A\mathcal{Y}d\xi \right)^{1/2} \left(\int_a^b (F_j - F)^* A(F_j - F)d\xi \right)^{1/2}$$

implies

$$\lim_{j\to\infty} G_j = \int_a^b \mathcal{Y}^* AF d\xi\,,$$

which is continuous in λ, we find

$$G = \int_a^b \mathcal{Y}^* AF d\xi\,, \quad a.e.$$

Thus if F vanishes near a and b,

$$\int_a^b F^* AF d\xi = \lim_{j\to\infty} \int_a^b F_j^* AF_j d\xi$$

$$= \lim_{j\to\infty} \int_{-\infty}^\infty G_j^* dPG_j$$

$$= \int_{-\infty}^\infty G^* dPG\,.$$

3. Finally, if F is arbitrary in $L_A^2(a,b)$, let

$$F_I = F\,, \quad x \in (a',b')$$
$$= 0\,, \quad \text{otherwise}\,.$$

Let

$$G_I = \int_a^b \mathcal{Y}^* A F_I d\xi = \int_{a'}^{b'} \mathcal{Y}^* A F d\xi \,.$$

Since if $I_1 \subset I_2$,

$$\int_{-\infty}^{\infty} (G_{I_1} - G_{I_2})^* dP (G_{I_1} - G_{I_2}) = \int_{I_2 - I_1} F^* A F d\xi \,,$$

$\{G_I\}$ is a Cauchy sequence as I approaches (a,b). Let

$$\lim_{I \to (a,b)} G_I = G \quad \text{in} \quad L_p^2(-\infty, \infty) \,.$$

Letting I approach (a,b) in the previous result, we get

$$\int_a^b F^* A F d\xi = \int_{-\infty}^{\infty} G^* dP G \,.$$

4. Since G_I approaches G in $L_p^2(-\infty, \infty)$,

$$\lim_{(a',b') \to (a,b)} \int_{-\infty}^{\infty} \left[G(\lambda) - \int_{a'}^{b'} \mathcal{Y}(\xi, \lambda)^* A F d\xi \right]^*$$

$$\times \, dP \left[G(\lambda) - \int_{a'}^{b'} \mathcal{Y}(\xi, \lambda)^* A F d\xi \right] = 0 \,. \qquad \square$$

XI.2.10. Theorem. *If $G(\lambda)$ is the limit of $\int_{a'}^{b'} \mathcal{Y}(\xi, \lambda)^* A(\xi) F(\xi) d\xi$ in $L_p^2(-\infty, \infty)$, then*

$$\int_{-\infty}^{\infty} \mathcal{Y}(x, \lambda) dP(\lambda) G(\lambda) = F(x)$$

in $L_A^2(a,b)$, that is

$$\lim_{I \to (-\infty, \infty)} \int_a^b \left[F - \int_I \mathcal{Y} dP G \right]^* A \left[F - \int_I \mathcal{Y} dP G \right] d\xi = 0 \,.$$

Proof. Let $I = (\mu, \nu)$ and

$$F_I = \int_I \mathcal{Y}(x, \lambda) dP(\lambda) G(\lambda) \,.$$

If $(a', b') \subset (a, b)$, then

$$\int_{a'}^{b'} [F - F_I]^* A F_I d\xi = \int_{a'}^{b'} [F - F_I]^* A \left[\int_I \mathcal{Y} dP G \right] d\xi$$

$$= \int_I \left[\int_{a'}^{b'} [F - F_I]^* A \mathcal{Y} d\xi \right] dP G \,.$$

Likewise

$$\int_{a'}^{b'} [F - F_I]^* A F d\xi = \int_{-\infty}^{\infty} \left[\int_{a'}^{b'} [F - F_I]^* A \mathcal{Y} d\xi \right] dPG \,.$$

Subtracting, we have

$$\int_{a'}^{b'} [F - F_I]^* A [F - F_I] d\xi = \int_{(-\infty,\infty)-I} \left[\int_{a'}^{b'} [F - F_I]^* A \mathcal{Y} d\xi \right] dPG \,.$$

Now, $\int_{a'}^{b'} \mathcal{Y}^* A [F - F_I] d\xi$ is the transform of a function in $L_A^2(a, b)$ which vanishes outside (a', b'). Consequently, the integral from a' to b' in brackets is in $L_p^2(-\infty, \infty)$. Applying Schwarz's inequality, we have

$$\left(\int_{a'}^{b'} [F - F_I]^* A [F - F_I] d\xi \right)^2$$

$$\leq \left(\int_{(-\infty,\infty)-I} \left[\int_{a'}^{b'} \mathcal{Y}^* A [F - F_I] d\xi \right]^* \right.$$

$$\left. dP \left[\int_{a'}^{b'} \mathcal{Y}^* A [F - F_I] d\xi \right] \right) \left(\int_{(-\infty,\infty)-I} G^* dPG \right) \,.$$

The first integral on the right is less than or equal to

$$\int_{a'}^{b'} [F - F_I]^* A [F - F_I] d\xi \,.$$

If this is inserted and cancelled,

$$\int_{a'}^{b'} [F - F_I]^* A [F - F_I] d\xi \leq \int_{(-\infty,\infty)-I} G^* dPG \,.$$

Let (a', b') approach (a, b). Then let I approach $(-\infty, \infty)$. The result is that $F = \lim_{I \to (-\infty,\infty)} F_I$, or

$$F(x) = \lim_{(\mu,\nu) \to (-\infty,\infty)} \int_{\mu}^{\nu} \mathcal{Y}(x, \lambda) dP(\lambda) G(\lambda)$$

in $L_A^2(a, b)$.

Theorem XI.2.9 may be extended to involve inner products by use of the polarization identity. The inner product form of Parseval's equality is

$$\int_{a}^{b} F_2(\xi)^* A(\xi) F_1(\xi) d\xi = \int_{-\infty}^{\infty} G_2(\lambda)^* dP(\lambda) G_1(\lambda) \,,$$

where

$$G_j(\lambda) = \int_a^b \mathcal{Y}(\xi, \lambda)^* A(\xi) F_j(\xi) d\xi, \qquad j = 1, 2.$$

Theorems XI.2.9 and XI.2.10 may be extended to represent the resolvent operator $(L - \lambda_0 I)^{-1}$ when λ_0 is not in the support of $dP(\lambda)$. Parseval's equality is

$$\int_a^b [(L - \lambda_0 I)^{-1} F(\xi)]^* A(\xi)[(L - \lambda_0 I)^{-1} F(\xi)] d\xi = \int_{-\infty}^{\infty} \frac{G(\lambda)^* dP(\lambda) G(\lambda)}{|\lambda - \lambda_0|^2}.$$

The resolvent expansion is

$$(L - \lambda_0 I)^{-1} F(x) = \int_{-\infty}^{\infty} \mathcal{Y}(x, \lambda) dP(\lambda) \frac{G(\lambda)}{\lambda - \lambda_0}. \qquad \square$$

XI.3 The Converse Problem

The preceding section began with choosing an F in $L_A^2(a, b)$, producing a G in $L_p^2(-\infty, \infty)$, and then showing that F could be recovered from G. In this section we begin with G, produce F, and then recover G.

Without the assumption that $JY' - BY = 0$, $AY = 0$ implies $Y = 0$, $L_p^2(-\infty, \infty)$ may be too large in the sense that $G \to F \to \widetilde{G}$, but $\widetilde{G}$ may be not equal G. $\widetilde{G}$ may be only in a subspace of $L_p^2(-\infty, \infty)$.

With the assumption which we made in the introduction (let $F = 0$), there is no difficulty.

XI.3.1. Lemma. *Let $G(\lambda)$ be in $L_p^2(-\infty, \infty)$. Let*

$$F_I(x) = \int_I \mathcal{Y}(x, \lambda) dP(\lambda) G(\lambda).$$

Then $\displaystyle\lim_{I \to (-\infty, \infty)} F_I(x)$ *exists in $L_A^2(a, b)$.*

Proof. Let $I_1 \subset I_2$. Then

$$F_{I_2} - F_{I_1} = \int_{I_2 - I_1} \mathcal{Y}(x, \lambda) dP(\lambda) G(\lambda)$$
$$= \int_{-\infty}^{\infty} \mathcal{Y}(x, \lambda) dP(\lambda) K_{I_2 - I_1}(\lambda) G(\lambda),$$

where

$$K_{I_2 - I_1}(\lambda) = 1, \qquad \lambda \in I_2 - I_1$$
$$= 0, \qquad \lambda \notin I_2 - I_1.$$

Let R be an arbitrary element of $L_A^2(a,b)$ that vanishes near a and b. Then

$$\int_a^b R^* A[F_{I_2} - F_{I-1}]d\xi = \int_a^b R^* A\left[\iint_{I_2 - I_1} \mathcal{Y}dPG\right]d\xi$$

$$= \int_{I_2 - I_1}\left[\int_a^b \mathcal{Y}^* ARd\xi\right]^* dPG$$

$$= \int_{I_2 - I_1} S^* dPG,$$

where S is the transform of R:

$$S = \int_a^b \mathcal{Y}^* ARd\xi.$$

We now let $R = F_{I_2} - F_{I_1}$ on (a', b'), but set $R = 0$ near a and b. Then

$$\int_a^b [F_{I_2} - F_{I_1}]^* A[F_{I_2} - F_{I_1}]d\xi = \int_{I_2 - I_1} S^* dPG$$

$$\leq \left(\int_{I_2 - I_1} S^* dPS\right)^{1/2}\left(\int_{I_2 - I_1} G^* dPG\right)^{1/2}$$

$$\leq \left(\int_{-\infty}^{\infty} S^* dPS\right)^{1/2}\left(\int_{I_2 - I_1} G^* dPG\right)^{1/2}.$$

But by Parseval's equality

$$\int_a^b [F_{I_2} - F_{I_1}]^* A[F_{I_2} - F_{I_1}]d\xi = \int_{-\infty}^{\infty} S^* dPS.$$

Take the square root, cancel with the inequality above, then square.

$$\int_a^b [F_{I_2} - F_{I_1}]^* A[F_{I_2} - F_{I_1}]d\xi \leq \int_{I_2 - I_1} G^* dPG.$$

Since the right side is independent of a' and b', let (a', b') approach (a, b). Then $F_{I_2} - F_{I_1}$ is in $L_A^2(a,b)$. As I approaches $(-\infty, \infty)$, the inequality shows that $\{F_I\}$ forms a Cauchy sequence in $L_A^2(a,b)$, and therefore

$$\lim_{I \to (-\infty, \infty)} F_I = F$$

in $L_A^2(a,b)$. $\square$

XI.3.2. Lemma. *Let $G(\lambda)$ be in $L_p^2(-\infty, \infty)$. Let*

$$F_I(x) = \int_I \mathcal{Y}(x, \lambda)dP(\lambda)(G(\lambda)), \quad F(x) = \lim_{I \to (-\infty, \infty)} \int_I \mathcal{Y}(x, \lambda)dP(\lambda)G(\lambda).$$

Let

$$\widetilde{G}(\lambda) = \int_a^b \mathcal{Y}(\xi, \lambda)^* A(\xi) F(\xi) d\xi \,,$$

and let

$$\widetilde{F}_I(x) = \int_I \mathcal{Y}(x, \lambda) dP(\lambda) \widetilde{G}(\lambda) \,.$$

Then

$$\lim_{I \to (-\infty, \infty)} \int_a^b [F_I - \widetilde{F}_I]^* A [F_I - \widetilde{F}_I] d\xi = 0 \,.$$

Proof. From Theorem XI.2.10, $F(x) = \lim\limits_{I \to (-\infty, \infty)} \widetilde{F}_I(x)$ in $L_A^2(a, b)$. But by defi-nition, $F(x) = \lim\limits_{I \to (-\infty, \infty)} F_I(x)$. The triangle inequality

$$\|F_I - \widetilde{F}_I\|_A \le \|F_I - F\|_A + \|F - \widetilde{F}_I\|_A$$

shows that as I approaches $(-\infty, \infty)$, $\|F_I - \widetilde{F}_I\|_A$ approaches 0.

At this point we have, given G, there is an F. F yields $\widetilde{G}$, which again yields F. And so the process stops. We continue to show that G and $\widetilde{G}$ coincide. $\qquad \square$

XI.3.3. Lemma. *Let λ_0 be a complex number with positive imaginary part, and let*

$$H_I(x, \lambda_0) = \int_I \mathcal{Y}(x, \lambda) dP(\lambda) \left[\frac{G(\lambda) - \widetilde{G}(\lambda)}{\lambda - \lambda_0} \right] \,.$$

Then for all fixed λ_0, $\lim\limits_{I \to (-\infty, \infty)} H_I(x, \lambda_0) = 0$.

Proof. H_I satisfies

$$J H_I' - (\lambda_0 A + B) H_I = A(x) \int_I \mathcal{Y}(x, \lambda) dP(\lambda) [G(\lambda) - \widetilde{G}(\lambda)] \,.$$

Further, H_I satisfies the boundary conditions at a and b. Because of the differential equation

$$H_I(x) = \int_a^b \mathbf{G}(\lambda_0, x, \xi) A(\xi) [F_I(\xi) - \widetilde{F}_I(\xi)] d\xi + \mathcal{Y}(x, \lambda_0) C \,,$$

where $\mathbf{G}$ is the Green's function for the singular boundary value problem. As I approaches $(-\infty, \infty)$, the integral approaches 0, so

$$H(x, \lambda_0) = \lim_{I \to (-\infty, \infty)} H_I(x, \lambda_0) = \mathcal{Y}(x, \lambda_0) C \,.$$

Applying the boundary conditions, we find

$$\begin{pmatrix} I_n & M_a \\ I_n & M_b \end{pmatrix} \begin{pmatrix} 0 & -I_n \\ I_n & 0 \end{pmatrix} C = 0 \,.$$

Since the coefficient of C is nonsingular, $C = 0$. $\qquad \square$

XI.3.4. Lemma. $G(\lambda) = \widetilde{G}(\lambda)$ *in* $L_p^2(-\infty, \infty)$.

Proof. Let K be a constant $2n \times 1$ matrix; let

$$Y_s(\lambda) = \int_e^s \mathcal{Y}(\xi, \lambda)^* A(\xi) K d\xi.$$

Then $Y_s(\lambda)$ is in $L_p^2(-\infty, \infty)$ since it is the transform of a function that vanishes near a and b (s may be greater or less than e). Then

$$\int_a^s K^* A(\xi) H_I(\xi, \lambda_0) d\xi = \int_I Y_s(\lambda)^* dP(\lambda) \left[\frac{G(\lambda) - \widetilde{G}(\lambda)}{\lambda - \lambda_0} \right].$$

Let I approach $(-\infty, \infty)$. Then

$$0 = \int_{-\infty}^\infty Y_s(\lambda)^* dP(\lambda) \left[\frac{G(\lambda) - \widetilde{G}(\lambda)}{\lambda - \lambda_0} \right].$$

We may assume without loss of generality that $G(\lambda) - \widetilde{G}(\lambda)$ is real. Consequently, if we take the imaginary part, we can see that

$$0 = \int_{-\infty}^\infty Y_s(\lambda) dP(\lambda)[G(\lambda) - \widetilde{G}(\lambda)] \left[\frac{\nu_0}{(\lambda - \mu_0)^2 + \nu_0^2} \right].$$

If we integrate with respect to μ_0 from α to β, and reverse the order of integration, we have

$$0 = \int_{-\infty}^\infty Y_s(\lambda)^* dP(\lambda)[G(\lambda) - \widetilde{G}(\lambda)] \left[\tan^{-1} \left(\frac{\beta - \lambda}{\nu_0} \right) - \tan^{-1} \left(\frac{\alpha - \lambda}{\nu_0} \right) \right].$$

Letting ν_0 approach 0, we get

$$0 = \int_\alpha^\beta Y_s(\lambda)^* dP(\lambda)[G(\lambda) - \widetilde{G}(\lambda)],$$

or

$$0 = \int_a^s \left[\int_\alpha^\beta K^* A(\xi) \mathcal{Y}(\xi, \lambda) dP(\lambda)[G(\lambda) - \widetilde{G}(\lambda)] \right] d\xi.$$

Differentiate with respect to s,

$$0 = \int_\alpha^\beta K^* A(x) \mathcal{Y}(x, \lambda) dP(\lambda)[G(\lambda) - \widetilde{G}(\lambda)].$$

Now apply an extension of the mean value theorem, remembering that the expression is 1×1 and $\mathcal{Y}$ is analytic in λ. We find for some λ_0 in $[\alpha, \beta]$

$$K^* A(x) \mathcal{Y}(x, \lambda_0) \int_\alpha^\beta dP(\lambda)[G(\lambda) - \widetilde{G}(\lambda)] = 0.$$

From this we conclude that

$$\int_\alpha^\beta dP(\lambda)[G(\lambda) - \widetilde{G}(\lambda)] = 0$$

for all α, β. We use this to build up integrals involving step functions, dense in $L_p^2(-\infty, \infty)$, that have as their limit

$$\int_{-\infty}^\infty [G(\lambda) - \widetilde{G}(\lambda)]^* dP(\lambda)[G(\lambda) - \widetilde{G}(\lambda)] = 0.$$

Hence $G = \widetilde{G}$ in $L_p^2(-\infty, \infty)$. $\qquad\Box$

We summarize.

XI.3.4. Theorem. *If $G(\lambda)$ is in $L_p^2(-\infty, \infty)$, there is a unique $F(x)$ in $L_A^2(a, b)$ such that*

$$F(x) = \int_{-\infty}^\infty \mathcal{Y}(x, \lambda) dP(\lambda) G(\lambda)$$

and

$$G(\lambda) = \int_a^b \mathcal{Y}(\xi, \lambda)^* A(\xi) F(\xi) d\xi.$$

XI.4 The Relation Between $\mathbf{M_a}$, $\mathbf{M_b}$ and $\mathbf{P}(\lambda)$

The matrices M_a, M_b can frequently be determined by a careful inspection of the solution used to determine appropriate L_A^2 solutions. More difficult is the determination of the spectral matrix $P(\lambda)$, since its existence follows from Helly's selection theorems. Fortunately, they are intimately connected.

XI.4.1. Theorem. *Let $H(x)$ be defined by*

$$\begin{aligned}
H(x) &= \chi_b(x, \lambda_0)[M_a(\lambda_0) - M_b(\lambda_0)]^{-1}[I_n, M_a(\lambda_0)], & e < x < b, \\
&= \chi_a(x, \lambda_0)[M_a(\lambda_0) - M_b(\lambda_0)]^{-1}[I_n, M_b(\lambda_0)], & a < x < e,
\end{aligned}$$

where $\nu = Im\, \lambda_0 \neq 0$. Then

$$\int_a^b H(\xi)^* A(\xi) H(\xi) d\xi = \frac{1}{2i\,Im\, \lambda_0}\, M(\lambda_0),$$

where $M = \begin{pmatrix} M_{11} & M_{12} \\ M_{21} & M_{22} \end{pmatrix}$ *and*

$$
\begin{aligned}
M_{11} &= (M_a - M_b)^{-1} - (M_a^* - M_b^*)^{-1} \\
&= 2i\,\mathrm{Im}\,(M_a - M_b)^{-1}\,, \\
M_{12} &= \frac{1}{2}(M_a - M_b)^{-1}(M_a - M_b) - \frac{1}{2}(M_a^* - M_b^*)^{-1}(M_a^* + M_b^*)\,, \\
M_{21} &= \frac{-1}{2}(M_a + M_b)(M_a - M_b)^{-1} + \frac{1}{2}(M_a^* + M_b^*)(M_a^* - M_b^*)^{-1} \\
&= -M_{12}^*\,, \\
M_{22} &= M_a(M_a - M_b)^{-1}M_b - M_b^*(M_a^* - M_b^*(M_a^* - M_b^*)^{-1}M_a^* \\
&= 2i\,\mathrm{Im}\,M_a(M_a - M_b)^{-1}M_b\,.
\end{aligned}
$$

It is tempting to try to write M_{12} and M_{21} as $2i$ times the imaginary part of something. For $n = 1$, the matrices involved are scalar and commute. Then

$$
M_{12} = 2i\mathrm{Im}\left[\frac{1}{2}(M_a - M_b)^{-1}(M_a + M_b)\right]\,,
$$

$$
M_{21} = 2i\mathrm{Im}\left[-\frac{1}{2}(M_a + M_b)(M_a - M_b)^{-1}\right]\,.
$$

This would be true in the general case if M_a and M_b commute. We have been unable to verify this.

Proof. An expression for $\int_{a'}^{b'} H^*AH d\xi$ was calculated in the proof of Theorem XI.2.8. We have only to let (a', b') approach (a, b) and expand the various matrix products. $\qquad\square$

XI.4.2. Theorem. *If λ_1 and λ_2 are real, then*

$$
P(\lambda_2) - P(\lambda_1) = \lim_{\nu \to 0+} \frac{1}{2\pi i} \int_{\lambda_1}^{\lambda_2} M(\mu + i\nu)d\mu\,.
$$

Proof. We have

$$
\int_{-\infty}^{\infty} |\lambda - \lambda_0|^{-2} dP(\lambda) = M(\lambda_0)/(2i\mathrm{Im}\,\lambda_0)\,.
$$

With $\lambda_0 = \mu + i\nu$, we integrate both sides with respect to μ from λ_1 to λ_2. Then let ν approach zero from above. $\qquad\square$

XI.5 The Spectral Resolution

We connect the results of Sections XI.2, XI.3, and XI.4 to the classic representation of the identity as an integral generated by a projection valued measure E_λ. Given

$$F(x) = \int_{-\infty}^{\infty} \mathcal{Y}(x,\lambda)dP(\lambda)G(\lambda)\,,$$

where

$$G(\lambda) = \int_{a}^{b} \mathcal{Y}^*(\xi,\lambda)A(\xi)F(\xi)d\xi\,,$$

we define

$$E_\lambda F(x) = \int_{-\infty}^{\lambda+} \mathcal{Y}(x,\lambda)dP(\lambda)G(\lambda)\,.$$

Considered as the limit of eigenfunction expansions, it is not difficult to show that E_λ is a projection, is continuous from above, and satisfies $E_{\lambda_1}E_{\lambda_2} = E_{\lambda_1}$ when $\lambda_1 \leq \lambda_2$, as well $E_{-\infty} = 0$, $E_\infty = I$. If we let $\{\lambda_j\}_{j=-\infty}^{\infty}$ be a partition of $(-\infty,\infty)$, $\lambda_i < \lambda_j$ if $i < j$, and

$$\Delta_j EF(x) = \int_{\lambda_j}^{\lambda_{j+1}} \mathcal{Y}(x,\lambda)dP(\lambda)G(\lambda)\,,$$

then $F(x) = \sum\limits_{j=-\infty}^{\infty} \Delta_j EF(x)$. As $\{\lambda_j\}_{j=-\infty}^{\infty}$ becomes finer, we may write

$$f(x) = \int_{-\infty}^{\infty} dE_\lambda F(x)$$

as the limit of the decomposition above.

If Y is in D, it has the representation

$$Y(x) = \int_{-\infty}^{\infty} \mathcal{Y}(x,\lambda)dP(\lambda)G(\lambda)\,,$$

where

$$G(\lambda) = \int_{a}^{b} \mathcal{Y}^*(\xi,\lambda)A(\xi)Y(\xi)d\xi\,.$$

Then

$$LY = \int_{-\infty}^{\infty} \lambda\mathcal{Y}(x,\lambda)dP(\lambda)G(\lambda),$$

which is equivalent to

$$LY(x) = \int_{-\infty}^{\infty} \lambda dE_\lambda Y(x)\,.$$

The resolvent operator also has the standard representation. If λ_0 is complex,

$$(L - \lambda_0 I)^{-1} F(x) = \int_{-\infty}^{\infty} \frac{1}{\lambda - \lambda_0}\, dE_\lambda F(x)\,.$$

Apparently, λ_0 is in the spectrum of L if and only if it is in the support of dE_λ or $dP(\lambda)$.

References

[1] F. V. Atkinson, **Discrete and Continuous Boundary Value Problems**, Academic Press, New York, 1964.

[2] F. Brauer, *Spectral theory for linear systems of differential equations*, Pacific J. Math. **10** (1960), 17–34.

[3] E. A. Coddington and N. Levinson, **Theory of Ordinary Differential Equations**, McGraw-Hill, New York, 1955.

[4] D. B. Hinton and J. K. Shaw, *Titchmarsh's λ-dependent boundary conditions for Hamiltonian systems*, Lecture Notes in Mathematics, Vol. 964, Springer-Verlag, Berlin, 1982, 318–326.

[5] ———, *Hamiltonian systems of limit-point or limit circle type with both endpoints singular*, J. Diff. Eq. **50** (1983), 444–464.

[6] A. M. Krall, **Applied Analysis**, D. Reidel, Dordrecht, Netherlands, 1987.

Chapter XII

Distributions

Our goal in the near future is to find and catagorize those boundary value problems which have orthogonal polynomial solutions, but first we must define what we mean by "orthogonal polynomials," and in order to do so we need some concepts from the theory of distributions.

We shall very briefly examine four sets of test functions D, S, P and E, and the linear functionals acting upon them D', S', P', E'. Of particular interest is the set P of functions of slow growth, which contains, among others, the polynomials. P has not been extensively studied. It was introduced in [4].

XII.1   Test Functions with Compact Support, D; Distributions Without Constraint, D$'$

We begin by examining the best known pair of test functions and distributions, D and D'. These serve as a role model for the other pairs S and S', P and P', E and E', to follow.

XII.1.1. Definition. A measurable function $f = f(x)$ is locally integrable if

$$\int_R |f(x)|\,dx$$

is finite for each bounded measurable set R in $(-\infty, \infty)$.

XII.1.2. Definition. The support of a function $f(x)$ is the set of all x such that $f(x) \neq 0$.

XII.1.3. Definition. A function $\phi(x)$ is called a test function of compact support if:

(1) ϕ is infinitely differentiable.

(2) ϕ has compact support. That is, there is a number N, depending on ϕ, such that $\phi(x) = 0$ when $|x| > N$.

We call the set of all test functions of compact support D.

As an example we present

$$\phi(x) = \begin{cases} 0, & |x| \geq a \\ \exp[(|x|^2 - a^2)^{-1}], & x \leq a. \end{cases}$$

It is evident that D is a linear space, that $\phi^{(k)}$ is in D when ϕ is, and that if f is infinitely differentiable, then $f\phi$ is also in D.

XII.1.4. Definition. A sequence $\{\phi_j\}_{j=1}^{\infty}$ converges to ϕ_0 in D if:

(1) All ϕ_j, $j = 1, \ldots$, and ϕ_0 vanish outside a common region.

(2) $\phi_j^{(k)}$ approaches $\phi_0^{(k)}$ as j approaches ∞, for all k.

XII.1.5. Definition. A distribution of rapid growth is a continuous linear functional on the linear space D. We call the space of such distributions D'.

We denote the action of f in D' on ϕ in D by $\langle f, \phi \rangle$.

There should be no confusion with this notation, which is similar to an inner product.

XII.1.6. Theorem. *Every locally integrable function f defines a distribution in D' by*

$$\langle f, \phi \rangle = \int_{-\infty}^{\infty} f \phi \, dx.$$

As a proof we need only to note that

$$\langle f, \phi \rangle < \sup_{x \in R_\phi} |\phi(x)| \int_{R_\phi} |f(x)| \, dx < \infty.$$

There are distributions different from these. They are important enough to merit a special name.

XII.1.7. Definition. A distribution of rapid growth is regular if it can be written in the form $\int_{-\infty}^{\infty} f \phi \, dx$. All other distributions are singular.

Just as with test functions D it is evident that D' is a linear space. Furthermore a topology can be induced on D'. Under this topology, it can be shown that the regular distributions are dense in D'. Hence every distribution is a limit of regular distributions.

XII.1.8. Examples. 1. Let R be a fixed region on the real line. Define the distribution H_R by

$$\langle H_R, \phi \rangle = \int_R \phi \, dx.$$

H_R is generated by the characteristic function of R:

$$\chi_R(x) = \begin{cases} 1\,, & x \text{ in } R, \\ 0\,, & x \text{ not in } R. \end{cases}$$

Thus

$$\langle H_R, \phi \rangle = \int_{-\infty}^{\infty} \chi_R \phi \, dx,$$

and is a regular distribution. It is called the Heaviside distribution after its inventor the British physicist Oliver Heaviside.

2. Let ξ be an arbitrary real number, and define the distribution δ_ξ by

$$\langle \delta_\xi, \phi \rangle = \phi(\xi)\,.$$

If ϕ_j approaches ϕ_0 in D, then clearly $\langle \delta_\xi, \phi_j \rangle$ approaches $\langle \delta_\xi, \phi_0 \rangle$, so δ_ξ is in D'. Frequently, δ_ξ is written symbolically as

$$\langle \delta_\xi, \phi \rangle = \int_{-\infty}^{\infty} \delta(x - \xi)\phi(x)dx,$$

where

$$\delta(x - \xi) = \begin{cases} 0\,, & x \neq \xi, \\ \infty\,, & x = \xi, \end{cases}$$

and $\delta(x - \xi)$ has the property

$$\int_R \delta(x - \xi)dx = \begin{cases} 1\,, & x \text{ in } R, \\ 0\,, & x \text{ not in } R. \end{cases}$$

Of course there is no such function $\delta(x - \xi)$, but the symbolism works quite well. δ_ξ is a perfectly valid singular distribution. It is called the Dirac distribution (or Dirac delta function), named after its inventor the British physicist Paul Dirac.

XII.1.9. Definition. Let f be a distribution and let g be infinitely differentiable. The distribution of rapid growth fg is defined by

$$\langle fg, \phi \rangle = \langle f, g\phi \rangle\,.$$

The infinite differentiability of g is needed. For if f and g were both generated by $1/|x|^{1/2}$, then fg would correspond to $1/|x|$. Unfortunately this is not a distribution in D' because of its singular behavior near 0.

If f and f' are locally integrable, then

$$\langle f', \phi \rangle = \int_{-\infty}^{\infty} f' \phi \, dx$$
$$= -\int_{-\infty}^{\infty} f \phi' \, dx = -\langle f, \phi' \rangle\,.$$

We can use this as a basis for a more general definition.

XII.1.10. Definition. The derivative of a distribution of rapid growth f is given by

$$\langle f', \phi \rangle = -\langle f, \phi' \rangle \,.$$

More generally

$$\langle f^{(k)}, \phi \rangle = (-1)^k \langle f, \phi^{(k)} \rangle \,, \quad k = 1, 2, \dots \,.$$

This is the most important characteristic of distributions. They inherit infinite differentiability from the test functions, even though they may not be differentiable in the ordinary sense.

XII.1.11. Examples. 1. δ'_ξ is calculated by

$$\langle \delta'_\xi, \phi \rangle = -\langle \delta_\xi, \phi' \rangle = -\phi'(\xi) \,.$$

2. If f is infinitely differentiable,

$$\langle (f \delta_\xi)', \phi \rangle = -\langle f \delta_\xi, \phi' \rangle = -\langle \delta_\xi, f \phi' \rangle = -f(\xi) \phi'(\xi) \,.$$

3. Likewise

$$\begin{aligned}
\langle f \delta'_\xi, \phi \rangle &= \langle \delta'_\xi, f \phi \rangle \\
&= -\langle \delta_\xi, (f \phi)' \rangle = -\langle \delta_\xi, f \phi' + f' \phi \rangle \\
&= -f(\xi) \phi'(\xi) - f'(\xi) \phi(\xi) \\
&= f(\xi) \langle \delta'_\xi, \phi \rangle - f'(\xi) \langle \delta_\xi, \phi \rangle \,.
\end{aligned}$$

Or $f \delta'_\xi = f(\xi) \delta'_\xi - f'(\xi) \delta_\xi$.

4. If L is a formal differential operator with real, infinitely differentiable coefficients defined by

$$Ly = \sum_{k=0}^{p} a_k(x) y^{(k)} \,,$$

with formal adjoint given by

$$L^+ z = \sum_{k=0}^{p} (-1)^k (a_k(x) z)^{(k)} \,,$$

then we may define the distributional operator Lf, f a distribution, by

$$\langle Lf, \phi \rangle = \langle f, L^+ \phi \rangle \,.$$

When f is sufficiently differentiable

$$\begin{aligned}
\langle Lf, \phi \rangle &= \int_{-\infty}^{\infty} Lf \, \phi \, dx, \\
&= \int_{-\infty}^{\infty} f \, L^+ \phi \, dx
\end{aligned}$$

after a number of integrations by parts. So the definition of Lf is consistent.

Finally we can extend the concept of support to distributions, again through test functions.

XII.1.12. Definition. A distribution f vanishes in a region R if $\langle f, \phi \rangle = 0$ for all test functions ϕ with support in R. f has support in R if $\langle f, \phi \rangle = 0$ for all test functions with support in the complement of R.

XII.2 Limits of Distributions

We would be remiss in our outline of distribution theory if we did not include the following. If a collection of distributions in D' depends upon a parameter, we may define a limit process involving that parameter just as with ordinary functions.

XII.2.1. Definition. Let the distribution f_α depend upon the real valued parameter α. Then $\lim\limits_{\alpha \to \alpha_0} f_\alpha = f_{\alpha_0}$ if

$$\lim_{\alpha \to \alpha_0} \langle f_\alpha, \phi \rangle = \langle f_{\alpha_0}, \phi \rangle$$

for all ϕ in D. Likewise $\lim\limits_{n \to \infty} f_n = f_0$ if

$$\lim_{n \to \infty} \langle f_n, \phi \rangle = \langle f_0, \phi \rangle$$

for all ϕ in D.

XII.2.2. An Example. We consider the function

$$f_\alpha(x) = 1/2\alpha, \quad c - \alpha < x < c + \alpha,$$
$$= 0, \qquad |x - c| > \alpha.$$

Then

$$\langle f_\alpha, \phi \rangle = \frac{1}{2\alpha} \int_{c-\alpha}^{c+\alpha} \phi(s)ds.$$

If α is small, then $\phi(s) \simeq \phi(c)$, so

$$\langle f_\alpha, \phi \rangle \simeq \left(\frac{1}{2\alpha} \right) \phi(c)\, 2\alpha = \phi(c).$$

Hence $\lim\limits_{\alpha \to 0} f_\alpha = \delta(x - c)$.

This can be generalized.

XII.2.3. Theorem. *Let f be a nonnegative integrable function satisfying* $\int_{-\infty}^{\infty} f(x)dx = 1$. *Let*

$$f_\alpha(x) = \frac{1}{\alpha}\, f(x/\alpha).$$

Then $\lim\limits_{\alpha \to 0} f_\alpha(x) = \delta(x)$.

We leave the proof as an exercise.

XII.2.4. Additional Examples. Apply the previous theorem to

$$f(x) = 1/[\pi(1+x^2)]\,,$$
$$f(x) = (1/\sqrt{\pi})e^{-x^2}\,.$$

XII.3   Test Functions of Rapid Decay, S; Distributions of Slow Growth, S′

We wish to enlarge our collection of test functions. In so doing the distributions, or continuous linear functionals on them, will be somewhat restricted.

XII.3.1. Definition. A function $\phi(x)$ belongs to S, the space of test functions of rapid decay, if:

(a) $\phi(x)$ is infinitely differentiable.

(b) $\phi(x)$, as well as all its derivatives, vanishes at $\pm\infty$ faster than the reciprocal of any polynomial. Thus for each j and k,

$$\lim_{x\to\pm\infty} x^\alpha \frac{d^k\phi}{dx^k} = 0\,, \quad j,k = 0,1,\ldots.$$

It is evident that $D \subset S$, but not the reverse. For example $\phi(x) = e^{-x^2}$ is in S, but not in D.

XII.3.2. Definition. A sequence $\{\phi_\alpha\}_{j=1}^{\infty}$ converges to ϕ_0 in S if for all $k,\ell = 0,1,\ldots,$

$$\lim_{j\to\infty}\ \sup_x \left| x^k \left[\frac{d^\ell \phi_\alpha}{dx^\ell}\right]\right| = 0\,.$$

This is slightly different from convergence in D.

XII.3.3. Definition. A distribution of slow growth is a continuous linear functional on the space S. The space of all continuous linear functionals on S is denoted by S'.

Since $D \subset S$, it follows that $S' \subset D'$.

XII.3.4. Definition. A function $f(x)$ is of slow growth if:

(a) $f(x)$ is locally integrable.

(b) There exist c, n, r such that

$$|f(x)| < c|x|^n,$$

when $|x| > r$.

XII.3.5. Theorem. *Every function, f, of slow growth generates a (regular) distribution of slow growth through the formula*

$$\langle f, \phi \rangle = \int_{-\infty}^{\infty} f(x)\phi(x)dx.$$

Proof. It is sufficient to show that if ϕ_α approaches 0 in S, then $\langle f, \phi_j \rangle$ also approaches 0.

For each j,

$$\int_{-\infty}^{\infty} f(x)\phi_\alpha(x)dx = \int_{-\infty}^{\infty} \frac{f(x)}{(1+x^2)^p}[1+x^2]^p \phi_\alpha(x)dx.$$

If p is large, then $|f(x)/[1+x^2]^p|$ is integrable. Thus

$$\left| \int_{-\infty}^{\infty} f(x)\phi_\alpha(x)dx \right| \le \sup_x |[1+x^2]^p \phi_\alpha(x)| \int_{-\infty}^{\infty} \frac{f(x)}{[1+x^2]^p}\, dx.$$

This approaches 0 as ϕ_α approaches 0 in S. $\qquad\qquad\square$

One very important reason for introducing the spaces S and S' is that if ϕ is in S, then its Fourier transform is also in S. Distributions f in S' inherit Fourier transforms from S through

$$\langle \mathcal{F}f, \phi \rangle = \langle f, \mathcal{F}\phi \rangle.$$

In higher dimensions this property is especially useful in solving partial differential equations. We cite [3] or [6] for further reference.

XII.4 Test Functions of Slow Growth, P; Distributions of Rapid Decay, P'

Since we wish to apply distribution theory to polynomials, we need a set of test functions which include them as a subset. Our next collection of test functions does this.

XII.4.1. Definition. A function $\phi(x)$ belongs to P, the space of test functions of slow growth, if:

(a) $\phi(x)$ is infinitely differentiable.

(b) $\lim\limits_{x \to \pm\infty} e^{-\alpha|x|}\phi^{(\ell)}(x)| = 0$ for all $\alpha > 0$ and integers $\ell \ge 0$.

We note that polynomials are in P. They are neither in D nor in S.

XII.4.2. Definition. A sequence ϕ_α converges to ϕ_0 in P if for all $\alpha > 0$ and integers $\ell \ge 0$,

$$\lim_{j \to \infty} \sup_x \left| e^{-\alpha|x|} \left[\frac{d^\ell \phi_j}{dx^\ell} - \frac{d^\ell \phi_j}{dx^\ell} \right] \right| = 0.$$

Here the roles of growth and decay found in the definition S are reversed.

XII.4.3. Definition. A distribution of rapid decay is a continuous linear functional on the space P. The space of all continuous linear functionals on P is denoted by P'.

XII.4.4. Definition. A function f is of rapid decay if:
  (a) $F(x)$ is locally integrable.
  (b) For all $a > 0$

$$\lim_{x \to \pm\infty} e^{-\alpha|x|} f(x) = 0 \,.$$

XII.4.5. Theorem. *Every function, f, of rapid decay generates a (regular) distribution of rapid decay through the formula*

$$\langle f, \phi \rangle = \int_{-\infty}^{\infty} f(x)\phi(x)dx \,.$$

Proof. Let $\alpha > 0$

$$|\langle f, \phi \rangle| = \left| \int_{-\infty}^{\infty} e^{|\alpha|x} f(x)\, e^{-\alpha|x|} \phi(x)\, dx \right| ,$$

$$\leq \sup_{x} |e^{-\alpha|x|} \phi(x)| \int_{-\infty}^{\infty} e^{-|x|} |e^{[\alpha+1]|x|} f(x)|\, dx.$$

This is clearly finite. If ϕ approaches 0 in P, then $\langle f, \phi \rangle$ does as well.

    We note that if $\mathcal{P}$ stands for the linear space of polynomials, then $\mathcal{P} \subset P$ and so distributions in P' are also distributions on $\mathcal{P}$. $\square$

XII.5   Test Functions Without Constraint; E; Distributions of Compact Support, E′

We would again be remiss in our outline of distributions if we did not at least mention E and E'. They, in some sense, round out our discussion. While they are not often used, in polynomial theory when the interval under consideration is finite, the weight functions associated with orthogonalizing the polynomials are in E'.

XII.5.1. Definition. A function $\phi(x)$ belongs to E, the space of unconstrained test functions, if $\phi(x)$ is infinitely differentiable.

    $\phi(x) = e^{x^2}$ serves as a classic example of such a test function.

XII.5.2. Definition. A sequence ϕ_j converges to ϕ_0 in E if for all compact sets C and all integers $\ell \geq 0$,

$$\lim_{j \to \infty} \sup_{x \in C} \left| \frac{d^\ell \phi_j}{dx^\ell} - \frac{d^\ell \phi_0}{dx^\ell} \right| = 0 \,.$$

XII.5.3. Definition. A distribution of compact support is a continuous linear functional on the space E. The space of all continuous linear functionals on E is denoted by E'.

XII.5.4. Theorem. *Let $f(x)$ be locally integrable and have compact support. The f generates a (regular) distribution of compact support through the formula*

$$\langle f, \phi \rangle = \int_{-\infty}^{\infty} f(x)\phi(x)dx.$$

Proof. Clearly this is well defined. If ϕ approaches 0 in E, then if the support of f is C,

$$|\langle f, \phi \rangle| < \sup_{x \in C} |\phi| \int_C |f(x)|dx,$$

which has limit 0. $\qquad\square$

XII.5.5. Remarks. We note that

$$D \subset S \subset P \subset E.$$

Thus

$$D' \supset S' \supset P' \supset E'.$$

There is a certain dichotomy in these sequences: $D \subset E', S \subset P', P \subset S', E \subset D'$ as well.

The delta function is in all the spaces D', S', P' and E', as are all its distributional derivatives. They have support only at $x = 0$. Multiplication of a distribution by "appropriate" infinitely differentiable functions is frequently permitted. One must be sure that the product of the function and a test function remains a test function. For instance in P', multiplication of a distribution f by a polynomial is permitted, since a test function ϕ in P, multiplied by a polynomial, is still a test function.

Derivatives of distributions in S', P' or E' are defined in the same way as in D':

$$\langle f^{(k)}, \phi \rangle = (-1)^k \langle f, \phi^{(k)} \rangle, \quad k = 1, 2, \ldots$$

are also defined in the same way provided the coefficients are appropriate.

When solving distributional differential equations, we first seek solutions in the largest space D', then see how far down the chain we can push it. When searching for polynomial weight functions, this is the rule.

XII.6 Distributional Differential Equations

Having built up an elaborate theory of distributions and their derivatives, we close with a discussion of solving some elementary differential equations in a distributional setting. We work in D', the largest distributional space in order to get the most general solution.

Before we begin we need a preliminary result.

XII.6.1. Theorem.   *The test function ϕ in D satisfies $\phi(0) = 0$ if and only if $\phi = x\psi$, where ψ is also a test function in D.*

Proof. If $\phi = x\psi$, the result is obvious. Conversely, let $\psi = \phi/x$. We must show that ψ is a test function in D. Support and differentiability except at $x = 0$ are no problem. We must show, however, that ψ is infinitely differentiable at $x = 0$.$\square$

XII.6.2. Lemma. *Let ϕ be a test function in D and let $\phi(0) = 0$. Let $\psi = \phi/x$. Then $\lim\limits_{x\to 0}\psi^{(n)}(x) = \phi^{(n+1)}(0)/(n+1)$.*

Proof. By Leibnitz's rule,

$$\psi^{(n)}(x) = \sum_{j=0}^{n} \binom{n}{j} (-1)^j \, j! \, x^{-1-j}\phi^{(n-j)}$$

$$= \left[\sum_{j=0}^{n} \frac{n!}{(n-j)!}(-1)^j x^{n-j}\phi^{(n-j)}\right]\bigg/ x^{n+1}\,.$$

Note that within the sum, the term with $j = n$ is not multiplied by a power of x. Since $\phi(0) = 0$ as x approaches 0, the quotient is indeterminate. Apply l'Hopital's rule.

$$\psi^{(n)}(x) \to \sum_{j=0}^{n} \frac{n!}{(n-j)!}(-1)^j \left[(n-j)x^{n-j-1}\phi^{(n-j)}(x)\right.$$

$$\left.+x^{n-j}\phi^{(n-j+1)}(x)\right]\bigg/(n+1)x^n\,.$$

Rewrite as two sums. In the first let $j = k - 1$. In the second let $j = k$.

$$\psi^{(k)}(x) \to \left[\sum_{k=1}^{n} \frac{n!}{(n-k)!}(-1)^{k-1}x^{n-k}\phi^{(n-k+1)}(x)\right.$$

$$\left.+ \sum_{k=0}^{n} \frac{n!}{(n-k)!}(-1)^k x^{n-k}\phi^{(n-k+1)}(x)\right]\bigg/(n+1)x^k\,.$$

This collapses, leaving

$$\psi^{(n)}(x) \to \phi^{(n+1)}(0)/(n+1)\,.$$

We now know that $\psi^{(n)}(x)$ has a limit as x approaches 0. We extend $\psi^{(n)}$ to $x = 0$ by defining it to be $\phi^{(n+1)}(0)/(n+1)$ at $x = 0$.

We must still show that $\psi^{(n)}$ is continuously differentiable at $x = 0$. We compute

$$\frac{\psi^{(n)}(x) - \psi^{(n)}(0)}{x - 0} = \left[\frac{\sum_{j=0}^{n} \frac{n!}{(n-j)!}(-1)^j x^{n-j} \phi^{(n-j)}(x)}{x^{n+1}} - \frac{\phi^{(n+1)}(0)}{(n+1)} \right] \Big/ x,$$

$$= \left[\sum_{j=0}^{n} \frac{(n+1)!}{(n-j)!}(-1)^j x^{n-j} \phi^{(n-j)}(x) - x^{n+1}\phi^{(n+1)}(0) \right] \Big/ (n+1)x^{n+2}.$$

Again, as $x \to 0$, this is indeterminate. Apply l'Hopital's rule.

$$\frac{\psi^{(n)}(x) - \psi^{(n)}(0)}{x - 0} \to$$

$$\left[\sum_{j=0}^{n} \frac{(n+1)!}{(n-j)!}(-1)^j \right.$$

$$\times \left[(n-j)x^{n-j-1}\phi^{(n-j)}(x) + x^{n-j}\phi^{(n-j+1)}(x) \right] - (n+1)x^n \phi^{(n+1)}(x)$$

$$\left. \times \left((n+1)(n+2)x^{n+1}\right)^{-1} \right].$$

Again split the sum apart, letting $j = k - 1$ in the first, and $j = k$ in the second. The sum collapses, leaving

$$\frac{\psi^{(n)}(x) - \psi^{(n)}(0)}{x - 0} \to \lim_{x \to 0} \left[\frac{\phi^{(n+1)}(x) - \phi^{(n+1)}(0)}{x - 0} \right] \Big/ (n+2).$$

Apply l'Hopital's rule.

$$\lim_{x \to 0} \frac{\phi^{(n)}(x) - \phi^{(n)}(0)}{x - 0} = \phi^{(n+2)}(0)/(n+2). \qquad \square$$

XII.6.3. Theorem. *The only distributional solution f in D' to the equation $xf = 0$ is $f = c\,\delta(x)$.*

Proof. Let ϕ_0 be an arbitrary, but fixed, test function satisfying $\phi_0(0) = 1$. Then for arbitrary ϕ in D,

$$\phi(x) = \phi(0)\phi_0(x) + [\phi(x) - \phi(0)\phi_0(x)].$$

The bracketed term vanishes at $x = 0$, hence

$$\phi(x) = \phi(0)\phi_0(x) + x\,\psi(x),$$

where ψ is also a test function in D.

We compute $\langle f, \phi \rangle$:

$$\langle f, \phi \rangle = \phi(0)\langle f, \phi_0 \rangle + \langle f, x\,\psi \rangle \,.$$

The second term equals $\langle xf, \psi \rangle$, and vanishes. Thus if $\langle f, \phi_0 \rangle = c$,

$$\langle f, \phi \rangle = c\,\phi(0) = \langle c\,\delta, \phi \rangle,$$

and $f = c\,\delta$.

We shall use this result extensively in order to find weight functionals for orthogonal polynomials. It is also relatively easy to show that if $x^n f = 0$, then $f = A_0\delta + A_1\delta' + \cdots + A_{n-1}\delta^{(n-1)}$. We invite the reader to write a proof. $\square$

XII.6.4. Theorem. *The only distributional solution to $f' = 0$ in D' is $f = c$, a constant.*

Proof. Note that if ψ is a test function, then

$$\phi(x) = \int_{-\infty}^{x} \psi(u)du$$

is also a test function if and only if

$$\int_{-\infty}^{\infty} \psi()du = 0 \,.$$

In that case $\phi' = \psi$.

The differential equation is equivalent to $\langle f', \phi \rangle = 0$, or $\langle f, \psi \rangle = 0$, where $\psi = \phi'$.

Let ϕ_0 be an arbitrary, but fixed, test function satisfying

$$\int_{-\infty}^{\infty} \phi_0(u)du = 1 \,.$$

Then write

$$\phi(x) = \phi_0(x) \int_{-\infty}^{\infty} \phi(u)du + \left[\phi(x) - \phi_0(x) \int_{-\infty}^{\infty} \phi(u)du \right] \,.$$

The term in brackets

$$\psi(x) = \phi(x) - \phi_0(x) \int_{-\infty}^{\infty} \phi(u)du$$

has the property $\int_{-\infty}^{\infty} \psi(u)du = 0$. Thus there exists a test function η such that $\eta' = \psi$, and

$$\langle f, \psi \rangle = \langle f, \eta' \rangle = -\langle f', \eta \rangle = 0 \,.$$

Compute $\langle f, \phi \rangle$. We find

$$\langle f, \phi \rangle = \langle f, \phi_0 \rangle \int_{-\infty}^{\infty} \phi(u) du\,.$$

If $c = \langle f, \phi_0 \rangle$, then

$$\langle f, \phi \rangle = \int_{-\infty}^{\infty} c\,\phi(u) du = \langle c, \phi \rangle\,,$$

and $f = c$.

This theorem has important implications. $\square$

XII.6.5. Theorem. *In each of the problems*

(1) $a_0 y^{(n)} + \cdots + a_n y = 0$,

where a_j, $j = 0, \ldots, n$, are infinitely differentiable and $a_0 \neq 0$, or

(2) $Y' + AY = 0$,

where Y is an n-dimensional vector, and A is an infinitely differentiable, non-singular $n \times n$ matrix, the distributional solution is identical to the classical solution.

Proof. We can use the standard embedding to write the scalar problem as a vector system. Hence it suffices to consider $Y' + AY = 0$.

Classically the system has a fundamental matrix $\mathcal{Y}$. We make the transformation $Y = \mathcal{Y}C$ to find $\mathcal{Y}C' = 0$. Since $\mathcal{Y}$ is invertible, $C' = 0$ and C is constant. Thus $Y = \mathcal{Y}C$, the classical solution. $\square$

XII.6.6. Theorem. *The only distributional solution to $f' = g$ in D', where g is continuous, is the classical solution.*

Proof. The differential equation is equivalent to

$$\langle f', \phi \rangle = \langle g, \phi \rangle\,,$$

or

$$\langle f, \phi' \rangle = -\langle y, \phi \rangle\,.$$

Using the decomposition of ϕ, developed previously,

$$\phi(x) = \phi_0(x) \int_{-\infty}^{\infty} \phi(u) du + \left[\phi(x) - \phi_0(x) \int_{-\infty}^{\infty} \phi(u) du \right]\,,$$

we have

$$\langle f, \phi \rangle = \langle f, \phi_0 \rangle \int_{-\infty}^{\infty} \phi(u) du + \langle f, \psi \rangle\,,$$

where ψ is the bracketed term above. ψ is the derivative of

$$\eta(x) = \int_{-\infty}^{x} \phi(u) du - \int_{-\infty}^{x} \phi_0(u) du \int_{-\infty}^{\infty} \phi(u) du\,.$$

Now $\langle f, \psi \rangle$ is known:

$$\langle f, \psi \rangle = -\langle g, \eta \rangle\,.$$

We define the distribution f_0 by setting

$$\langle f_0, \phi \rangle = \langle g, -\eta \rangle\,.$$

Then

$$\langle f, \phi \rangle = \langle f, \phi_0 \rangle \int_{-\infty}^{\infty} \phi(u)du + \langle f_0, \phi \rangle$$
$$= \langle c_f, \phi \rangle + \langle f_0, \phi \rangle\,,$$

where $c_f = \langle f_0, \phi \rangle$.

Now let y represent a classical solution of $y' = g$. The same operations yield

$$\langle y, \phi \rangle = \langle c_y, \phi \rangle + \langle f_0, \phi \rangle\,,$$

where $c_y = \langle y, \phi_0 \rangle$. Then

$$\langle f, \phi \rangle = \langle y, \phi \rangle + \langle (c_f - c_y), \phi \rangle,$$

or $f = y + c$, which is the classical general solution. $\square$

XII.6.7. Theorem. *In each of the problems*
   (1) $a_0 y^{(n)} + \cdots + a_n y = f$,
   where a_j, $j = 0, \ldots, n$, are infinitely differentiable and $a_0 \neq 0$, or
   (2) $Y' + AY = F$,
   where Y and F are n-dimensional vectors F is continuous, and A is an infinitely differentiable, nonsingular $n \times n$ matrix, the distributional solution is identical to the classical solution.

We leave the proof to the reader.

Distributional, or generalized, solutions fall into three categories.

   (1) The solution corresponds to a function which is sufficiently differentiable in the ordinary sense. These are classical solutions.

   For example, the solution to $f' = 0$ in D' is $f = c$. It is also in S', but not in P' or E' unless $c = 0$.

   (2) The solution corresponds to an ordinary function which is not differentiable, but satisfies the equation as a distribution. These are weak solutions.

   For example the solution to $xf' = 0$ is $f = c_1 + c_2 H(x)$, where $H(x)$ is the Heaviside function

$$H(x) = 1, \quad x \geq 0,$$
$$= 0, \quad x < 0.$$

Here $H' = \delta(x)$ and $x H' = x \delta = 0$. f is in D' and S', but not in P' or E' unless $c_1 = 0$ and $c_2 = 0$.

(3) The solution is a singular distribution and does not correspond to an ordinary function. These are called generalized solutions.

For example $x^2 f' = 0$ has a general solution $f = c_1 + c_2 H(x) + c_3 \delta(x)$. Here $f' = c_2 \delta(x) + c_2 \delta'(x)$ and $x^2 f' = 0$. f lies in D' and S'. It lies in P' and E' only when $c_1 = 0$, $c_2 = 0$.

References

[1] H. Bremermann, **Distributions, Complex Variables and Fourier Transforms**, Addison-Wesley, Reading, Mass., 1965.

[2] I. M. Gelfand and G. E. Shilov, **Generalized Functions 1**, Academic Press, New York, 1964.

[3] A. M. Krall, **Applied Analysis**, D. Reidel, Dordrecht, Netherlands, 1987.

[4] R. D. Morton and A. M. Krall, *Distributional weight functions and orthogonal polynomials*, SIAM J. Math. Anal. **9** (1978), 604–626.

[5] L. Schwartz, **Théorie des Distributions**, Hermann, Paris, 1966.

[6] I. Stakgold, **Boundary Value Problems of Mathematical Physics, vol. II**, Macmillan, New York, 1965.

[7] A. H. Zemanian, **Distribution Theory and Transform Analysis**, McGraw-Hill, New York, 1965.

Chapter XIII

Orthogonal Polynomials

We plan to examine collections of orthogonal polynomials satisfying second, fourth
and higher order differential equations in detail. However, since they have a great
deal in common, we develop that common ground here.

XIII.1   Basic Properties of Orthogonal Polynomials

XIII.1.1. Definition. Let $\{\mu_n\}_{n=0}^{\infty}$ be an infinite collection of real numbers satisfying

$$\Delta_n = \begin{vmatrix} \mu_0 & \cdots & \mu_n \\ \vdots & & \vdots \\ \mu_n & \cdots & \mu_{2n} \end{vmatrix} \neq 0 \ , \quad n = 0, 1, \ldots .$$

The collection $\{\mu_n\}$ is called a collection of moments.

XIII.1.2. Definition. Let $\{\mu_n\}$ be a collection of moments. Then the Tchebycheff
polynomials $\{p_n\}_{n=0}^{\infty}$ are defined by $p_0 = 1$,

$$p_n = \begin{vmatrix} \mu_0 & \cdots & \mu_n \\ \vdots & & \vdots \\ \mu_{n-1} & \cdots & \mu_{2n-1} \\ 1 & \cdots & x^n \end{vmatrix} \Big/ \Delta_{n-1} \ , \quad n = 1, 2, \ldots .$$

These polynomials can be considered as test functions of slow growth, lying
in P, and we shall do so. They have the beneficial property of containing only
finitely many terms, and so frequently linear functionals may be applied to them,
even if they may not be in P'.

XIII.1.3. Definition. Let w be a linear functional containing $\{x^n\}_{n=0}^{\infty}$, and hence
all polynomials, in its domain. Then w is called a weight functional for $\{p_n\}_{n=0}^{\infty}$ if

$$\langle w, x^n \rangle = \mu_n \ , \quad n = 0, 1, \ldots,$$

where $\{\mu_n\}_{n=0}^{\infty}$ is a collection of moments and $\{p_n\}_{n=0}^{\infty}$ is its corresponding set of orthogonal polynomials.

XIII.1.4. Theorem. *If w is a weight functional for $\{p_n\}_{n=0}^{\infty}$, then the collection $\{p_n\}_{n=0}^{\infty}$ is an orthogonal set under the inner product*

$$\{p_n, p_m\} = \langle w, p_n p_m \rangle, \quad n, m = 0, 1, \ldots .$$

Proof. Let $m < n$. For each such m,

$$\langle w, p_n x^m \rangle = \begin{vmatrix} \mu_0 & \cdots & \mu_n \\ \vdots & & \\ \mu_{n-1} & \cdots & \mu_{2n-1} \\ \mu_m & & \mu_{m+n} \end{vmatrix} / \Delta_{n-1} . \qquad \square$$

The bottom row is the same as one of those above. Hence $\langle w, p_n x^m \rangle = 0$. Consequently, p_n and p_m have inner product 0 for $m < n$.

Further, if $m = n$, then

$$\langle w, p_n^2 \rangle = \langle w, p_n x^n \rangle$$

$$= \begin{vmatrix} \mu_0 & \cdots & \mu_n \\ \vdots & & \vdots \\ \mu_{n-1} & \cdots & \mu_{2n-1} \\ \mu_n & \cdots & \mu_{2n} \end{vmatrix} / \Delta_{n-1}$$

$$= \Delta_n / \Delta_{n-1} \neq 0 .$$

If $\Delta_n > 0$ for all n, the inner product can be thought of as defining a pre-Hilbert space. If not, then some of the polynomials have negative norm squares. If the number is finite, the setting is ultimately a Pontrjagin space. If infinite, a Krein space.

XIII.1.5. Theorem. *There exists a weight functional w, defined on the Tchebycheff polynomials $\{p_n\}_{n=0}^{\infty}$.*

Proof. We justify this statement in two ways. First Boas [2] has shown that, given any collection of moments $\{\mu_n\}_{n=0}^{\infty}$, there exist infinitely many weights w such that $\langle w, x^n \rangle = \mu_n$, $n = 0, 1, \ldots$. This result has been refined by Duran [4], who shows that w may be chosen in S. What one wishes classically, of course, is a function of bounded variation with the same properties. Such functions are sometimes extremely hard to find.

Second, we can exhibit a linear functional, defined on polynomials, which generates the moments $\{\mu_n\}_{n=0}^{\infty}$, [8]. It is

$$w = \sum_{j=0}^{\infty} (-1)^j \mu_j \delta^{(j)}(x)/(j!) .$$

So

$$\langle w, x^n \rangle = \sum_{j=0}^{\infty} (-1)^j \mu_j \langle \delta^{(j)}, x^n \rangle / (j!) \, .$$

The only nonzero term in the series is the term with $j = n$. Hence

$$\begin{aligned}
\langle w, x^n \rangle &= (-1)^n \mu_n \langle \delta^{(n)}, x^n \rangle / (n!) \\
&= (-1)^n \mu_n (-1)^n \langle \delta, n! \rangle / (n!) \\
&= \mu_n \, .
\end{aligned}$$

$\square$

XIII.1.6. Theorem. *The Tchebycheff polynomials $\{p_n\}_{n=0}^{\infty}$ satisfy a three term recurrence relation in x*

$$p_{n+1}(x) = (x + B_n)p_n(x) - C_n p_{n-1}(x) \, ,$$

where if $p_n(x) = x^n - S_n x^{n-1} \cdots$, then

$$B_n = -S_{n+1} + S_n, \quad and \quad C_n - \Delta_n \Delta_{n-2} / \Delta_{n-1}^2 \, .$$

Proof. Clearly

$$p_{n+1} = (x + B_n)p_n + \sum_{i=0}^{n-1} \alpha_i p_i \, .$$

Multiply by p_m, $m < n - 1$, and apply w. The result is

$$\langle w, p_m p_{n+1} \rangle = \langle w, (x p_m) \rangle + B_n \langle w, p_m p_n \rangle + \sum_{i=0}^{n-1} \alpha_i \langle w, p_m p_i \rangle \, ,$$

or

$$0 = 0 + 0 + \alpha_m \langle w, p_m^2 \rangle.$$

Thus $\alpha_m = 0$, $m = 0, \ldots, m - 1$.

We let $C_n = -\alpha_{n-1}$ to derive the relation

$$p_{n+1} = (x + B_n)p_n - C_n p_{n-1} \, .$$

Comparing the coefficients of x^n within, we find $B_n = -S_{n+1} + S_n$. To find c_n, multiply by p_{n-1} and apply w.

$$0 = \langle w, (x p_{n-1})p_n \rangle + B_n \langle w, p_{n-1} p_n \rangle - C_n \langle w, p_{n-1}^2 \rangle \, .$$

We can replace $x p_{n-1}$ by p_n. Hence

$$0 = \langle w, p_n^2 \rangle - C_n \langle w, p_{n-1}^2 \rangle \, .$$

Since

$$\begin{aligned}
\langle w, p_n^2 \rangle &= \Delta_n / \Delta_{n-1} \, , \\
C_n &= (\Delta_n / \Delta_{n-1}) / (\Delta_{n-1} / \Delta_{n-2}) = \Delta_n \Delta_{n-2} / \Delta_{n-1}^2 \, .
\end{aligned}$$

$\square$

At this point the development of orthogonal polynomials can go in many directions. We refer the reader to [1], [3], [10], [11] and [12] for further information.

We examine the connections between orthogonal polynomials, ordinary differential equations, the differential equations' symmetry factors (functions which, when used to multiply the differential equations by, result in a symmetric differential expression), and moments.

XIII.2   Orthogonal Polynomials, Differential Equations, Symmetry Factors and Moments

We examine when the collection of Tchebycheff polynomials $\{p_n\}_{n=0}^{\infty}$ satisfies a collection of differential equations of the form

$$\sum_{i=1}^{N} A_i(x)p_n^{(i)} = \lambda_n w(x)p_n(x) , \qquad n = 0, 1, \ldots,$$

where the only coefficient which varies in n is λ_n. By successive substitutions of $p_0, p_1, \ldots$, it is easy to show that the differential equation may be rewritten as

$$\sum_{i=0}^{N} a_i(x)p_n^{(i)} = \lambda_n p_n , \quad \text{where} \quad a_i(x) = \sum_{j=0}^{i} \ell_{ij}x^j ,$$

and

$$\lambda_n = \ell_{00} + \ell_{11}n + \ell_{22}n(n-1) + \cdots + \ell_{NN}n(n-1)\cdots(n-N+1) .$$

The results to follow which characterize the various relations were first investigated by the author's father, H. L. Krall (1907–1994), in [5]. Further improvements were made by K. H. Kwon, L. L. Littlejohn and R. H. Yoo [7]. We present here the version by L. L. Littlejohn and D. Race [8]. Because of the formidability of the derivations, we shall later give simpler versions of the results which apply to the second and fourth order differential equations, so the reader need not despair.

For notational brevity, let

$$\ell y = \sum_{i=0}^{N} a_i(x)y^{(i)} , \quad \text{where} \quad a_i(x) = \sum_{j=0}^{i} \ell_{ij}x^j .$$

XIII.2.1. Definition. The expression $w\ell$ is symmetric over polynomials if $w\ell y = (w\ell)^*y$, its formal Lagrange adjoint, when $w\ell$ is applied to polynomials set in the inner product place generated by the inner product $\{y, z\}$, where

$$\{y, z\} = \langle w, yz \rangle ,$$

discussed in the previous section.

XIII.2.2. Theorem. *The collection $\{p_n\}_{n=0}^{\infty}$ of Tchebycheff polynomials satisfies a collection of differential equations $\ell p_n = \lambda_n p_n$ if and only if $w\ell$ is symmetric over polynomials.*

Proof. Assume $w\,\ell$ is symmetric. Compute ℓp_n. This is a polynomial of degree n. So

$$\ell p_n = \sum_{i=0}^{N} a_i p_n^{(i)} = \sum_{j=0}^{n} c_j p_j \,.$$

Hence

$$\left\langle w, \left(\sum_{j=0}^{n} c_j p_j\right) p_k \right\rangle = c_k \langle w, p_k^2 \rangle$$

for fixed k. Further, we have

$$c_k \langle w, p_k^2 \rangle = \langle (w\ell)p_n, p_k \rangle,$$

which, by symmetry equals

$$\langle p_n, (w\ell)p_k \rangle = \left\langle w, \sum_{j=0}^{N} a_j p_k^j p_n \right\rangle .$$

This is zero if $k < n$. That is $c_k = 0$, $k < n$. We therefore have

$$\ell p_n = c_0 p_n \,.$$

Letting $c_0 = \lambda_n$, we have

$$\sum_{i=0}^{N} a_i(x) p_n^{(i)} = \lambda_n p_n \,.$$

Conversely, suppose $\{p_n\}_{n=0}^{\infty}$ satisfies a collection of differential equations $\ell p_n = \lambda_n p_n$. Then

$$\langle w, (\ell p_n)p_m \rangle = \left\langle w, \sum_{i=0}^{N} a_i p_n^{(i)} p_m \right\rangle = \sum_{i=0}^{N} \langle p_m a_i w, p_n^{(i)} \rangle = \sum_{n=0}^{N} (-1)^i \langle (p_m a_i w)^{(i)}, p_n \rangle$$

$$= \left\langle \sum_{i=0}^{N} (-1)^\ell (w a_i p_m)^{(i)}, p_n \right\rangle = \langle (w\ell)^* p_m, p_n \rangle .$$

Since $\ell p_n = \lambda_n p_n$ and $\ell p_m = \lambda_m p_m$, we have

$$(\lambda_n - \lambda_m)\langle p_n, p_m \rangle = \langle w, \lambda_n p_n p_m \rangle - \langle w, p_n \lambda_m p_m \rangle$$
$$= \langle w, (\ell p_n)p_m \rangle - \langle w, p_n (\ell p_m) \rangle$$
$$= \langle (w\ell)p_n, p_m \rangle - \langle (w\ell)^* p_n, p_m \rangle$$
$$= \langle [(w\ell) - (w\ell)^*]p_n, p_m \rangle .$$

Since p_n and p_m are mutually orthogonal, this is zero, and $w\ell = (w\ell)^*$ over polynomials. $\qquad\square$

XIII.2.3. Corollary. *If the collection of Tchebycheff polynomials $\{p_n\}_{n=0}^{\infty}$ satisfies a collection of differential equations $\ell p_n = \lambda_n p_n$, then the order of ℓ is even, $N = 2M$.*

Proof. Comparing the coefficients of the highest derivatives in $w\ell$ and in $(w\ell)^*$, we find $w\,a_N = (-1)^N w\,a_N$. Thus $N = 2M$.

In the actual study of orthogonal polynomials satisfying differential equations, one usually begins with only the differential equations

$$\ell p_n = \lambda_n p_n\,, \qquad n = 0, 1, \ldots,$$

and does not necessarily have an orthogonalizing weight functional w readily available in any recognizable form. It is essential to find conditions which will enable the computation of w. □

XIII.2.4. Definition. Let $\{p_n\}_{n=0}^{\infty}$ be a collection of Tchebycheff polynomials satisfying the differential equations

$$\ell p_n = \sum_{i=1}^{2M} a_i \lfloor_n^{(i)} = \lambda_n p_n\,, \quad \text{where} \quad a_i = \sum_{j=0}^{i} \ell y\, x^j\,,$$

and

$$\lambda_n = \ell_{00} + \ell_{11} n + \ell_{22} n(n-1) + \cdots + \ell_{2M,2M}\, n(n-1)\cdots(n - 2M + 1)\,.$$

If w is a weight functional for $\{p_n\}_{n=0}^{\infty}$, then w is called a symmetry factor for ℓ.

Consider

$$w\,\ell\, y = \sum_{i=0}^{2M} w\, a_i(x)\, y^{(i)}\,.$$

The formal adjoint is

$$(w\,\ell)^* y = \sum_{i=0}^{2M} (-1)^i (w\, a_i\, y)^{(i)} = \sum_{i=0}^{2M} (-1)^i \left[\sum_{j=0}^{i} \binom{i}{j} (w\, a_i)^{(i-j)} y^{(j)} \right]$$

$$= \sum_{j=0}^{2M} (-1)^i \left[\sum_{i=0}^{i} \binom{i}{j} (w\, a_j)^{(j-i)} y^{(i)} \right]$$

$$= \sum_{i=0}^{2M} \left[\sum_{j=i}^{2M} (-1)^j \binom{j}{i} (w\, a_j)^{(j-i)} \right] y^{(i)}\,.$$

$w\,\ell = (w\,\ell)^*$ in a formal sense if and only if

$$S_{i+1} = \sum_{j=i}^{2M} (-1)^j \binom{j}{i} (w\, a_j)^{(j-i)} - w\, a_i = 0\,,$$

$i = 0, \ldots, 2M.$

This can be improved upon. Only half of the terms, those with even index, must be required to vanish. The others, those with odd index, then do so automatically. The proof is surprisingly difficult and requires the use of the Bernoulli numbers, which are generated by the expansion

$$x/(e^x - 1) = 1 - x/2 + \sum_{i=1}^{\infty} \frac{B_{2i}}{(2i)!}\, x^{2i}\,.$$

Here $B_0 = 1$, $B_1 = -\frac{1}{2}$ and $B_{2i+1} = 0$, $i = 1, 2, \ldots$.

XIII.2.5. Theorem. *Let $\{x_{2i-1}\}_{i=1}^{\infty}$ and $\{y_{2i-1}\}_{i=1}^{\infty}$ be solutions to*

$$x_{2r-1} + \sum_{i=1}^{r} \binom{2r-1}{2i-1} x_{2i-1} = 1\,,$$

$$\sum_{i=1}^{r} \binom{2r}{2i-1} y_{2i-1} = 1\,, \qquad r = 1, 2, \ldots;$$

then $x_{2r-1} = y_{2r-1} = r^{-1}(2^{2r} - 1)B_{2r}$, $r = 1, \ldots, \infty$, where $\{B_{2r}\}_{r=1}^{\infty}$ are the Bernoulli numbers.

Proof. (See [6].) Let

$$\tanh\left(\frac{1}{2}x\right) = \sum_{i=1}^{\infty} c_{2i-1} x^{2i-1} \big/ (2i-1)!\,.$$

Then

$$\sum_{i=0}^{\infty} x^{2r}/(2r)! = \cosh x - 1 = \tanh\left(\frac{1}{2}x\right) \sinh x$$

$$= \sum_{i=1}^{\infty} \frac{c_{2i-1} x^{2i-1}}{(2i-1)!} \sum_{s=0}^{\infty} \frac{x^{2s+1}}{(2s+1)!}$$

$$= \sum_{i=1}^{\infty} \sum_{s=0}^{\infty} \frac{c_{2i-1} x^{2(i+s)}}{(2i-1)!(2s+1)!}\,.$$

Let $s = r - i$ and eliminate s. The sum equals

$$\sum_{i=1}^{\infty} \sum_{r=1}^{\infty} \frac{c_{2i-1} x^{2r}}{(2i-1)!(2r-2i+1)!}$$

$$= \sum_{r=1}^{\infty} \left[\sum_{i=1}^{r} \frac{c_{2i-1}}{(2i-1)!(2r-2i+1)!} \right] x^{2r}\,,$$

since here $2r - 2c + 1 \geq 0$ implies $r + \frac{1}{2} \geq i$.

Equate coefficients, recalling the expansion of $\cosh x - 1$,

$$\frac{1}{(2r)!} = \sum_{i=1}^{r} \frac{c_{2i-1}}{(2i-1)!(2r-2i+1)!} \,, \quad r = 1, \ldots,$$

or

$$1 = \sum_{i=1}^{r} \binom{2r}{2i-1} c_{2i-1}, \quad r = 1, \ldots .$$

Thus $y_{2i-1} = c_{2i-1}$, $i = 1, \ldots$.

Likewise,

$$\sum_{r=1}^{\infty} \frac{x^{2r-1}}{(2r-1)!} = \sinh x = \tanh\left(\frac{1}{2}x\right) + \tanh\left(\frac{1}{2}x\right)\cosh x$$

$$= \sum_{i-1}^{\infty} \frac{c_{2i-1}x^{2i-1}}{(2i-1)!} + \sum_{i=1}^{\infty} \frac{c_{2i-1}x^{2i-1}}{(2i-1)!} \sum_{s=0}^{\infty} \frac{x^{2s}}{(2s)!}$$

$$= \sum_{r=1}^{\infty} \frac{c_{2r-1}x^{2r-1}}{(2r-1)!} + \sum_{i=1}^{\infty} \sum_{s=0}^{\infty} \frac{c_{2i-1}x^{2i+2s-1}}{(2i-1)!(2s)!}$$

$$= \sum_{r=1}^{\infty} \frac{c_{2r-1}x^{2r-1}}{(2r-1)!} + \sum_{i=1}^{\infty} \sum_{r=1}^{\infty} \frac{c_{2i-1}x^{2r-1}}{(2i-1)!(2i)!} \,,$$

where, in the second sum we have let $s = r - i$ and eliminated s. The sums equal

$$\sum_{r=1}^{\infty} \frac{c_{2r-1}x^{2r-1}}{(2r-1)!} + \sum_{r=1}^{\infty} \left[\sum_{i=1}^{r} \frac{c_{2i-1}}{(2i-1)!(2r-2i)!}\right] x^{2r-1} \,,$$

since $2r - 2i \geq 0$ and $r \geq i$.

Again equate coefficients

$$\frac{1}{(2r-1)!} = \frac{c_{2r-1}}{(2r-1)!} + \sum_{i=1}^{r} \frac{c_{2i-1}}{(2i-1)!(2r-2i)!} \,,$$

or

$$1 = c_{2r-1} + \sum_{i=1}^{r} \binom{2r-1}{2i-1} c_{2i-1} \,, \quad r = 1, \ldots .$$

Thus $x_{2i-1} = c_{2i-1}$, $i = 1, \ldots$.

Now since

$$\tanh\left(\frac{1}{2}x\right) = \sum_{i=1}^{\infty} \frac{(2^{2i}-1)}{(2i-1)!i} B_{2i}\, x^{2i-1} \,,$$

$$x_{2i-1} = y_{2i-1} = (2^{2i}-1)B_{2i}/i, \quad i = 1, \ldots .$$

If we make the substitution $x_{2r-1} = y_{2r-1} = (2r-1)!c_{r-1} = (2r-1)!d_{r-1}$, then

$$\frac{1}{(2i+1)!} = \sum_{r=0}^{i} \frac{c_r}{(2i-2r)!} + c_i\,, \quad i = 0, 1, \ldots, \infty\,,$$

$$\frac{1}{(2i+2)!} = \sum_{r=0}^{i} \frac{d_r}{(2i-2r+1)!}\,, \quad i = 0, 1, \ldots, \infty\,,$$

and

$$c_r = d_r = 2 \left[\frac{2^{2r+2} - 1}{(2r+2)!}\right] B_{2r+2}\,, \quad r = 0, 1, \ldots\,. \qquad \square$$

XIII.2.6. Theorem. *For $k = 0, 1, \ldots, 2M$,*

$$S_{2M-2k-1} = \sum_{j=0}^{k} \frac{2(2^{2j+2} - 1)}{j+2)!} B_{2j+2} P(2M - 2k - 1 + 2j, 2j + 1) S_{2m-2k+2j}^{(2j+1)}\,,$$

where $P(n, k) = n(n-1) \cdots (n - k + 1)$.

Proof. (see [8].)

$$S_{2M-2k-1} - \sum_{J=0}^{k} c_j P(2M - 2k - 1 + 2j, 2j + 1)(wa_{2M-2k-2+j})^{(j)}$$

$$= \sum_{j=1}^{2k+2} (-1)^j \binom{2M - 2k - 2 + j}{2M - 2k - 2} (wa_{2M-2k-2+j})^{(j)}$$

$$- \sum_{j=0}^{k} \left[\sum_{u=2M-2k+2j-1}^{2M} (-1)^u c_j P(2M - 2k - 1 + 2j, 2j + 1) \binom{u}{2M - 2k + 2j - 1} \right.$$

$$\left. \times (wa_u)^{(u-2M+2k+2)} \right]$$

$$+ \sum_{j=0}^{k} c_j P(2M - 2k - 1 + 2j, 2j + 1)(wa_{2M-2k+2j-1})^{(2j+1)}$$

$$= -\sum_{i=0}^{k} \left[\binom{2M - 2k + 2i - 1}{2M - 2k + 2r - 1} - \sum_{r=0}^{i} P(2M - 2k + 2r - 1, 2r + 1) \right.$$

$$\times \binom{2M - 2k + 2i - 1}{2M - 2k + 2r - 1} c_r - c_i P(2M - 2k + 2i - 1, 2i + 1) \left. \right] (wa_{2M-2k+2i-1})^{(2i+1)}$$

$$+ \sum_{i=0}^{k} \left[\left(\binom{2M - 2k + 2i}{2M - 2k + 2r} \right) - \sum_{r=0}^{i} P(2M - 2k + 2r - 1, 2r + 1) \right.$$

$$\times \binom{2M - 2k + 2i}{2M - 2k + 2r - 1} c_r \left. \right] (wa_{2M-2k+2i})^{(2i+2)}$$

$$= -\sum_{i=0}^{k} \frac{(2M-2k+2i-1)!}{(2M-2k-2)!} \left[\frac{1}{(2i+1)!} - \sum_{r=0}^{k} \frac{c_r}{(2i-2r)!} - c_i \right] (w a_{2M-2k+2i})^{(2i+2)}$$

$$+ \sum_{i=0}^{k} \frac{(2M-2k+2i)!}{(2M-2k-2)!} \left[\frac{1}{(2i+2)!} - \sum_{k=0}^{i} \frac{c_r}{(2i-2r+1)!} - c_i \right] (w a_{2M-2k+2i})^{(2i+2)}$$

$$= 0,$$

when $c_r = \frac{2(2^{2r+2}-1)}{(2r+2)!} B_{2r+2}$. $\qquad\qquad\qquad\qquad\qquad\qquad\square$

XIII.2.7. Theorem. $S_{k+1} = 0$, $k = 0, 1, \ldots, 2M - 1$ *if and only if* $S_{2k+2} = 0$, $k = 0, 1, \ldots, M - 1$.

Proof. (See [8].) Clearly if $S_{k+1} = 0$, $k = 0, \ldots, 2M - 1$, then those with even index are 0.

Conversely, if those of even index are 0, then all are by Theorem XIII.2.6. $\square$

XIII.2.8. Theorem. *The expression*

$$\ell y = \sum_{i=0}^{2M} a_i(x) y^{(i)}$$

has w as a symmetry factor if and only if

$$S_{2k+2} = \sum_{j=0}^{2M-2k-1} (-1)^{j+1} \binom{2k+j+1}{2k+1} (w\, a_{2k+j+1})^{(j)} - w\, a_{2k+1} = 0,$$

$k = 0, 1, \ldots, M - 1$.

We remark that it is possible to replace S_{2k+1}, $k = 0, 1, \ldots, M - 1$ by

$$W_{k+1} = \sum_{i=2k+1}^{2M} (-1)^i \binom{i-k-l}{k} (w\, a_i)^{i-2k-1}, \quad k = 0, 1, \ldots, n - 1.$$

The proof may be found in [8, Theorem 5.3]. It is quite intricate and very tedious, since it really gives no additional information. The terms $w_{k+1} = 0$, $k = 0, \ldots, n - 1$, applied to x^n to yield moment relations, were first exhibited by H. L. Krall [5]. The proof given there is unbelievable. It involves several pages of very intricate sums, some of them quadrouple sums. In contrast, the procedure presented here, due to Littlejohn and Race [8], is quite mild.

Finally we show the equivalence of the symmetry factor equations and various moment relations.

XIII.2.9. Theorem. *The collection $\{p_n\}_{n=0}^{\infty}$ of Tchebycheff polynomials satisfies a collection of differential equations*

$$\sum_{i=0}^{2M} \left[\sum_{j=0} \ell_{ij} x^j \right] p_n = \lambda_n p_n,$$

where

$$\lambda_n = \ell_{00} + \ell_{11} n + \cdots + \ell_{2M,2M} n(n-1)\cdots(n-2M+1),$$

if and only if the moments $\{\mu_n\}_{n=0}^{\infty}$, associated with $\{p_n\}_{n=0}^{\infty}$, satisfy

$$\sum_{j=0}^{2M-2k-1} \binom{2k+j+1}{2k+1} P(n,j) \sum_{i=0}^{2k+j+1} \ell_{2k+j+1,i}\mu_{n-j+i} + \sum_{i=0}^{2k+1} \ell_{2k+1,i}\mu_{n+i} = 0,$$

$k = 0, 1, \ldots, M-1$, $n = 0, 1, \ldots$.

Proof. If $\{p_n\}_{n=0}^{\infty}$ satisfies a collection of differential equations $\ell p_n = \lambda_n p_n$, then there is a w so that $w\ell$ is formally symmetric and the coefficients of ℓ, $a_i = \sum_{j=0}^{i} \ell_{ij} x_j$ satisfy the symmetry equations of the previous theorem. Apply S_{2k+2} to x^n, $k = 0, 1, \cdot, M-1$.

$$\langle S_{2k+2}, x^n \rangle$$

$$= \sum_{j=0}^{2M-2k-1} (-1)^{j+1} \binom{2k+j+1}{2k+1} \langle w\, a_{2k+j+1})^{(j)}, x^n \rangle - \langle w\, a_{2k+1}, x^n \rangle$$

$$= \sum_{j=0}^{2M-2k-1} (-1)^{j+1} \binom{2k+j+1}{2k+1} (-1)^j P(n,j) \langle w\, a_{2k+j+1}, x^{n-j} \rangle$$

$$\quad - \langle w\, a_{2k+1}, x^n \rangle$$

$$= - \sum_{j=0}^{2M-2k-1} \binom{2k+j+1}{2k+1} P(n,j) \left\langle w \sum_{i=0}^{2k+j+1} \ell_{2k+j+1,i} x^i, x^{n-j} \right\rangle$$

$$\quad - \left\langle w \sum_{i=0}^{2k+1} \ell_{2k+1,i} x^i, x^n \right\rangle$$

$$= - \left[\sum_{j=0}^{2M-2k-1} \binom{2k+j+1}{2k+1} P_n(n,j) \sum_{i=0}^{2k+j+1} \ell_{2k+j+1,i}\mu_{n-j+i} \right.$$

$$\left. + \sum_{i=0}^{2k+1} \ell_{2k+1,i}\mu_{n+i} \right]$$

$$= 0.$$

We have used here $\langle w\, \ell_{pq} x^i, x^r \rangle = \ell_{pq}\mu_{i+r}$.

Conversely if the moment equations are satisfied, then the symmetry equations are satisfied over polynomials and ℓ is a formally symmetric differential operator over polynomials.

While the moment equations seem very complicated, they are, in fact, quite useful as the next chapters will illustrate.

We note that we have three equivalent statements:

1. The collection of Tchebycheff polynomials $p_0 = 1$,

$$p_n = \begin{vmatrix} \mu_0 & \cdots & \mu_n \\ \vdots & & \\ \mu_{n-1} & \cdots & \mu_{2n-1} \\ 1 & \cdots & x^n \end{vmatrix} \Big/ \Delta_{n-1} ,$$

where

$$\Delta_n = \begin{vmatrix} \mu_0 & \cdots \mu_n \\ \vdots & \\ \mu_n & \cdots & \mu_{2n} \end{vmatrix} \neq 0 ,$$

$n = 1, \ldots$, satisfies a formally symmetric (over polynomials) differential equation

$$\sum_{i=0}^{2M} a_i(x) p_n^{(i)} = \lambda_n p_n ,$$

where $a_0 = \ell_{00}$,

$$a_i(x) = \sum_{j=0}^{i} \ell y \, x^j ,$$

and

$$\lambda_n = \ell_{00} + \ell_{11} n + \ell_{22} n(n-1) + \cdots + \ell_{2M,2M} n(n-1) \cdots (n - 2M + 1) .$$

2. The coefficients $\{a_i\}_{i=0}^{2M}$ satisfy the symmetry equations

$$S_{2k+2} = \sum_{j=0}^{2M-2k-1} (-1)^{j+1} \binom{2k+j+1}{2k+1} (w\, a_{2k+j+1})^{(j)} - w\, a_{2k+1} = 0 ,$$

$k = 0, 1, \ldots, M - 1$.

3. The moments $\{\mu_i\}_{i=0}^{\infty}$ satisfy the moment equations

$$\sum_{j=0}^{2M-2k-1} \binom{2k+j+1}{2k+1} P(n,j) \sum_{i=0}^{2k+j+1} \ell_{2k+j+1,i}\mu_{n-j+i} + \sum_{i=0}^{2k+1} \ell_{2k+1,i}\mu_{n+i} = 0 .$$

Finally, in closing, we note that the weight function w satisfies

$$S_{2M} = 2[M(w\, a_{2M})' - (w\, a_{2M-1})] = 0$$

over polynomials. Usually $S_{2M} = 0$ and

$$w = C \, \exp\left[\frac{1}{M} \int^x \frac{a_{2M-1}}{a_{2M}} \, dx \right] \Big/ a_{2M}$$

or w equals the expression plus some singular distribution integrals. But in the case of the Bessel polynomials "0" can be *any* polynomial annihilator, and not necessarily 0. (See [8].) $\qquad\square$

References

[1] L. C. Andrews, **Special Functions for Engineers and Applied Mathematicians**, Macmillan, New York, 1985.

[2] R. P. Boas, *The Stieltjes moment problem for functions of bounded variation*, Bull. Amer. Math. Soc. **45** (1939), 399–404.

[3] T. S. Chihara, **An Introduction to Orthogonal Polynomials**, Gordon and Breach, New York, 1978.

[4] A. J. Duran, *The Stieltjes moment problem for rapidly decreasing functions*, Proc. Amer. Math. Soc. **107** (1989), 731–741.

[5] H. L. Krall, *Certain differential equations for Tchebycheff polynomials*, Duke Math. J. **4** (1938), 705–718.

[6] ______, *Self-adjoint differential expressions*, Amer. Math. Monthly **67** (1960), 867–878.

[7] K. H. Kwon, L. L. Littlejohn and B. H. Yoo, *Characterizations of orthogonal polynomials satisfying certain differential equations*, SIAM J. Math. Anal. **1** (1993), 10–24.

[8] L. L. Littlejohn and D. Race, *Symmetric and symmetrisable differential expressions*, Proc. London Math Soc. **60** (1990), 344–364.

[9] R. D. Morton and A. M. Krall, *Distributional weight functions and orthogonal polynomials*, SIAM J. Math. Anal. **9** (1978), 604–626.

[10] E. D. Rainville, **Special Functions**, Macmillan, New York, 1960.

[11] J. A. Shohat and J. D. Tamarkin, **The Problem of Moments**, Amer. Math. Soc., Providence, RI, 1943.

[12] G. Szego, **Orthogonal Polynomials**, Amer. Math. Soc., Providence, RI, 1939.

Chapter XIV

Orthogonal Polynomials Satisfying Second Order Differential Equations

We begin by repeating the results of the previous chapter while restricting ourselves to differential equations of the second order. We then discuss the four classical sets of orthgonal polynomials satisfying a collection of differential equations of second order, both formally, then in an L^2 setting as eigenfunctions for a differential operator. Subcases are also exhibited. Finally we examine the one enigmatic case, the Bessel polynomials.

XIV.1 The General Theory

We again consider an infinite collection of moments $\{\mu_n\}_{n=0}^\infty$, satisfying

$$\Delta_n = \begin{bmatrix} \mu_0 & \cdots & \mu_n \\ \vdots & & \vdots \\ \mu_n & \cdots & \mu_{2n} \end{bmatrix} \neq 0, \qquad n = 0, 1, \ldots,$$

and define the Tchebycheff polynomials by setting $p_0 = 1$,

$$p_n = \begin{bmatrix} \mu_0 & \cdots & \mu_n \\ \vdots & & \vdots \\ \mu_n & \cdots & \mu_{2n} \\ 1 & \cdots & x^n \end{bmatrix} \Big/ \Delta_{n-1}, \qquad n = 1, 2, \ldots.$$

They are mutually orthogonal with respect to the distributional weight function

$$w = \sum_{j=0}^\infty (-1)^j \mu_j \delta^{(j)}(x)/j!,$$

as well as others, including (at least) one given by a function of bounded variation.

Likewise a three term recurrence relation

$$p_{n+1}(x) = (x + B_n)p_n(x) - C_n p_{n-1}(x) = 0$$

also exists. The proofs are identical to those of the previous chapter.

We assume that the polynomials $\{p_n\}_{n=0}^{\infty}$ satisfy a collection of differential equations of the form

$$lp_n = (l_{22}x^2 + l_{21}x + l_{20})p_n'' + (l_{11}x + l_{10})p_n' = \lambda_n p_n,$$

where

$$\lambda_n = l_{11}n + l_{22}n(n - 1), \qquad n = 0, 1, \dots.$$

It is easy to see that the differential expression wl is symmetric over polynomials. Again the proof is unchanged.

To find the weight function we require that wl be symmetric. Thus $wl = (wl)^*$, or

$$(wa_2)y'' + (wa_1)y' = (wa_2 y)'' - (wa_1 y)',$$

where

$$a_2 = l_{22}x^2 + l_{21}x + l_{20},$$
$$a_1 = l_{11}x + l_{10},$$

and w acts as a distribution over polynomials. The constraint is equivalent to

$$S_0 = (wa_2)' - (wa_1) = 0$$

over polynomials. Solved in P', this essentially yields

$$w = C\left\{\exp\left[\int \frac{a_1}{a_2}\,dx\right]\right\}\Big/a_2.$$

In every case but the Bessel polynomials (to come) this actually gives the appropriate weight.

The moments, $\{\mu_n\}_{n=0}^{\infty}$, satisfy

$$n[l_{22}\mu_{n+1} + l_{21}\mu_n + l_{20}\mu_{n-1}] + [l_{11}\mu_{n+1} + l_{10}\mu_n] = 0,$$

$n = 0, 1, \dots.$ These are easily solved for μ_n, $n = 0, 1, \dots.$

It is apparent that the second derivative coefficient $l_{22}x^2 + l_{21}x + l_{20}$ has two different zeros, one zero, no zeros or a double zero.

In the first case, a linear transformation of x gives

$$(1 - x^2)p_n'' + (B - A - [2 + A + B]x)p_n' + n(n + A + B + 1)p_n = 0,$$

the Jacobi equation.

In the second case, a linear transformation of x gives

$$xp_n'' + (\alpha + 1 - x)p_n' + np_n = 0,$$

the generalized Laguerre equation.

In the third case, a linear transformation of x gives

$$p_n'' - 2xp_n' + 2np_n = 0,$$

the Hermite equation.

Finally, in the fourth case, a linear transformation of x gives

$$x^2 p_n'' + (ax + b)p_n' - n(n + a + 1)p_n = 0,$$

the Bessel polynomial equation.

These cases have been discussed extensively [2], [17], [22], [23], all with slightly differing points of view. We examine each case in more detail.

XIV.2 The Jacobi Polynomials

The Jacobi polynomials $\{p_n^{(A,B)}(x)\}_{n=0}^\infty$ satisfy the differential equation

$$ly = -[(1 - x)^{A+1}(1 + x)^{B+1}y']'/(1 - x)^A(1 + x)^B = n(n + A + B + 1)y.$$

They are generated by moments $\{\mu_n\}_{n=0}^\infty$ satisfying

$$(A + B + n + 2)\mu_{n+1} + (A - B)\mu_n - n\mu_{n-1} = 0.$$

This can be solved by first replacing x by $1 + x$ or $1 - x$ and finding the moment relations about -1 or 1, then translating back.

$$\mu_n = \sum_{j=0}^n \binom{n}{j} (-1)^j 2^j (A + 1)_j / (A + B + 2)_j,$$

or

$$\mu_n = \sum_{j=0}^n \binom{n}{j} (-1)^{n-j} 2^j (B + 1)_j / (A + B + 2)_j,$$

where

$$(a)_j = a(a + 1) \cdots (a + j - 1).$$

The weight equation is

$$(1 - x^2)w' - [(B - A) - (A + B)x]w = 0.$$

If $A > -1$ and $B > -1$, this is solved in P' the distributions of rapid decay. We find

$$w = (1 - x)^A(1 + x)^B[\alpha H(1 - x) + \beta H(1 + x) + \gamma],$$

where $H(x)$ is the Heaviside function,

$$\begin{aligned} H(x) &= 1, \quad x \geq 0, \\ &= 0, \quad x < 0, \end{aligned}$$

and α, β, γ are constants. In order for w to vanish at $\pm\infty$, we must have $\alpha = 1$, $\beta = 1$, $\gamma = -1$. w is then the classic weight function

$$w = (1 - x)^A (1 + x)^B, \quad -1 \leq x \leq 1,$$
$$= 0 \qquad , \qquad x < -1 \text{ or } 1 < x.$$

When either A or B is less than -1, a Cauchy regularization of the weight given above is required [26]. If $-N - 1 < \alpha < -N$ and $-M - 1 < \beta < -M$, where M and N are positive integers, the action of w in P' on test functions ϕ in P is given by

$$\langle w, \phi \rangle = \left(\frac{\Gamma(A+B+2)}{\Gamma(A+1)\Gamma(B+1)2^{A+B+1}} \right)$$

$$\times \left[\int_0^1 \left((1-x)^A \left\{ (1+x)^B \phi(x) - \sum_{j=0}^{N-1} \frac{[(1+x)^A \phi(x)]^{(j)}}{j!} \Bigg|_{x=1} \right. (-1)^j (1-x)^j \right\} dx \right.$$

$$+ \int_{-1}^0 (1+x)^B \left\{ (1-x)^A \phi(x) - \sum_{j=0}^{N-1} \frac{[(1-x)^B \phi(x)]^{(j)}}{j!} \Bigg|_{x=-1} (1+x)^j \right\} dx$$

$$\left. + \sum_{j=0}^{N-1} \frac{(-1)^j [(1+x)^B \phi(x)]^{(j)}|_{x=1}}{j!(A+j+1)} + \sum_{j=0}^{M-1} \frac{[(1-x)^A \phi(x)]^{(j)}|_{x=-1}}{j!(B+j+1)} \right],$$

(see [26]).

This inner product and the Pontrjagin space it generates have not been carefully examined. (See [3].)

There is another approach, however. The substitutions

$$y = (1-x)^{-A} z \qquad , \quad \text{when } A < 1,\ B > -1,$$
$$y = (1+x)^{-B} z \qquad , \quad \text{when } A > 1,\ B < 1,$$
$$y = (1-x)^{-A}(1+x)^{-B} z \quad , \quad \text{when } A < 1,\ B < 1,$$

transform the situation into a new problem in which the new $A > -1$ and/or new $B > -1$, [7]. Thus we will only consider the case $A > -1$, $B > -1$.

The Jacobi polynomials $\{P_n^{(A,B)}(x)\}_{n=0}^{\infty}$ have series representations

$$P_n^{(A,B)}(x) = \sum_{k=0}^{n} \binom{n+A}{n-k}\binom{n+B}{n-k}\left(\frac{x-1}{2}\right)^k \left(\frac{x+1}{2}\right)^k$$

$$= \sum_{k=0}^{n} \binom{n+A}{n-k}\binom{n+k+A+B}{k}\left(\frac{x-1}{2}\right)^k$$

$$= \sum_{k=0}^{n} (-1)^{n-k} \binom{n+B}{n-k}\binom{n+k+A+B}{k}\left(\frac{x+1}{2}\right)^k$$

(see [1]), which are almost never used. They satisfy a three term recurrence relation

$$2n(A + B + n)(A + B + 2n - 2)P_n^{(A,B)}(x)$$

$$= (A + B + 2n - 1)[A^2 - B^2 + x(A + B + 2n)(A + B + 2n - 2)P_{n-1}^{(A,B)}(x)$$

$$-2(A + n - 1)(B + n - 1)(A + B + 2n)P_{n-2}^{(A,B)}(x),$$

which may be manipulated to yield the norm square

$$\int_{-1}^{1} P_n^{(A,B)}(x)^2 (1 - x)^A (1 + x)^B \, dx = \frac{2^{A+B+1}\Gamma(A + n + 1)\Gamma(B + n + 1)}{(A + B + 2n + 1)n!\Gamma(A + B + n + 1)} \, .$$

The differential equation is easily manipulated to show

$$\int_{-1}^{1} P_n^{(A,B)}(x)P_m^{(A,B)}(x)(1 - x)^A (1 + x)^B = 0, \qquad n \neq m \, .$$

As references we cite [1], [7], [11], [24], [27], [28].

In order to discuss a differential operator in $L^2(-1, 1; (1 - x)^A(1 + x)^B)$, we need to consider boundary conditions. Solutions to $ly = 0$ are $y_1 = 1$ and $y_2 = Q_0^{(A,B)}(x)$, the Jacobi function of the second kind [29]. Boundary values at ± 1 are, therefore Wronskians with these two functions. For y in the domain of the maximal operator they are

$$B_{1,1}(y) = \lim_{x \to 1} (1 - x)^{A+1}(1 + x)^{B+1}[-y'(x)] \, ,$$

$$B_{1,2}(y) = \lim_{x \to 1} (1 - x)^{A+1}(1 + x)^{B+1}[Q_0^{(A,B)}(x)'y(x) - Q_0^{(A,B)}(x)y'(x)] \, ,$$

$$B_{-1,1}(y) = \lim_{x \to -1} (1 - x)^{A+1}(1 + x)^{B+1}[-y'(x)] \, ,$$

$$B_{-1,2}(y) = \lim_{x \to -1} (1 - x)^{A+1}(1 + x)^{B+1}[Q_0^{(A,B)}(x)'y(x) - Q_0^{(A,B)}(x)y'(x)] \, .$$

The Jacobi polynomials satisfy $B_{1,1}(y) = 0$ and $B_{-1,1}(y) = 0$ for all $A > -1$, $B > -1$.

An examination of the operator l shows that $x = 1$ is in the limit circle case when $-1 < A < 1$, hence both $B_{1,1}$ and $B_{1,2}$ exist. When $1 \leq A$, l is in the limit point case. $B_{1,1}$ is an annihilator boundary value (always 0), while $B_{1,2}$ may not exist. The situation is similar at $x = -1$.

With these preliminaries, we are now in a position to define the self-adjoint Jacobi operator in $L^2(-1, 1; (1 - x)^A(1 + x)^B)$.

XIV.2.1. Definition. We denote by D_L these elements y in $L^2(-1, 1; (1 - x)^A(1 + x)^B)$ satisfying:

(1) y is differentiable a.e. on $(-1, 1)$.

(2) $(1 - x)^{A+1}(1 + x)^{B+1}y'(x)$ is differentiable a.e. on $(-1, 1)$ and ly is in $L^2(-1, 1; (1 - x)^A(1 + x)^B)$.

(3) $B_{1,1}(y) = 0$ and $B_{-1,1}(y) = 0$.

We define the Jacobi operator L by setting $Ly = ly$ for all y in D_L.

Clearly the Jacobi polynomials $\{P_n^{A,B}(x)\}_{n=0}^{\infty}$ are eigenfunctions for L with eigenvalues $\{\lambda_n = n(n + A + B + 1)\}_{n=0}^{\infty}$.

XIV.2.2. Theorem.   *The Jacobi polynomials $\{P_n^{A,B}(x)\}_{n=0}^{\infty}$ form a complete orthogonal set in $L^2(-1, 1, (1 - x)^A(1 + x)^B)$.*

There are several proofs of this. All seem to get bogged down in obscure technical details. We present here a variation on an idea found in [13]. Assume that there is an element f in $L^2(-1, 1; (1 - x)^A(1 + x)^B)$ which is orthogonal to $\{P_n^{A,B}\}_{n=1}^{\infty}$. Then f is orthogonal to $\{x^n\}_{n=0}^{\infty}$ as well, and, for arbitrary λ, $e^{i\lambda x}$. (Convergence here is uniform.) Therefore

$$\int_{-1}^{1} f(x)e^{i\lambda x}(1 - x)^A(1 + x)^B dx = 0$$

for all λ. This is the Fourier transform of a function in $L^1(-\infty, \infty)$, and is 0. So the function $f(x)(1-x)^A(1+x)^B$ is 0 a.e. on $(-1, 1)$. We see, therefore, that $f = 0$ a.e. on $(-1, 1)$ and also as an element in $L^2(-1, 1; (1 - x)^A(1 + x)^B)$.

Knowing that $\{P_n^{(A,B)}\}_{n=0}^{\infty}$ forms a complete orthogonal set enables us to bypass the rather difficult theory concerning the spectral resolution of the operator L. If f is now an arbitrary element of $L^2(-1, 1; (1 - x)^A(1 + x)^B)$, then $f(x) = \sum_{n=0}^{\infty} c_n P_n^{(A,B)}(x)$, where

$$c_n = \frac{\int_{-1}^{1} f(x)\, P_n^{(A,B)}(x)(1 - x)^A(1 + x)^B dx}{\left(\frac{2^{A+B+1}\Gamma(A+n+1)\Gamma(B+n+1)}{(A+B+2n+1)n!\Gamma(A+B+n+1)}\right)}.$$

If f is in D_L, then

$$Lf = \sum_{n=0}^{\infty} \lambda_n c_n P_n^{(A,B)}(x),$$

where

$$\lambda_n = n(n + A + B + 1).$$

The two expansions above can be rewritten as Stieltjes integrals. If we let

$$\phi_n(x) = P_n^{(A,B)}(x) \Big/ \left(\frac{2^{A+B+1}\Gamma(A + n + 1)\Gamma(B + n + 1)}{(A + B + 2n + 1)\, n!\, \Gamma(A + B + n + 1)}\right)^{\frac{1}{2}},$$

then

$$f(x) = \sum_{n=0}^{\infty} \langle f, \phi_n \rangle \phi_n(x), \quad \text{and} \quad Lf(x) = \sum_{n=0}^{\infty} \lambda_n \langle f, \phi_n \rangle \phi_n(x),$$

where $\langle \cdot, \cdot \rangle$ stands for the inner product in $L^2(-1, 1; (1-x)^A (1+x)^B)$. Denote by $P_n f$, the term $\langle f, \phi_n \rangle \phi_n(x)$. Define $P(\lambda)$ by setting $P(0-) = 0$ and having $P(\lambda)$ jump by P_n at $\lambda_n = n(n + A + B + 1)$.

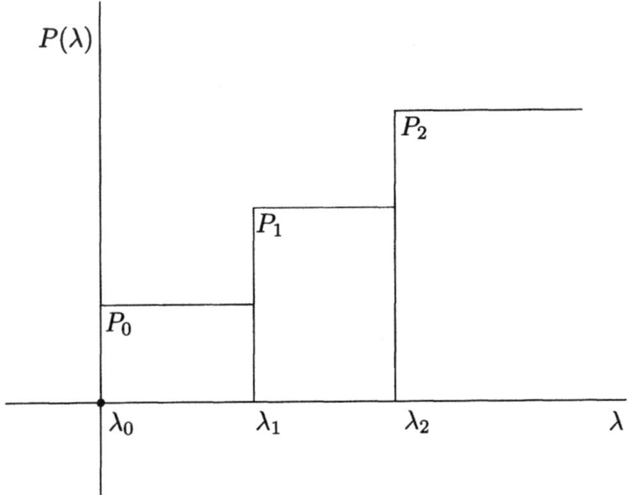

We then have

$$f(x) = \int_{0-}^{\infty} dP(\lambda) f(x), \qquad f \text{ in } L^2(-1, 1; (1-x)^A (1+x)^B),$$

and

$$Lf(x) = \int_{0-}^{\infty} \lambda dP(\lambda) f(x), \qquad f \text{ in } D_L.$$

It is clear the $P(\lambda)$ is a projection valued measure:

$$P(\lambda) f(x) = \sum_{\lambda_n \leq \lambda} P_n f(x).$$

XIV.3 The Legendre Polynomials

If $A = B = 0$, the Jacobi case of the previous section is reduced to a discussion of the Legendre polynomials $\{P_n(x)\}_{n=0}^{\infty}$ set in $L^2(-1, 1)$, and their expansions. The Legendre polynomials $\{P_n\}_{n=0}^{\infty}$ satisfy the differential equations

$$ly = -((1 - x^2) y')' = \lambda_n y,$$

where

$$\lambda_n = n(n + 1), \qquad n = 0, 1, \ldots.$$

The moment equation is

$$(n+2)\mu_{n+1} - n\mu_{n-1} = 0.$$

If we set $\mu_0 = 1$, $\mu_1 = 0$, then

$$\mu_{2n} = 1/(2n+1),$$
$$\mu_{2n+1} = 0, \qquad n = 0, 1, \ldots.$$

The weight equation is

$$(1 - x^2)w' = 0,$$

which has a distributional solution

$$w(x) = \alpha H(1 - x) + \beta H(1 + x) + \gamma.$$

We choose $\alpha = 1$, $\beta = 1$ and $\gamma = -1$ so that $w = 0$ if $x < -1$ or $x > 1$.

$$w(x) = 1, \qquad -1 \le x \le 1,$$
$$0, \qquad x < -1 \text{ or } 1 < x.$$

The Legendre polynomials, $\{P_n(x)\}_{n=0}^{\infty}$, satisfy a three term recurrence relation

$$nP_n(x) = (2n - 1)\, x\, P_{n-1}(x) - (n - 1)P_{n-2}(x),$$

which may be used to show that

$$\int_{-1}^{1} P_n(x)^2 dx = 2/(2n + 1), \qquad n = 0, 1, \ldots.$$

The differential equation may be used to show that

$$\int_{-1}^{1} P_n(x)P_m(x)dx = 0, \qquad m \ne n.$$

In order to discuss a differential operator in $L^2(-1, 1)$, we use the solutions of $ly = 0$, $y_1 = 1$ and $y_2 = \frac{1}{2}\ln[(1 + x)/(1 - x)]$. Both are in $L^2(-1, 1)$ and so the limit circle case holds at both ± 1. We set

$$B_{11}(y) = \lim_{x \to 1}(1 - x^2)(-y'(x)),$$
$$B_{12}(y) = \lim_{x \to 1}(1 - x^2)\left(\frac{1}{1 - x^2}\,y(x) - \frac{1}{2}\ln\left(\frac{1 + x}{1 - x}\right)y'(x)\right),$$
$$B_{-1,1}(y) = \lim_{x \to 1}(1 - x^2)(-y'(x)),$$
$$B_{-1,2}(y) = \lim_{x \to 1}(1 - x^2)\left(\frac{1}{1 - x^2}\,y(x) - \frac{1}{2}\ln\left(\frac{1 + x}{1 - x}\right)y'(x)\right),$$

for elements in the domain of the maximal operator.

The Legendre polynomials satisfy $B_{1,1}(y) = 0$, $B_{-1,1}(y) = 0$. Therefore,

XIV.3.1. Definition. We denote by D_L those elements y in $L^2(-1,1)$ satisfying:

(1) y is differentiable a.e. on $(-1,1)$.

(2) $(1-x^2)y'(x)$ is differentiable a.e. on $(-1,1)$ and $ly = -((1-x^2)y')'$ is in $L^2(-1,1)$.

(3) $B_{1,1}(y) = 0$ and $B_{-1,1}(y) = 0$.

We define the Legendre operator L by setting $Ly = ly$ for all y in D_L.

XIV.3.2. Theorem. *The Legendre polynomials $\{P_n(x)\}_{n=0}^{\infty}$ form a complete orthogonal set in $L^2(-1,1)$.*

The proof is a repetition of Theorem XIV.2.2, and is omitted.

With completeness, we find that if f is in $L^2(-1,1)$, then

$$f(x) = \sum_{n=0}^{\infty} c_n P_n(x), \quad \text{where} \quad c_n = \frac{2n+1}{2} \int_{-1}^{1} P_n(x)f(x)dx.$$

If f is in D_L, then

$$f(x) = \sum_{n=0}^{\infty} (n(n+1))c_n \, P_n(x).$$

These expansions can be written as Stieltjes integrals, similar to those in the previous section.

There are other, closely related examples which we leave to the reader as exercises. If $A = B$, the Jacobi polynomials are called the ultraspherical polynomials or Gegenbauer polynomials [27, p. 277].

If $A = B = -\frac{1}{2}$, the polynomials are the Tchebycheff polynomials of the first kind. If $A = B = \frac{1}{2}$, the polynomials are the Tchebycheff polynomials of the second kind.

Associated with each are boundary conditions, a differential operator on an L^2 space, and various eigenfunction expansions or spectral resolutions.

XIV.4 The Generalized Laguerre Polynomials

The generalized Laguerre polynomials $\{L_n^{(\alpha)}(x)\}_{n=0}^{\infty}$ satisfy the differential equations

$$ly = -(x^{\alpha+1}e^{-x}y')'/x^\alpha e^{-x} = ny.$$

Hence the moments $\{\mu_n\}_{n=0}^{\infty}$ satisfy

$$\mu_{n+1} - (\alpha + n + 1)\mu_n = 0.$$

If we set $\mu_0 = 1$, then

$$\mu_n = (\alpha+1)(\alpha+2)\cdots(\alpha+n) = (\alpha+1)_n.$$

So long as α is not $-1, -2, \ldots$, this may be rewritten as

$$\mu_n = \Gamma(\alpha + n + 1)/\Gamma(\alpha + 1).$$

The weight equation is

$$xw' + (x - a)w = 0,$$

where 0 is "0" over polynomials.

Rewritten as

$$x^{\alpha+1}e^{-x}(x^{-\alpha}e^x w)' = 0,$$

it is easy to see that it has the distributional solution

$$w = x^\alpha e^{-x}[AH(x) + B].$$

In order to have w vanish rapidly at $\pm\infty$, we choose $A = 1$, $B = 0$. Hence

$$w = x^\alpha e^{-x} \quad , \qquad x \geq 0,$$
$$= 0 \qquad , \qquad x < 0.$$

When $\alpha < -1$ and α is not a negative integer, a Cauchy regularization of w may be used instead. Acting on polynomial test functions, or test functions of slow growth ϕ

$$\langle w, \phi \rangle = \frac{(-1)^j}{\Gamma(\alpha + j + 1)} \int_0^\infty x^{\alpha+j}(e^{-x}\phi(x))^{(j)}\,dx,$$

when $-j - 1 < \alpha < -j$. (See [26].) The Pontrjagin space generated by the regularization has been examined briefly, but needs more study. [3], [16], [19].

Like the Jacobi poynomials, there is another approach. When $\alpha < -1$, the substitution $y = x^{-\alpha}z$ in the equation

$$-xy'' - (\alpha + 1 - x)y' = ny$$

results in

$$-xz'' - (-\alpha + 1 - x)z' = (n + \alpha)z$$

where $-\alpha > -1$. Although n is replaced by $n + a$ (the eigenparameter is shifted), an analysis of the case $\alpha < -1$ easily follows [6], [8].

The Laguerre polynomials $\{L_n^{(\alpha)}\}_{n=0}^\infty$ have a series representation

$$L_n^{(\alpha)}(x) = \sum_{k=0}^n \frac{(-1)^k(1+\alpha)_n x^k}{k!(n-k)!(1+\alpha)_k},$$

where $(\alpha + 1)_k = (\alpha + 1)(\alpha + 2)\cdots(\alpha + k)$, or

$$L_n^{(\alpha)}(x) = \Gamma(\alpha + n + 1)\sum_{k=0}^n \frac{(-1)^k x^k}{k!(n-k)!\Gamma(\alpha + k + 1)}$$

as well, when α is not a negative integer.

They satisfy a three term recurrence relation

$$nL_n^{(\alpha)}(x) = (2n - 1 + \alpha - x)L_{n-1}^{(\alpha)}(x) - (n - 1 + \alpha)L_{n-2}^{(\alpha)}(x).$$

The norm square

$$\int_0^\infty L_n^{(\alpha)}(x)^2 x^\alpha e^{-x} dx = \Gamma(\alpha + 1 + n)/n!$$

may be derived from the recurrence relation. That is,

$$\int_0^\infty L_n^{(\alpha)}(x)L_m^{(\alpha)}(x)x^\alpha e^{-x} dx = 0, \quad n \neq m,$$

may be derived in a number of ways [11], [27].

In order to discuss a differential operator in $L^2(0, \infty; x^\alpha e^{-x})$ we need to consider boundary conditions. Solutions to $ly = 0$ are $y_1 = 1$ and

$$y_2 = \int_1^x \frac{e^\xi}{\xi^{\alpha+1}}\, d\xi.$$

At $x = 0$ we have for elements y in the maximal domain

$$B_1(y) = \lim_{x \to 0} -x^{\alpha+1}e^{-x}y'(x),$$

and

$$B_2(y) = \lim_{x \to 0} -x^{\alpha+1}e^{-x}\left[y(x)\left(\frac{e^x}{x^{\alpha+1}}\right) - y'(x)\int_1^x \frac{e^\xi}{\xi^{\alpha+1}}\, d\xi\right].$$

$x = 0$ is in the limit circle case (look at the indicial roots of $ly = 0$) when $-1 < \alpha < 1$. Both $B_1(y)$ and $B_2(y)$ serve as boundary values. When $\alpha \geq 1$, only $B_1(y) = 0$ exists and is an annihilator.

In either case, the Laguerre polynomials satisfy $B_1(y) = 0$. ∞ is in the limit point case [4, p.229] and so we have the automatic or annihilator boundary condition

$$B_\infty(y) = -\lim_{x \to \infty} x^{\alpha+1}e^{-x}y'(x) = 0.$$

XIV.4.1. Definition. We denote by D_L, those elements y in $L^2(0, \infty; x^\alpha e^{-x})$ satisfying:

(1) y is differentiable a.e. on $(0, \infty)$.

(2) $x^{\alpha+1}e^{-x}y'(x)$ is differentiable a.e. on $(0, \infty)$ and ly is in $L^2(0, \infty, x^\alpha e^{-x})$.

(3) $\lim_{x \to 0+} -x^{\alpha+1}e^{-x}y'(x) = 0.$

We define the Laguerre operator L by setting $Ly = ly$ for all y in D_L.

Clearly the Laguerre polynomials $\{L_n^{(\alpha)}(x)\}_{n=0}^\infty$ are eigenfunctions for L, with eigenvalues $\{\lambda_n = n\}_{n=0}^\infty$.

XIV.4.2. Theorem. *The Laguerre polynomials $\{L_n^{(\alpha)}(x)\}_0^\infty$ form a complete orthogonal set in $L^2(0,\infty;x^\alpha e^{-x})$.*

Here, too, there are several unsatisfactory proofs found in the literature. We again use a modification of a technique found in [13], as corrected by [12].

Assume that f in $L^2(0,\infty;x^\alpha e^{-x})$ is orthogonal to the Laguerre polynomials, $\{L_n^{(\alpha)}(x)\}_{n=0}^\infty$. Then f is orthogonal to $\{x^n\}_{n=0}^\infty$ as well. Thus

$$\int_0^\infty e^{-x}x^{\alpha+n}f(x)dx = 0 \quad , \quad n = 0,1,\dots .$$

Consider now the function

$$G(\lambda) = \int_0^\infty e^{i\lambda x}e^{-x}x^\alpha f(x)dx.$$

This is the L^1 Fourier transform of $e^{-x}x^\alpha f(x)H(x)$. Further

$$
\begin{aligned}
|G(\lambda)| &= \left| \int_0^\infty e^{i\lambda x}e^{-x}x^\alpha f(x)dx \right| \\
&\leqq \int_0^\infty e^{-\operatorname{Im}\lambda x}e^{-x}x^\alpha|f(x)|dx \\
&\leqq \left(\int_0^\infty e^{-2(\operatorname{Im}\lambda+\frac{1}{2})x}x^\alpha dx \right)^{\frac{1}{2}} \left(\int_0^\infty e^{-x}x^\alpha f(x)^2 dx \right)^{\frac{1}{2}},
\end{aligned}
$$

using Schwarz's inequality. We see immediately that if $\operatorname{Im}\lambda+\frac{1}{2} > 0$, then $G(\lambda)$ is well defined and analytic in λ. $G(\lambda)$ is analytic in the half plane $\operatorname{Im}\lambda > -\frac{1}{2}$. Thus

$$G(\lambda) = \sum_{n=0}^\infty G^{(n)}(0)\,\lambda^n/n!\,.$$

But $G(0) = 0$. Further

$$
\begin{aligned}
\frac{d^nG}{d\lambda^n}(0) &= \int_0^\infty (ix)^n e^{i\lambda x}e^{-x}x^\alpha f(x)dx\Big|_{\lambda=0} \\
&= 0
\end{aligned}
$$

as well. So $G(\lambda)$ is identically 0.

This implies that $f(x) = 0$ a.e. and in $L^2(0,\infty;x^\alpha e^{-x})$, and $\{L_n^{(\alpha)}(x)\}_{n=0}^\infty$ is a complete orthogonal set.

We can again bypass the rather difficult theory concerning the spectral resolution of the operator L. If f is an arbitrary element of $L^2(0,\infty;x^\alpha e^{-x})$, then

$$f(x) = \sum_{n=0}^\infty c_n L_n^{(\alpha)}(x)\,, \quad \text{where} \quad c_n = \frac{\int_0^\infty f(x)L_n^{(\alpha)}(x)x^\alpha e^{-x}dx}{\Gamma(\alpha+1+n)/n!}\,.$$

If f is in D_L, then

$$Lf = \sum_{n=0}^{\infty} \lambda_n c_n L_n^{(\alpha)}(x), \qquad \text{where} \qquad \lambda_n = n.$$

The two expansions above can be rewritten as Stieltjes integrals. If we let

$$\phi_n(x) = L_n^{(\alpha)}(x) / \left[(\alpha + 1 + n)/n! \right]^{\frac{1}{2}},$$

then

$$f(x) = \sum_{n=0}^{\infty} \langle f, \phi_n \rangle \phi_n(x), \qquad \text{and} \qquad Lf(x) = \sum_{n=0}^{\infty} \lambda_n \langle f, \phi_n \rangle \phi_n(x),$$

where $\langle \cdot, \cdot \rangle$ stands for the inner product in $L^2(0, \infty; x^\alpha e^{-x})$. Denote by $P_n f$ the term $\langle f, \phi_n \rangle \phi_n(x)$. Define $P(\lambda)$ by setting $P(0-) = 0$ and having $P(\lambda)$ jump by P_n at $\lambda_n = n$. We then have

$$f(x) = \int_{0-}^{\infty} \lambda \, dP(\lambda) \, f(x), \quad f \text{ in } D_L.$$

It is clear that $P(\lambda)$ is a projection valued measure:

$$P(\lambda)f(x) = \sum_{\lambda_n \leq \lambda} P_n f(x).$$

When $\alpha = 0$, we recover the ordinary Laguerre poynomials. Here the Gamma function may be replaced, since

$$\Gamma(n + 1) = n!$$

XIV.5 The Hermite Polynomials

The Hermite polynomials $\{H_n(x)\}_{n=0}^{\infty}$ satisfy the differential equations

$$ly = -(e^{-x^2} y')'/e^{-x^2} = 2n\, y.$$

The moments $\{\mu_n\}_{n=0}^{\infty}$ satisfy

$$2\mu_{n+1} - n\, \mu_{n-1} = 0.$$

If we set $\mu_0 = 1$, $\mu_1 = 0$, then

$$\mu_{2n} = (2n)!/2^{2n} n!,$$
$$\mu_{2n+1} = 0 \quad, \quad n = 0, 1, \ldots.$$

The weight equation is

$$w' + 2xw = 0 \, ,$$

which has the solution

$$w = c^{-x^2}, \quad -\infty < x < \infty \, .$$

The Hermite polynomials $\{H_n\}_{n=0}^{\infty}$ have a series representation

$$H_n(x) = \sum_{k=0}^{\left[\frac{n}{2}\right]} \frac{(-1)^k n! (2x)^{n-2k}}{k!(n-2k)!} \, , \quad n = 0, 1, \ldots \, ,$$

where $\left[\frac{n}{2}\right]$ stands for the integer part of $n/2$.

The polynomials satisfy a three term recurrence relation

$$H_n(x) = 2x H_{n-1}(x) - 2(n-1) H_{n-2}(x) \, .$$

This may be used to calculate the norm square

$$\int_{-\infty}^{\infty} H_n(x)^2 e^{-x^2} \, dx = 2^n n! \sqrt{\pi} \, .$$

The differential equation easily shows that

$$\int_{-\infty}^{\infty} H_n(x) H_m(x) e^{-x^2} \, dx = 0 \, , \quad n \neq m \, .$$

An application of Levinson's criterion [4, p. 229] shows that both $\pm\infty$ are limit points. Using the solution $H_0 = 1$ of $ly = 0$ to generate the boundary conditions

$$B_{\infty}(y) = -\lim_{x \to \infty} e^{-x^2} y'(x),$$

$$B_{-\infty}(y) = -\lim_{x \to -\infty} e^{-x^2} y'(x),$$

we have two automatic (annihilator) conditions which are satisfied by all y in the maximal domains.

XIV.5.1. Definition. We denote by D_L, those elements y in $L^2(-\infty, \infty; e^{-x^2})$ satisfying:

(1) y is differentiable a.e. on $(-\infty, \infty)$.

(2) $e^{-x^2} y'(x)$ is differentiable a.e. on $(-\infty, \infty)$ and ly is in $L^2(-\infty, \infty; e^{-x^2})$.

We define the Hermite operator L by setting $Ly = ly$ for all y in D_L.

Clearly the Hermite polynomials $\{H_n(x)\}_{n=0}^{\infty}$ are eigenfunctions for L with eigenvalues $\{\lambda_n = 2n\}_{n=0}^{\infty}$.

XIV.5.2. Theorem. *The Hermite polynomials $\{H_n(x)\}_{n=0}^{\infty}$ form a complete orthogonal set in $L^2(-\infty, \infty; e^{-x^2})$.*

For the last time we use the technique found in [13].

Assume f in $L^2(-\infty, \infty; e^{-x^2})$ is orthogonal to the Hermite polynomials $\{H_n(x)\}_{n=0}^{\infty}$. Then f is orthogonal to $\{x^n\}_{n=0}^{\infty}$ as well. Thus

$$\int_{-\infty}^{\infty} x^n f(x) e^{-x^2} dx = 0, \quad n = 0, 1, \ldots.$$

Consider the function

$$H(\lambda) = \int_{-\infty}^{\infty} e^{i\lambda x} f(x) e^{-x^2} dx.$$

This is the L^1 Fourier transform of $f(x)e^{-x^2}$. Further

$$|H(\lambda)| \leq \int_{-\infty}^{\infty} e^{-\operatorname{Im} \lambda x} |f(x)| e^{-x^2} dx$$

$$\leqq \left(\int_{-\infty}^{\infty} e^{-2\operatorname{Im} \lambda x} e^{-x^2} dx \right)^{\frac{1}{2}} \left(\int_{-\infty}^{\infty} |f(x)|^2 e^{-x^2} dx \right)^{1/2},$$

using Schwarz's inequality. We conclude, therefore, that $H(\lambda)$ is an entire analytic function of λ. Thus

$$H(\lambda) = \sum_{n=0}^{\infty} H^{(n)}(0) \lambda^n / n!.$$

But $H(\lambda)$ has derivatives of all orders which vanish at $\lambda = 0$. Thus $H(\lambda) = 0$ identically. This implies $f(x) = 0$ a.e. in $L^2(-\infty, \infty; e^{-x^2})$ as well.

We bypass the Sturm–Liouville spectral theory to write when f is in $L^2(-\infty, \infty; e^{-x^2})$

$$f(x) = \sum_{n=0}^{\infty} C_n H_n(x),$$

where

$$H_n(x) = \frac{\int_{-\infty}^{\infty} H_n(x) f(x) e^{-x^2} dx}{2^n n! \sqrt{\pi}}.$$

If f is in D_L, then

$$Lf = \sum_{n=0}^{\infty} \lambda_n C_n H_n(x),$$

where $\lambda_n = 2n$.

The expansions above can be written as Stieltjes integrals. If we let

$$\phi_n(x) = H_n(x)/[2^n n! \sqrt{\pi}]^{\frac{1}{2}},$$

then

$$f(x) = \sum_{n=0}^{\infty} \lambda_n \langle F, \phi_n \rangle \phi_n(x), \quad \text{and} \quad Lf(x) = \sum_{n=0}^{\infty} \lambda_n \langle F, \phi_n \rangle \phi_n(x),$$

where $\langle \cdot, \cdot \rangle$ stands for the inner product in $L^2(-\infty, \infty; e^{-x^2})$. Denote by $P_n f$ the term $\langle f, \phi_n \rangle \phi_n(x)$. Define $P(\lambda)$ by setting $P(0-) = 0$ and having $P(\lambda)$ jump by P_n at $\lambda_n = 2n$. We then have

$$f(x) = \int_{0-}^{\infty} dP(\lambda) f(x), \quad f \text{ in } L^2(-\infty, \infty; e^{-x^2}),$$

and

$$Lf(x) = \int_{0-}^{\infty} \lambda dP(\lambda) f(x), \quad f \text{ in } D_L.$$

It is clear that $P(\lambda)$ is a projection valued measure:

$$P(\lambda) f(x) = \sum_{\lambda_n \leq \lambda} P_n f(x).$$

XIV.6   The Generalized Hermite Polynomials

Like the ordinary Laguerre polynomials $\{L_n(x)\}_{n=0}^{\infty}$, which have a generalization, the generalized Laguerre polynomials $\{L_n^{(\alpha)}(x)\}_{n=0}^{\infty}$, the ordinary Hermite polynomials $\{H_n(x)\}_{n=0}^{\infty}$ have one as well, the generalized Hermite polynomials, $\{H_n^{(\mu)}(x)\}_{n=0}^{\infty}$. The difference between the two is that, while the generalized Laguerre polynomials satisfy a single differential equation, that is, are eigenfunctions for a unique differential operator, the even and odd degree generalized Hermite polynomials $\{H_{2n}^{(\mu)}(x)\}_{n=0}^{\infty}$ and $\{H_{2n+1}^{(\mu)}(x)\}_{n=0}^{\infty}$ satisfy different differential equations, that is, they are eigenfunctions for *two* differential operators, properly defined on $(0, \infty)$, rather than $(-\infty, \infty)$. Nonetheless the entire set $\{H_n^{(\mu)}(x)\}_{n=0}^{\infty}$ share some of the usual properties associated with the other polynomials, [18], [20].

The series representations are

$$H_{2n}^{(\mu)}(x) = \sum_{k=0}^{n} \binom{n + \mu - \frac{1}{2}}{n - k} (-1)^{n-k} 2^{2n} n! x^{2k} / k!,$$

and

$$H_{2n+1}^{(\mu)}(x) = \sum_{k=0}^{n} \binom{n + \mu + \frac{1}{2}}{n - k} (-1)^{n-k} 2^{2n+1} n! x^{2k+1} / k!.$$

They satisfy

$$x H_{2n}^{(\mu)\prime\prime} + 2(\mu - x^2) H_{2n}^{(\mu)\prime} + [2(2n)x] H_{2n}^{(\mu)} = 0,$$

and

$$x H_{2n+1}^{(\mu)\prime\prime} + 2(\mu - x^2) H_{2n+1}^{(\mu)\prime} + [2(2n + 1) - 2\mu x^{-1}] H_{2n+1}^{(\mu)} = 0.$$

They form a complete orthogonal set on $(-\infty, \infty)$. (Orthogonality of two different polynomials of even or odd degree follows from the differential equations. Orthogonality of an even and an odd degree polynomial follows from antisymmetry on $(-\infty, \infty)$. Completeness is proved in the same way as the ordinary Hermite polynomials.) Orthogonality guarantees the existence of a three term recurrence relation [20]

$$H_{n+1}^{(\mu)}(x) = 2xH_n^{(\mu)}(x) - 2(n + \theta_n)H_{n-1}^{(\mu)}(x)\,,$$

where θ_n is the same variable factor in the last term of the differential equations

$$\begin{aligned}\theta_n &= 0\,, & n &= 2m\,,\\ &= 2\mu\,, & n &= 2m+1\,, \quad m = 0, 1, \ldots\,.\end{aligned}$$

The three term recurrence relation may be used to compute

$$\int_{-\infty}^{\infty} H_m^{(\mu)}(x)^2 |x|^{2\mu} e^{-x^2}\, dx = 2^{2n} \left[\frac{n}{2}\right]!\,\Gamma\left(\left[\frac{n+1}{2}\right] + \mu + \frac{1}{2}\right)\,,$$

where $[\,\cdot\,]$ denotes the greatest integer within.

In order to consider boundary value problems for these polynomials, however, it is necessary to consider the even polynomials and the odd polynomials separately, and to set them in $L^2(0, \infty; x^{2\mu} e^{-x^2})$. Note that both differential equations are singular at 0 as well as ∞.

XIV.6.1   The Generalized Hermite Polynomials of Even Degree

The classic orthogonality condition becomes

$$\int_0^{\infty} H_{2n}^{(\mu)}(x) H_{2m}^{(\mu)}(x) x^{2\mu} e^{-x^2}\, dx = 2^{4n-1} n!\,\Gamma(n + \mu + 1)\delta_{mn}\,,$$

when $\mu > -\frac{1}{2}$. The differential equation

$$ly = -\frac{1}{2}(x^{2\mu} e^{-x^2} y')'/(x^{2\mu} e^{-x^2}) = (2n)y\,.$$

If it is put in Liouville normal form by setting $y = x^{-\mu} e^{x^2/2} u$, ly is transformed into

$$\tilde{l}u = -\frac{1}{2}(u'' - [\mu(\mu - 1)x^{-2} - (2\mu - 1) + x^2]u) = (2n)u$$

on $L^2(0, \infty; 1)$. Infinity is established to be in the limit point case [4; p. 229]. The substitution $t = x^2$ transforms the original differential equation into

$$t\frac{d^2y}{dt^2} + \left(\frac{1}{2} + \mu - t\right)\frac{dy}{dt} + ny = 0\,.$$

If $\alpha = \mu - \frac{1}{2}$, this is

$$t\frac{d^2y}{dt^2} + (1 + \alpha - t)\frac{dy}{dt} + ny = 0,$$

the generalized Laguerre equation. It is well known that

$$y = H_{2n}^{(\mu)}(x) = (-1)^n 2^{2n} n! L_n^{(\mu - \frac{1}{2})}(t).$$

Recalling information from the Laguerre problem, we find that the polynomial generating boundary condition is

$$\lim_{t \to 0} t^{\alpha+1} e^{-t} \frac{dy}{dt} = 0.$$

This is equivalent to

$$\lim_{x \to 0} x^{2\mu} e^{-x^2} \frac{dy}{dx} = -\lim_{x \to 0} x^{2\mu} e^{-x^2} W[y, 1](x) = 0$$

in the Hermite case. The even degree Hermite polynomials satisfy this boundary condition at $x = 0$.

Again by comparing the Hermite and Laguerre equations we determine that l is in the limit circle case when $-\frac{1}{2} < \mu < \frac{3}{2}$. Examination of Frobenius roots establishes the same conclusion.

XIV.6.1.1. Definition. We denote by D_L those elements y in $L^2(0, \infty; x^{2\mu} e^{-x^2})$ satisfying:

(1) y is differentiable a.e. on $(0, \infty)$.

(2) $x^{2\mu} e^{-x^2} y'$ is differentiable a.e. on $(0, \infty)$ and

$$ly = -\frac{1}{2}(x^{2\mu} e^{-x^2} y')'/(x^{2\mu} e^{-x^2})$$

is in $L^2(0, \infty; x^{2\mu} e^{-x^2})$.

(3) $\lim_{x \to 0} x^{2\mu} e^{-x^2} y'(x) = 0$ (automatic if $\mu \leq \frac{1}{2}$ or $\frac{3}{2} \leq \mu$).

We define the generalized even Hermite operator L by setting $Ly = ly$ for all y in D_L.

XIV.6.1.2. Theorem. *The generalized Hermite polynomials* $\{H_{2n}^{(\mu)}\}_{n=0}^{\infty}$, $\mu > -\frac{1}{2}$, *form a complete orthogonal set in* $L^2(0, \infty; x^{2\mu} e^{-x^2})$.

We leave the proof to the reader.

We bypass the spectral theory concerning L by recognizing that its resolution is again an eigenfunction expansion.

If f is an arbitrary element of $L^2(0, \infty; x^{2\mu}e^{-x^2})$, then

$$f(x) = \sum_{n=0}^{\infty} c_{2n} H_{2n}^{(\mu)}(x)\,, \quad \text{where} \quad c_{2n} = \frac{\int_0^{\infty} f(\xi) H_{2n}^{(\mu)}(\xi) \xi^{2\mu} e^{-\xi^2}\, d\xi}{2^{4n-1} n! \Gamma(n + \mu + 1)}\,.$$

If f is in D_L, then

$$Lf(x) = \sum_{n=0}^{\infty} 2n\, c_{2n}\, H_{2n}^{(\mu)}(x)\,.$$

These formulas can be rewritten as Stieltjes integrals with respect to a spectral measure, just as in the case of the ordinary Hermite polynomials.

XIV.6.2 The Generalized Hermite Polynomials of Odd Degree

The classic orthogonality condition becomes

$$\int_0^{\infty} H_{2n+1}^{(\mu)}(x) H_{2m+1}^{(\mu)}(x) x^{2\mu} e^{-x^2}\, dx = 2^{4n+1} n! \Gamma\left(n + \mu + \frac{3}{2}\right) \delta_{mn}\,,$$

where $\mu > -\frac{3}{2}$. The differential equation becomes

$$my = -\frac{1}{2}((x^{2\mu}y')' - 2\mu\, x^{2\mu-2} e^{-x^2} y)/(x^{2\mu} e^{-x^2})(x^{2\mu} = (2n+1)y\,.$$

If it is put in Liouville normal form by setting $y = x^{-\mu} e^{-x^2/2} u$, $m\,y$ is transformed into

$$\tilde{m}u = -\frac{1}{2}(u'' - [\mu(\mu+1)x^{-2} - (2\mu+1) + x^2]u) = (2n+1)u$$

on $L^2(0, \infty)$. Infinity is established to be in the limit point case [4; p. 229]. No boundary condition is required at ∞.

The substitutions $t = x^2$, $y = \sqrt{t}z$ transform the original differential equation into

$$t\frac{d^2 z}{dt^2} + \left(\mu + \frac{3}{2} - t\right)\frac{dz}{dt} + nz = 0\,.$$

Letting $\alpha = \mu + \frac{1}{2}$, this is

$$t\frac{d^2 z}{dt^2} + (1 + \alpha - t)\frac{dz}{dt} + nz = 0\,,$$

the generalized Laguerre equation, It is well known that

$$y = H_{2n+1}^{(\mu)}(x) = c_n \sqrt{t} L_n^{(\mu + fr12)}(t)\,,$$

where $c_n = (-1)^n 2^{2n+1} n!$. Recalling information from the Laguerre problem, we find that the polynomial generating boundary condition is

$$\lim_{t \to 0} t^{\alpha+1} e^{-t} \frac{dz}{dt} = 0\,.$$

This is equivalent to

$$\lim_{x\to 0} x^{2\mu}e^{-x^2}\left(x\frac{dy}{dx}-y\right) = -\lim_{x\to 0} x^{2\mu}e^{-x^2}W[y,x] = 0\,.$$

The odd degree Hermite polynomials satisfy the boundary condition at $x=0$.

The Laguerre boundary conditions, needed when $-1 < \alpha < 1$, tell us that the Hermite boundary condition is needed when $-\frac{3}{2} < \mu < \frac{1}{2}$. Examination of Frobenius roots establishes the same conclusion.

XIV.6.2.1. Definition. We denote by D_M those elements y in $L^2(0,\infty; x^{2\mu}e^{-x^2})$ satisfying:

(1) y is differentiable a.e. on $(0,\infty)$.

(2) $x^{2\mu}e^{-x}y'$ is differentiable a.e. on $(0,\infty)$ and

$$my = -\frac{1}{2}((x^{2\mu}e^{-x^2}y')' - 2\mu\,x^{2\mu-2}e^{-x^2}y)/;(x^{2\mu}e^{-x^2})$$

is in $L^2(0,\infty; x^{2\mu}e^{-x^2})$.

(3) $\lim_{x\to 0} x^{2\mu}e^{-x^2}(xy'(x) - y(x)) = 0$ (automatic if $-\frac{3}{2} < \mu$, or $\frac{1}{2} < \mu$).

We define the generalized odd Hermite operator M by setting $My = my$ for all y in D_M.

XIV.6.2.2. Theorem. *The generalized Hermite polynomials $\{H^{(\mu)}_{2n+1}\}_{n=0}^{\infty}$ form a complete orthogonal set in $L^2(0,\infty; x^{2\mu}e^{-x^2})$.*

We leave the proof to the reader.

We bypass the spectral theory concerning M by recognizing that its resolution is again an eigenfunction expansion.

If F is an arbitrary element of $L^2(0,\infty; x^{2\mu}e^{-x^2})$, then

$$f(x) = \sum_{n=0}^{\infty} C_{2n+1}H^{(\mu)}_{2n+1}(x)\,,$$

where

$$c_{2n+1} = \frac{\int_0^{\infty} f(\xi)H^{(\mu)}_{2n+1}(\xi)\xi^{2\mu}e^{-\xi^2}\,d\xi}{2^{4n+1}n!\,\Gamma(n+\mu+\frac{3}{2})}\,.$$

If f is in D_M, then

$$Mf(x) = \sum_{n=0}^{\infty}(2n+1)c_{n+1}H^{(\mu)}_{2n+1}(x)\,.$$

These results can be rewritten as Stieltjes integrals with respect to a spectral measure, just as in the case of the ordinary Hermite polynomials.

XIV.7 The Bessel Polynomials

We now consider the great enigma concerning orthogonal polynomials. The Bessel polynomials. Discovered several times, they were finally formally recognized in 1949 when H. L. Krall and Orrin Frink [21] published the first detailed study concerning them. They satisfy the differential equation

$$ly = x^2 y'' + (ax + b)y' = n(n + a - 1)y.$$

The moments satisfy

$$(n + a - 1)\mu_n + b\mu_{n-1} = 0,$$

so if $\mu_0 = -b$, $\mu_n = (-b)^{n+1}/(a)_n$, $n = 0, 1, \ldots$, where $(a)_n = a(a+1)\cdots(1+n-1)$.

The weight equation is

$$x^2 w' - ((a - 2)x + b)w = 0,$$

which may be rewritten as

$$x^a e^{-b/x}(x^{2-a} e^{b/x} w)' = 0.$$

The solution to this equation

$$w = x^{a-2} e^{-b/x},$$

which seems obvious, is *not* a weight function, however. While it vanishes at $x = 0$ from the right, it does not in general vanish anywhere else. Even if $a < 2$, so it vanishes at both 0 and ∞, it goes to 0 at ∞ too slowly to serve.

There is a δ function distribution which works in a combinatory way, [26]. It is given by

$$w = \sum_{n=0}^{\infty} b^{n+1} \delta^{(n)}(x)/n!(a)_n.$$

It has never been connected to any classical function, however.

H. L. Krall and Orrin Frink [21] found a complex weight function which served to orthogonalize the Bessel polynomials when the path of integration contained the origin in its interior. When a is an integer, $w = x^{a-2} e^{-b/x}$ also fills this role.

A partial solution to the weight problem was found in 1992 by Kim, Kwon and Hahn [14]. When $a = b = 2$, they noticed that in the weight equation

$$x^2 x' - 2w = 0,$$

the term 0 does not actually need to be 0, but need only annihilate polynomials. For instance, the function

$$g(x) = 0, \qquad x \le 0,$$
$$= e^{-x^{\frac{1}{4}}} \sin x^{\frac{1}{4}} \, x \ge 0,$$

has this property, as we shall show shortly (see [10] [29]). Solving

$$x^2 w' - 2w = e^{-x^{\frac{1}{4}}} \sin x^{\frac{1}{4}}$$

on $[0, \infty)$ results in

$$w = e^{-2/x} \int_x^\infty e^{2/t} t^{-2} e^{-t^{\frac{1}{4}}} \sin t^{\frac{1}{4}} dt$$

on $(0, \infty)$. A brief reflection verifies that w is continuous on $[0, \infty)$ (even at 0+) and $|w|$ decreases like an exponential function of the form $e^{-t^{\frac{1}{4}}/2}$, so that all its moments are finite.

To see that

$$\int_0^\infty t^n e^{-t^{\frac{1}{4}}} \sin t^{\frac{1}{4}} dt = 0, \qquad n = 0, 1, \dots,$$

let $t^{\frac{1}{4}} = x$. The result is

$$\mu_n = 4 \int_0^\infty e^{-x} \sin x \, x^{4n+3} dx$$
$$= 4\mathrm{Im} \int_0^\infty e^{-(1-i)x} x^{4n+3} dx \,.$$

This is easily evaluated by integration by parts:

$$\mu_n = \mathrm{Im} \, [(4n+3)! \cos(n+1)\pi/2^{2n+2}]$$
$$= 0 \,.$$

To see that the correct moments are generated, note that on $[0, \infty)$

$$\int_0^\infty [x^2 w' - 2w] x^n dx = \int_0^\infty g(x) x^n dx = 0 \,.$$

Thus

$$\langle x^2 w', x^n \rangle - 2\langle w, x^n \rangle = 0 \,,$$
$$-\langle w, (x^{n+2})' \rangle - 2\langle w, x^n \rangle = 0 \,,$$
$$-(n+2)\langle w, x^{n+1} \rangle - 2\langle w, x^n \rangle = 0 \,.$$

Hence the moments satisfy

$$\mu_{n+1} = -2\mu_n/(n+2) \,,$$

the same relation exhibited earlier. If $\mu_0 = -2$, $\mu_n = (-2)^{n+1}/(n+1)!$.

The differential equation in general can be reduced to the case

$$x^2 y'' + (ax + w)y' = n(n + a - 1)y$$

by setting $x = bz/2$. With a arbitrary and $b = 2$, the solution to

$$x^2 w' - ((a-2)x + 2)w = g(x)$$

is

$$w = -x^{a-2} e^{-2/x} \int_x^\infty \frac{e^{2/t} e^{-t^{\frac{1}{4}}} \sin t^{\frac{1}{4}}}{t^a} \, dt.$$

There is a problem. w might also generate 0 moments. It is therefore necessary to show that

$$\mu_0 = \int_0^\infty w(t)dt \neq 0$$

in order to avoid a vacuous situation. Moroni [25] has shown that for $a > 12 \left(\frac{2}{\pi}\right)^4 \simeq 1.97$, μ_0 is indeed nonzero. For $a \leq 12 \left(\frac{2}{\pi}\right)^4$ The problem remains open.

Han and Kwon [9] have developed a Krein space setting for the Bessel polynomial operator. It is rather abstract, however, and does not resemble at all the L^2 settings in existence for the other polynomial sets.

References

[1] L. C. Andrews, **Special Functions for Engineers and Applied Mathematicians**, Macmillan, New York, 1985.

[2] S. Bochner *Über Sturm-Liouvillesche Polynomsysteme*, Math. Zeit. **29** (1929), 730–736.

[3] J. Bognar, **An Indefinite Inner Product Spaces**, Springer-Verlag, New York, 1974.

[4] E. A. Coddington and N. Levinson, **Theory of Ordinary Differential Equations**, Macmillan, New York, 1955.

[5] W. D. Evans, W. N. Everitt, A. M. Krall, K. H. Kwon and L. L. Littlejohn, *A solution to the general Bessel moment problem*, World Scientific Publishing Co., (WSSIAA 1 (1992), 205–220.

[6] M. Hajmirzahmad, *The spectral resolution of Laguerre operators in right definite and left definite spaces*, Ph.D. Dissertation, The Pennsylvania State University, University Park, PA 16802, 1990.

[7] ______, *Jacobi polynomial expansions*, J. Math. Anal. Appl. **181** (1994), 35–61.

[8] ______, *Laguerre polynomial expansions*, J. Comp. Appl. Math. **59** (1995), 25–37.

[9] S. S. Han and K. H. Kwon, *Spectral analysis of Bessel polynomials in Krein space*, Quaes. Math. **14** (1991), 327–335.

[10] C. Hermite and T. J. Stieltjes, **Correspondence D'Hermite et Stieltjes**, Gautier-Villars, 2 (1905), 337.

[11] D. Jackson, **Fourier Series and Orthogonal Polynomials**, Math. Assn. Amer., 1941.

[12] A. Jirari, *On the completeness of orthogonal polynomials*, Master's Thesis, the Pennsylvania State University, University Park, PA 16802, 1989.

[13] J. Keener, **Principles of Applied Mathematics**, Addison Wesley, Reading, Mass., 1988.

[14] S. S. Kim, K. H. Kwon and S. S. Hahn, *Orthogonalizing weights of Tchebycheff sets of polynomials*, Bull. London Math. Soc. **24** (1992), 361–367.

[15] A. M. Krall, *Orthogonal polynomials through moment generating functionals*, SIAM J. Math. Anal. **9** (1978), 604–626.

[16] ———, *Laguerre polynomials in indefinite inner product spaces*, J. Math. Anal. Appl. **70** (1979), 267–279.

[17] ———, *Chebycheff sets of polynomials which satisfy an ordinary differential equation*, SIAM Review **22** (1980), 436–441.

[18] ———, *On the generalized Hermite polynomials* $\{H_n^{(\mu)}\}_{n=0}^{\infty}$, $\mu < -\frac{1}{2}$, Indiana J. Math. **30** (1981), 73–78.

[19] ———, *On boundary values for the Laguerre operator in indefinite inner product spaces*, J. Math. Anal. Appl. **85** (1982), 406–408.

[20] ———, *Spectral analysis for the generalized Hermite polynomials*, Trans. Amer. Math. Soc. **344** (1994), 155–172.

[21] H. L. Krall and O. Frink, *A new class of orthogonal polynomials: The Bessel polynomials*, Trans. Amer. Math. Soc. **65** (1949), 100–115.

[22] K. H. Kwon and L. L. Littlejohn, *Classification of classical orthogonal polynomials*, submitted for publication.

[23] P. Lesky *Die Charakterisierung der klassischen orthogonalen Polynome durch Sturm-Liouvillesche Differentialgleichungen*, Arch. Rational Mech. Anal. **10** (1962), 341–351.

[24] L. L. Littlejohn and A. M. Krall, *Orthogonal polynomials and singular Sturm-Liouville systems, I*, Rocky Mt. J. Math. **16** (1986), 435–479.

[25] P. Maroni, *An integral representation for the Bessel form*, J. Comp. Appl. Math. **57** (1995), 251–260.

[26] R. D. Morton and A. M. Krall, *Distributional weight functions for orthogonal polynomials*, SIAM J. Math. Anal. **9** (1978), 604–626.

[27] E. D. Rainville, **Special Functions**, Macmillan, New York, 1960.

[28] G. Szego, **Orthogonal Polynomials**, American Mathematical Society, Providence, RI, 1939.

[29] D. V. Widder, **The Laplace Transform**, Princeton Univ. Press, Princeton, 1941, 125–126.

Chapter XV

Orthogonal Polynomials Satisfying Fourth Order Differential Equations

Since they are rather important and quite accessible, we repeat the general theoretical facts concerning weights, moments and polynomials pertaining to fourth order differential equations. We then briefly discuss the squares of the differential equations of the second order, giving a number of easily derived examples of fourth order problems. This is followed by three new orthogonal polynomial sets satisfying fourth order differential equations, but which do not satisfy second order differential equations.

XV.1 The General Theory

We again consider an infinite collection of moments $\{\mu_n\}_{n=0}^{\infty}$, satisfying

$$\Delta_n = \begin{vmatrix} \mu_0 & \cdots & \mu_n \\ \vdots & & \vdots \\ \mu_n & \cdots & \mu_{2n} \end{vmatrix} \neq 0 , \quad n = 0, 1, \ldots,$$

and define the Tchebycheff polynomials by setting $p_0 = 1$,

$$p_n = \begin{vmatrix} \mu_0 & \cdots & \mu_n \\ \vdots & & \vdots \\ \mu_{n-1} & & \\ 1 & \cdots & x^n \end{vmatrix} / \Delta_n , \quad n = 1, 2, \ldots .$$

They are mutually orthogonal with respect to the distributional weight function

$$w = \sum_{n=0}^{\infty} (-1)^j \mu_j \delta^{(j)}(x)/j! ,$$

as well as others, including (at least) one given by a function of bounded variation.

Likewise a three term recurrence relation

$$p_{n+1}(x) = (x + B_n)p_n(x) - c_n p_{n-1}(x) = 0$$

also exists. The proofs are identical to those found in Chapter XIII.

We assume that the polynomials satisfy a collection of differential equations of the form

$$\begin{aligned}
lp_n = {}& (l_{44}x^4 + l_{43}x^3 + l_{42}x^2 + l_{41}x + l_{40})p_n^{(iv)} \\
& + (l_{33}x^3 + l_{32}x^2 + l_{31}x + l_{30})p_n''' \\
& + (l_{22}x^2 + l_{21}x + l_{20})p_n'' \\
& + (l_{10}x + l_{00})p_n' = \lambda_n\, p_n\,,
\end{aligned}$$

where

$$\lambda_n = l_{11}n + l_{22}n(n-1) + l_{33}n(n-1)(n-2) + l_{44}n(n-1)(n-2)(n-3), \quad n = 0, 1, \dots .$$

It is easy to see that the differential expression $w\,l$ is symmetric over polynomials. The proof is identical to that of Chapter XIII.

To find the weight equations, we require that $w\,l$ be symmetric. Thus

$$\begin{aligned}
(w\,a_4)y^{iv} &+ (w\,a_3)y''' + (w\,a_2)y'' + (w\,a_1)y' \\
&= (w\,a_4 y)^{(iv)} - (w\,a_3 y)''' + (w\,a_2 y)'' - (w\,a_1 y)'\,,
\end{aligned}$$

where

$$\begin{aligned}
a_4 &= l_{44}x^4 + l_{43}x^3 + l_{42}x^2 + l_{41}x + l_{40}\,, \\
a_3 &= l_{33}x^3 + l_{32}x^2 + l_{31}x + l_{30}\,, \\
a_2 &= l_{22}x^2 + l_{21}x + l_{20}\,, \\
a_1 &= l_{11}x + l_{10}\,,
\end{aligned}$$

and w acts as a distribution over polynomials.

The requirement that $w\,l$ be symmetric is equivalent to

$$S_2 = 2(w\,a_1) - 2(w\,a_2)' + 3(w\,a_e)'' - r(w\,a_r)''' = 0$$

and

$$S_4 = 2(w\,a_3) - 4(w\,a_4)' = 0\,.$$

$S_4 = 0$ determines w:

$$w = c\left\{\exp\left[\int \frac{a_3}{2a_4}\, dx\right]\right\}/a_4\,.$$

$S_2 = 0$ is a compatibility condition.

Applying $S_2 = 0$ and $S_4 = 0$ as distributions to x^n results in the moment equations

$$(m - 1)(m - 2)(m - 3)[l_{44}\mu_m + l_{43}\mu_{m-1} + l_{42}\mu_{m-2} + l_{41}\mu_{m-3} + l_{40}\mu_{m-4}]$$
$$+ (m - 1)(m - 2)[l_{33}\mu_m + l_{32}\mu_{m-1} + l_{31}\mu_{m-2} + l_{30}\mu_{m-3}]$$
$$+ (m - 1)[l_{22}\mu_m + l_{21}\mu_{m-1} + l_{20}\mu_{m-2}]$$
$$+ [l_{11}\mu_m + l_{10}\mu_{m-1}] = 0 \quad , \quad m \geq 1,$$

and

$$2(m - 3)[l_{44}\mu_m + l_{43}\mu_{m-1} + l_{42}\mu_{m-2} + l_{41}\mu_{m-3} + l_{40}\mu_{m-4}]$$
$$+ [l_{33}\mu_m + l_{32}\mu_{m-1} + l_{31}\mu_{m-2} + l_{30}\mu_{m-3}] = 0 \quad , \quad m \geq 3.$$

H. L. Krall [8] showed that there are seven sets of orthogonal polynomials satisfying fourth order differential equations. Each of the second order examples, the Jacobi, Laguerre, Hermite and Bessel polynomial differential equations may be squared to yield a fourth order equation. In addition there are three new differential equation sets with orthogonal polynomial solutions. Because of many close relations to the previous examples, they are called the Legendre-type, the Laguerre-type and the Jacobi-type polynomials.

We shall list these examples with their various properties.

XV.2 The Jacobi Polynomials

We recall that when $A > -1$, $B > -1$, the Jacobi operator of second order is.

$$L_2 y = -((1 - x)^{A+1}(1 + x)^{B+1}y')'/(1 - x)^A(1 + x)^B .$$

Its square is

$$L_4 y = \{[(1 - x)^{A+2}(1 + x)^{B+2}y'']''$$
$$- (2 + A + B)[(1 - x)^{A+1}(1 + x)^{B+1}y']'\}/(1 - x)^A(1 + x)^B .$$

The moment relations deduce to

$$(A + B + n + 2)\mu_{n+1} + (A - B)\mu_n - (m - 1)\mu_{n-1} = 0, \, n \geq 0,$$

the same as in XIV.2. The moments are also the same.

The weight equations also give

$$w = (1 - x)^A(1 + x)^B \quad , \quad -1 \leq x \leq 1,$$
$$= 0 \quad\quad\quad , \quad x < -1 \text{ or } 1 < x.$$

While the cases $A \leq -1$ and/or $B \leq -1$ have not been discussed via the transformation given in XIV, the results are no doubt the same.

At $x = 1$ the problem is limit 4 if $-1 < A < 1$, limit 3 if $1 \le A < 3$, and limit 2 if $3 \le A$. If we let

$$q_2 = (1 - x)^{A+1}(1 + x)^{B+1},$$
$$q_1 = -(2 + A + B)(1 - x)^{A+1}(1 + x)^{B+1},$$

the coefficients in the fourth order operator, and

$$=(q_2 y'')'\bar{z} - (q_2 y'')\bar{z}' - (q_2 \bar{z}'')'y + (q_2 \bar{z}'')y'$$
$$+ (q_1 y')\bar{z} - q_1 \bar{z}'y,$$

the fourth order Wronskian, then boundary conditions at $x = 1$ are

$$\lim_{x \to 1} [y, P_0^{(A,B)}] = 0 \quad , \quad \text{(automatic if } 1 \le A\text{)},$$
$$\lim_{x \to 1} [y, P_1^{(A,B)}] = 0 \quad , \quad \text{(automatic if } 3 \le A\text{)}.$$

The same situation occurs at $x = -1$, with constraints on B instead of A.

The Legendre, Gegenbauer and Tchebycheff operators are again special cases with $A = B = 0$; $A = B$; or $A = B = -\frac{1}{2}$, or $A = B = \frac{1}{2}$.

XV.3   The Generalized Laguerre Polynomials

The second order Laguerre operator is

$$L_2 y = -(x^{\alpha+1} e^{-x} y')'/x^\alpha e^{-x}, \qquad a > -1.$$

Its square is

$$L_4 y = \{[x^{\alpha+2} e^{-x} y'']'' - [x^{\alpha+1} e^{-x} y']'\}/(x^\alpha e^{-x}).$$

The moment relations reduce to

$$\mu_{n+1} - (\alpha + n + 1)\mu_n = 0.$$

The moments are

$$\mu_n = (\alpha + 1)(\alpha + 2) \cdots (\alpha + n) = (\alpha + 1)_n,$$

the same as in Chapter XIV.

The weight equations also give

$$w = x^\alpha e^{-x} \quad , \quad 0 \le x < \infty,$$
$$= 0 \quad , \qquad x < 0.$$

The case $\alpha < -1$ can also be transformed to an equivalent $\alpha > -1$ by the transformation used in Chapter XIV.

Infinity is in the limit 2 case. No boundary condition is required [10].

0 is in the limit 4 case, $-1 < \alpha < 1$; the limit 3 case, $1 \leq \alpha < 3$; and in the limit 2 case of $3 \leq \alpha$. If we let

$$q_2 = x^{\alpha+2}e^{-x},$$
$$q_1 = -x^{\alpha+1}e^{-x},$$

and

$$\begin{aligned}
&= (q_2 y'')'\bar{z} - (q_2 y'')\bar{z}' - (q_2\bar{z}'')'y + (q_2\bar{z}'')y' \\
&\quad + (q_1 y')\bar{z} - q_1\bar{z}'y,
\end{aligned}$$

the fourth order Wronskian, then boundary conditions at $x = 0$ are

$$\lim_{x \to 1}[y, L_0^{(\alpha)}](x) = 0 \qquad , \quad (\text{automatic if } 1 \leq \alpha),$$

$$\lim_{x \to 1}[y, L_1^{(\alpha)}](x) = 0 \qquad , \quad (\text{automatic if } 3 \leq \alpha).$$

If $\alpha = 0$, the ordinary Laguerre operator of fourth order

$$L_4 y = \{[x^2 e^{-x} y'']'' - [xe^{-x}y']'\}/e^{-x}$$

is limit 4 at 0. The same boundary conditions (both) are required at 0.

XV.4 The Hermite Polynomials

The fourth order Hermite operator is

$$L_4 y = \{[e^{-x^2}y'']'' - 2[e^{-x^2}y']'\}/e^{-x^2}$$

on $(-\infty, \infty)$. Both $\pm\infty$ are limit 2, and no boundary conditions are required.

We invite the reader to derive the weight and moment equations and to verify that the weight function and moments are the same as in the second order problem.

XV.5 The Legendre-Type Polynomials

Discovered by H. L. Krall in 1940 [7], [8], these, the first of three sets of orthogonal polynomials satisfying fourth order differential equations, but not second order equations, remained virtually hidden by their obscure place of publication for some forty years. When the author needed some examples of fourth order problems in the late 1970's to serve as examples, he recalled various conversations with his father concerning these polynomials, and resurrected them [4]. They are a bit different from the three previously mentioned cases in that the weight measure has Stieltjes jumps at ± 1.

The Legendre-type operator is

$$L_4 y = [(1 - x^2)^2 y'']'' + 4[(\alpha(x^2 - 1) - 2)y']',$$

set in $L^2(-1, 1; w)$, where

$$w = \frac{1}{2}[\delta(x + 1) + \delta(x - 1)] + \frac{\alpha}{2}[H(x + 1) - H(x - 1)],$$

where $\alpha > 0$, $\delta(x)$ is the Dirac delta function and $H(x)$ is the Heaviside function. w is the distributional solution of the weight equations

$$s_2 = (x^2 - 1)w''' + 12x(x^2 - 1)w'' + [(24 - 4\alpha)x^2 + 4\alpha]w' = 0,$$

and

$$s_4 = (x^2 - 1)^2 w' = 0,$$

which vanishes at $\pm\infty$.

The moment equations are

$$(n + 1)\mu_n - 2(n - 1)\mu_{n-2} + (n - 3)\mu_{n-4} = 0 \quad , \quad n \geq 3,$$
$$(n - 1 + \alpha)(n + 1)\mu_n - (n + 1 + \alpha)(n - 1)\mu_{n-2} = 0 \quad , \quad n \geq 1.$$

These are easily solved to give

$$\mu_{2n} = (\alpha + 2n + 1)/(2n + 1)!,$$
$$\mu_{2n+1} = 0 \quad , \quad n = 0, 1, \ldots,$$

which gives the distributional series

$$w = \sum_{n=0}^{\infty} \frac{(\alpha + 2n + 1)}{(2n + 1)!} \delta^{2n}(x)$$

for w. This is easily connected to the other expression for w through the Fourier transform [4].

Solving the differential equations

$$L_4 y = \lambda_n y$$

for polynomial solutions gives the polynomials themselves. They are

$$P_n^{(\alpha)}(x) = \sum_{k=0}^{[n/2]} \frac{(-1)^k (2n - 2k)!(\alpha + \frac{1}{2}(n)(n - 1) + 2k)x^{n-2k}}{2^n k!(n - k)!(n - 2k)!} \quad , \quad n = 0, 1, \ldots .$$

They can also be found by using the generating function

$$\left(\alpha - x\frac{\partial}{\partial x} + \frac{1}{2}t\frac{\partial^2}{\partial t^2}t \right)(1 - 2xt + t^2)^{-\frac{1}{2}} = \sum_{n=0}^{\infty} P_n^{(\alpha)}(x)t^n .$$

The polynomials satisfy a three term recurrence relation

$$P_{n+1}^{(\alpha)}(x) = (A_n x + B_n)P_n^{(\alpha)}(x) - C_n P_{n-1}^{(\alpha)}(x),$$

where ·

$$A_n = \frac{(2n+1)(\alpha + \frac{1}{2}(n)(n+1))}{(n+1)(\alpha + \frac{1}{2}(n-1)(n))},$$

$$B_n = 0,$$

and

$$C_n = \frac{n(\alpha + \frac{1}{2}(n+1)(n+2))}{(n+1)(\alpha + \frac{1}{2}(n-1)(n))}.$$

Of course orthogonality of the polynomials with respect to w follows from Green's formula and the differential equations. Likewise the norm-squares of the polynomials follow from the use of the recurrence relations. They may be summarized by

$$\int_{-1}^{1} P_n^{(\alpha)} P_m^{(\alpha)}(x) w\, dx = \frac{\alpha(\alpha + \frac{1}{2}n(n-1))(\alpha + \frac{1}{2}(n+1)(n+2))}{(2n+1)} \delta_{mn}.$$

We shall highlight the appropriate differential operator in $L^2(-1,1;w)$ for which these polynomials form a complete set of eigenfunctions. We cite [1] and [2] for full details concerning the operator and its domain.

Lagrange's formula has the appearance of

$$\bar{z}L_4 y - y\overline{L_4 z} = [(x^2-1)^2(\bar{z}y''' - y\bar{z}''') + 4x(x^2-1)(\bar{z}y'' - y\bar{z}'')$$
$$+(4\alpha(x^2-1)-8)(\bar{z}y' - y\bar{z}''') - (x^2-1)^2(\bar{z}'y'' - y'\bar{z}'')]'.$$

Consequently, if the y and z terms are well behaved, when integration is performed over $[-1,1]$, the terms on the right side involving (x^2-1) all vanish. This yields a Green's formula

$$\int_{-1}^{1} [\bar{z}L_4 y - yL_4\bar{z}]\left(\frac{\alpha}{2}\right) dx = -4\alpha(\bar{z}y' - y\bar{z}')\Big|_{-1}^{1}.$$

Further since $L_4 y(1) = 8\alpha\, y'(1)$, and $L_4 y(-1) = -8\alpha y'(-1)$,

$$[\bar{z}L_4 y - yL_4\bar{z}]\left[\frac{1}{2}\right](1) = 4\alpha(\bar{z}y' - y\bar{z}')(1),$$

and

$$[\bar{z}L_4 y - yL_4\bar{z}]\left[\frac{1}{2}\right](-1) = -4\alpha(\bar{z}y' - y\bar{z}')(-1).$$

If these expressions are added, the result is

$$\int_{-1}^{1} [\bar{z}L_4 y - yL_4\bar{z}]w(x)dx = 0.$$

It is apparent this holds for polynomials. It can be shown it holds more generally [1], [2].

If the method of Frobenius is applied about ± 1, it is found that the indicial roots are $-1, 0, 1, 2$. There are, therefore, three square summable solutions at ± 1. The limit 3 case holds at both ± 1. This implies that one boundary condition is necessary.

It is apparent that $L_4 y(\pm 1) = \pm 8\alpha y'(\pm 1) = \lambda y(\pm 1)$ is that boundary condition, and so the appropriate λ-dependent boundary value problem is

$$L_4 y = \lambda y \,,$$
$$8\alpha y'(1) = \lambda y(1) \,,$$
$$-8\alpha y'(-1) = \lambda y(-1) \,,$$

set in $L^2(-1, 1; \frac{\alpha}{2})$.

When $\alpha > 0$, an alternative can be found in which the operator has a fixed domain: We consider $H = L^2(-1, 1; \frac{\alpha}{2}) \otimes R \otimes R$. With Y in H given by $Y = y(x), y_1, y_{-1})^T$, Z in H given by $Z = (z(x), z_1, z_{-1})^T$, the inner product in H is

$$\langle Y, Z \rangle = \int_{-1}^{1} y(x)\bar{z}(x) \left(\frac{\alpha}{2}\right) dx + \frac{1}{2}y_1\bar{z}_1 + \frac{1}{2}y_{-1}\bar{z}_{-1} \,.$$

The operator A (for which the Legendre-type polynomials form a complete set of eigenfunctions) is defined as follows.

XV.5.1. Definition. We denote by D_A those elements $Y = (y(x), y_1, y_{-1})^T$ satisfying the following:

(1) $y(x)$ is in $L^2(-1, 1)$.

(2) y', y'', y''' exist and y''' is absolutely continuous.

(3) $L_4 y$ exits a.e. and is in $L^2(-1, 1)$.

(4) $y_1 = y(1)$.

(5) $y_{-1} = y(-1)$.

We define the Legendre-type operator A by setting

$$AY = \begin{pmatrix} L_4 y \\ L_4 y(1) \\ L_4 y(-1) \end{pmatrix} = \begin{pmatrix} L_4 y \\ 8\alpha y'(1) \\ -8\alpha y'(-1) \end{pmatrix}$$

for all Y in D_A.

XV. 5.2. Theorem. *The operator A is self-adjoint.*

Proof. The symmetry of A follows from Green's formula. To show self-adjointness, let Y in D_A have a first component which vanishes near ± 1.

Then $AY = (L_4 y, 0, 0)^T$. Let Z be in D_{A^*}. Then

$$\langle Y, A^* Z \rangle = \langle AY, Z \rangle = \int_{-1}^{1} z(x) L_4 y(x) \left(\frac{\alpha}{2} \right) \, dx.$$

Standard techniques [Theorem IV.4.5] can now be used to show that in $(-1, 1)$ the first component of $A^* Z$ is $L_4 z$.

Now let $y(x)$ vanish only near $x = -1$. Then

$$\langle AY, Z \rangle - \langle Y, A^* Z \rangle = -\frac{1}{2} y(1)(8\alpha z'(1) - A^* Z(1)) = 0.$$

Since $y(1)$ is arbitrary, $A^* Z(1) = 8\alpha z'(1)$.

Likewise $A^* Z(-1) = -8\alpha z'(-1)$.

This implies that $D_A = D_{A^*}$. Since the forms of A and A^* are the same, $A = A^*$.

The eigenfunctions of A, $\{(P_n^{(\alpha)}(x), P_n^{(\alpha)}(1), P_n^{(\alpha)}(-1))^T\}_{n=0}^{\infty}$ can be used to express arbitrary elements of H in an eigenfunction expansion. $\square$

XV.5.3. Theorem. *Let H_0 be the subspace of H spanned by*

$$\{(P_n^{(\alpha)}(x), P_n^{(\alpha)}(1), P_n^{(\alpha)}(-1))^T\}_{n=0}^{\infty},$$

and let $F = (f(x), f(1), f(-1))^T$ in H be orthogonal to H_0.
Let $X^n = (X^n, 1^n(-1)^n)^T$. Then $\langle F, X^n \rangle = 0$ for all $n = 0, 1, \ldots$.

Proof. X^n consists of linear combinations of the eigenfunctions.

We note that H can be identified with $L^2(-1, 1; w)$ where w was given earlier:

$$w = \frac{1}{2}[\delta(x + 1) + \delta(x - 1)] + \frac{\alpha}{2}[H(x + 1) = H(x - 1)]. \qquad \square$$

XV.5.4. Lemma. *The measure ϕ given by $\phi(E) = \int_E f(x)w(x)dx$ is of bounded variation if f is in $L^2(-1, 1; w)$.*

Proof. This follows from Schwarz's inequality since the total variation of ϕ satisfies

$$|\phi|(-1, 1) \leq \int_{-1}^{1} |f| w \, dx,$$

$$\leq \left(\int_{-1}^{1} |f|^2 w \, dx \right)^{\frac{1}{2}} \left(\int_{-1}^{1} |1|^2 w \, dx \right)^{\frac{1}{2}}. \qquad \square$$

XV.5.5. Lemma. *The Fourier transform of ϕ, $\Phi(\lambda)$ is 0.*

Proof.

$$\Phi(\lambda) = \frac{1}{\sqrt{2\pi}} \int_{-1}^{1} e^{-idx} f(x)w(x)dx$$

$$= \lim_{n\to\infty} \sum_{n=0}^{N} \frac{(-i\lambda)^n}{\sqrt{2\pi}n!} \int_{-1}^{1} f(x)x^n w(x)dx \,,$$

where the limit in N is uniform in x. Since the integrals are all 0, so is Φ. $\qquad\square$

XV.5.6. Theorem. *If F is orthogonal to H_0, then $F = 0$.*

Upon inverting the Fourier transform Φ, we see that the measure ϕ is identically 0. This implies that $f(x) = 0$ a.e. in $(-1,1)$ and $f_1 = 0$, $f_{-1} = 0$. Thus $F = 0$ in H.

XV.5.7. Corollary. *If $F = (f(x), f_1, f_{-1})^T$ is in H, then*

$$F = \sum_{n=0}^{\infty} c_n(P_n^{(\alpha)}(x), P_n^{(\alpha)}(1), P_n^{(\alpha)}(-1))^T,$$

where the Fourier coefficients c_n are given by

$$c_n = \frac{[\int_{-1}^{1} f(x)P_n^{(\alpha)}(x)\frac{\alpha}{2}dx + \frac{1}{2}f_1 P_n^{(\alpha)}(1) + \frac{1}{2}f_{-1}P_n^{(\alpha)}(-1)]}{\int_{-1}^{1} P_n^{(\alpha)}(x)^2 w \, dx}.$$

The generating function shows that $P_n^{(\alpha)}(1) = \alpha$, $P_n^{(\alpha)}(-1) = (-1)^n\alpha$.

XV.6   The Laguerre-Type Polynomials

These polynomials were also discovered by H. L. Krall in 1940 [8]. They are the second set of polynomials satisfying fourth order differential equations, but not second order differential equations. They also remained virtually hidden by their obscure place of publication for over forty years, and also serve as an example of eigenfunctions associated with a differential operator. In this case the weight measure is e^{-x} from 0 to ∞, but also has a Stieltjes jump at 0, [4].

The Laguerre-type operator is

$$L_4 y = [(x^2 e^{-x}y'')'' - (([2R+2]x+2)e^{-x}y')']/e^{-x},$$

set in $L^2(0, \infty; w)$, where

$$w = (1/R)\delta(x) + e^{-x}H(x) \quad , \quad R > 0 \,.$$

Here again $\delta(x)$ is the Dirac delta function, and $H(x)$ is the Heaviside function. w is the distributional solution of

$$S_2 = x^2 w''' + 6x\,w'' + [-x^2 + (2R+6)x + y]w' + [2Rx + 6]w = 0$$

and

$$S_4 = x^2\,w' + x^2 w = 0\,,$$

which vanishes at $\pm\infty$.

The moment equations are

$$\mu_{n-1} - (n-1)\mu_{n-2} = 0 \quad , \quad n \geq 3\,,$$
$$(2R + n + 1)\mu - (2R + [2R+6][n-1] + [n-1][n-2]\mu_{n-1}$$
$$+ (2[n-1][n-2])\mu_{n-2} = 0 \quad , \quad n \geqq 1.$$

These are easily solved to give

$$\mu_0 = (R+1)/R\,,$$
$$\mu_n = n! \quad , \quad n = 1, 2, \ldots\,,$$

which gives the distributional series

$$w = (1/R)\delta(x) + \sum_{n=0}^{\infty} (-1)^n \delta^{(n)}(x)$$

for w. This is easily connected to the other expression for w through the Fourier transform [4].

Solving the differential equations

$$L_4 y = \lambda_n y$$

for the polynomials gives

$$R_n(x) = \sum_{k=0}^{n} \frac{(-1)^k}{(k+1)!} \binom{n}{k} [k(R+n+1) + R]\,x^n\,, \qquad n = 0, 1, \ldots\,.$$

They can also be found by using the generating function

$$\frac{[R(1-t^2) - xt]}{(1-t)^3} \exp\left(\frac{-xt}{1-t}\right) = \sum_{n=0}^{\infty} R_n(x)t^k\,.$$

The polynomials satisfy a three term recurrence relation

$$R_{n+1}(x) = (A_n x + B_n)R_n - C_n\,R_{n-1}\,,$$

where

$$A_n = -[R + n + 1]/[(n+1)(R+n)]\,,$$
$$B_n = \frac{[(2n+1)R^2 + 4(n+1)(n)R + (2n+1)(n)(n+1)]}{(n+1)(R+n)^2}\,,$$

and

$$C_n = \frac{[(n)(R+n+1)^2]}{(n+1)(R+n)^2} .$$

Again orthogonality of these polynomials with respect to w follows from Green's formula and the differential equations. The norm-squares of the polynomials follow from the use of the recurrence relations, i.e.,

$$\int_0^\infty R_n(x)\, R_m(x)\, w\, dx = (R+n+1)(R+n)\delta_{mn} .$$

The appropriate differential operator in $L^2(0,\infty; w)$ for which these polynomials form a complete set of eigenfunctions may be described briefly as follows. We cite [3] for details concerning the operator and its domain.

Lagrange's formula has the appearance of

$$\bar{z}L_4y - y\overline{L_4z} = [(x^2e^{-x}y'')' - (x^2e^{-x})y''\bar{z}' - (x^2e^{-x}\bar{z}'')'y + (x^2e^{-x})\bar{z}''y'$$
$$+([2R+2]x+2)e^{-x}(y'\bar{z}-y\bar{z}')]'$$

Consequently, if y and z are well behaved, when integration is performed over $[0,\infty)$, the terms involving x^2 and x vanish at 0. Race [10] has shown that all terms vanish at $x = \infty$ (it is in the limit 2 case), and so Green's formula is

$$\int_0^\infty [\bar{z}L_4y - y\overline{L_4z}]e^{-x}dx = 2(y'\bar{z} - y\bar{z}')(0).$$

Since $L_4y(0) = -2Ry'(0)$,

$$\frac{1}{R}\,\bar{z}\,L_4y(0) = -2\,\bar{z}(0)y'(0)\,, \quad \text{and} \quad \frac{1}{R}\,yL_4z(0) = -2\,y(0)\,\bar{z}'(0).$$

If these are added to Green's formula, it becomes

$$\int_0^\infty [\bar{z}\,L_4(y) - y\,L_4\bar{z}]\,w\,dx = 0.$$

It is apparent this holds for polynomials. It can be shown that it holds more generally [3].

If the method of Frobenius is applied at $x = 0$, it is found to be in the limit 3 case, while Race [10] has shown that ∞ is in the limit 2 case, i.e., we need not apply boundary conditions at ∞.

The boundary condition $[y, x^2](0) = 0$ holds automatically.

$$= [(x^2e^{-x}y'')'\bar{z} - (x^2e^{-x})y''\bar{z}' - (x^2e^{-x}\bar{z}'')'y + (x^2e^{-x})\bar{z}''y'$$
$$+([2R+2]x+2)e^{-x}(y'\bar{z} - y\bar{z}')]$$

is the term on the right side of Lagrange's formula.

We use the other boundary terms $[y_0, 1](0)$ and $[y, x](0)$ to define a λ-dependent boundary condition

$$R[y, 1](0) = (\lambda/2)[y, x](0).$$

Since $[y, 1](0) = -2y'(0)$, $[y, x](0) = 2y(0)$ and $L_4 y(0) = -2Ry'(0)$, the boundary condition and differential equations are equivalent to

$$L_4 y = \lambda y, \qquad 0 < x < \infty,$$
$$L_4 y = \lambda y, \qquad x = 0,$$

or $L_4 y = \lambda y$ in $L^2(0, \infty; w)$.

An alternative notation is preferred. We consider $H = L^2(0, \infty; e^{-x}) \otimes R$, where Y and Z in H are given by

$$Y = (y(x), y_0)^T \quad , \quad Z = (z(x), z_0)^T,$$

and the inner product of Y and Z is

$$\langle Y, Z \rangle = \int_0^\infty y(x)\overline{z}(x)e^{-x}dx + \frac{1}{R}y_0\overline{z}_0 .$$

The operator A (for which the Laguerre-type polynomials form a complete set of eigenfunctions) is defined as follows.

XV.6.1. Definition. We denote by D_A those elements $Y = (y(x), y_0)^T$ satisfying the following:

(1) $y(x)$ is in $L^2(0, \infty; e^{-x})$.

(2) y', y'', y''' exist and y''' is absolutely continuous on $(0, \infty)$.

(3) $L_4 y$ exists a.e. and is in $L^2(0, \infty; e^{-x})$.

(4) $y_0 = \frac{1}{2}[y, x](0)$.

We define the Laguerre-type operator A by setting

$$AY = \begin{pmatrix} L_4 y \\ R[y, 1](0) \end{pmatrix}$$

for all Y in D_A.

XV.6.2. Theorem. *The operation A is self-adjoint.*

Proof. Symmetry follows from Green's formula. To show self-adjointness, let Y in D_A have a first component which vanishes near 0. Then $AY = (L_4 y, 0)^T$. Let Z be in D_A^*. Then

$$\langle Y, A^* Z \rangle = \langle AY, Z \rangle = \int_0^\infty \overline{z}(x)\, Ly(x)e^{-x}dx.$$

Standard techniques [9] now show that the first component of A^*Z is $L_4 Z$. Now let Y in D_A not necessarily vanish at 0. Then

$$\langle AY, Z \rangle - \langle Y, A^*Z \rangle = -y(0)[2\overline{z}'(0) + \frac{1}{R}\overline{(A^*Z)}(0)]$$
$$= 0.$$

Since $y(0)$ is arbitrary,
$$A^*Z(0) = -2R\, z(0).$$

This shows $D_{A^*} = D_A$ and $A^*Z = AZ$.

The eigenfunctions of A, $\{(R_n(x), R)^T\}_{n=0}^\infty$ can be used to express arbitrary elements of H in an eigenfunction expansion. $\qquad\qquad\square$

We denote by H_0 the subspace of H spanned by $\{(R_n(x), R)^T\}_{n=0}^\infty$.

XV.6.3. Theorem. *Let $F = (f(x), f_0)^T$ be orthogonal to H_0. Then $F = 0$.*

The proof is similar to that given for the Legendre-type polynomials.

XV.6.4. Corollary. *If $F = f(x), f_0)$ is in H, then*

$$F = \sum_{n=0}^\infty C_n (R_n\ R)^T,$$

where the Fourier coefficients are given by

$$C_n = \frac{[\int_0^\infty f(x) R_n(x) e^{-x} dx + \frac{1}{R} f_0 R]}{\int_0^\infty R_n(x)^2\, w\, dx}.$$

The formula for $R_n(x)$ with $x = 0$ shows that $R_n(0) = R$.

XV.6.5. Corollary. *If f is in $L^2(0\infty; e^{-x})$, then*

$$f(x) = \sum_{n=0}^\infty c_n\, R_n(x),$$

where

XV.7   The Jacobi-Type Polynomials

These represent the third set of semiclassical polynomials found by H. L. Krall in 1940 [8]. Here we have a continuous weight function $(1-x)^\alpha$ on $0 \leq x \leq 1$, $\alpha > -1$, together with a Stieltjes jump of $\frac{1}{M}$ at 0, $M > 0$.

The Jacobi-type operator is

$$L_4 y = \{([(1-x)^{\alpha+4} - 2(1-x)^{\alpha+3} + (1-x)^{\alpha+2}]y'')''$$
$$+ ([(2\alpha + 2 + 2M)(1-x)^{\alpha+2} - (2\alpha + 4 + 2M)(1-x)^{\alpha+1}]y')'\}/(1-x)^\alpha,$$

set in $L^2(0, 1; w)$, where

$$w = \frac{1}{M}\delta(x) + [H(x) - H(x - 1)](1 - x)^\alpha\,,$$

$\delta(x)$ is the Dirac delta function and $H(x)$ is the Heaviside function. w is the distributional solution of the weight equations

$$\begin{aligned}
S_2 =\,&2([\alpha + 2][2\alpha + 2 + 2M]x - 2M)w \\
&- 2(x([\alpha^2 + 9\alpha + 14 + 2M]x - [6\alpha + 12 + 2M])w)' \\
&+ 3(2x(x - 1)([\alpha + 4]x - 2)w)'' \\
&- 4((x^2 - x)^2 w)''' = 0\,,
\end{aligned}$$

and

$$S_4 = (x^2 - x)(\alpha x - 2)w - (x^2 - x)^2 w' = 0\,,$$

which vanishes at $\pm\infty$.

The moment equations are

$$(n + \alpha + 1)\mu_n - (2n + \alpha)\mu_{n-1} + (n - 1)\mu_{n-2} = 0\,, \qquad n \geq 3\,,$$

and

$$\begin{aligned}
(n + \alpha + 1)\mu_n - n\,\mu_{n-1} &= 0\,, \qquad n \geq 2\,, \\
(\alpha + 1 + M)(\alpha + 2)\mu_1 - M\,\mu_0 &= 0\,.
\end{aligned}$$

These are easily solved to give

$$\begin{aligned}
\mu_0 &= (\alpha + 1 + M)/M(\alpha + 1)\,, \\
\mu_1 &= 1/(\alpha + 1)(\alpha + 2)\,, \\
\mu_n &= n!/(\alpha + 1)_{n+1}\,, \quad n = 1, 2, \ldots.
\end{aligned}$$

The distributional weight function

$$w = \frac{1}{M}\delta(x) + \sum_{n=0}^{\infty}(-1)^n \delta^{(n)}(x)/(\alpha + 1)_{n+1}$$

is obtained from these and is equivalent to the form of w given earlier through use of the Fourier transform, [4].

Solving the differential equations

$$L_4 y = \lambda_n y$$

for polynomial solutions gives

$$S_n = \sum_{k=0}^{\infty} \frac{(-1)^{n-k}\binom{n}{k}(\alpha + 1)_{n+k}(k[n + a][n + 1] + [k + 1]M)x^k}{(k + 1)!(\alpha + 1)_n}\,,$$

$n = 0, 1, \ldots.$ (Do not confuse this S_n with the moment equations $S_2 = 0$, $S_4 = 0$.)

They can also be found by using the generating function defined as follows:
Let

$$F = ((1-x)\frac{\partial}{\partial x} + t^2\frac{\partial^2}{\partial t^2} + (\alpha+2)t\frac{\partial}{\partial t} + M),$$

and

$$G_0(x,t) = \rho^{-1}(2/[1-t+\rho])^\alpha = \sum_{n=0}^{\infty} P_n^{(\alpha,0)}(x)t^n,$$

where $P_n^{(\alpha,0)}$ are the classical Jacobi polynomials. Then

$$F \cdot G_0(x,t) = \sum_{n=0}^{\infty} S_n(x)t^n$$

is a generating function.

There is an interesting relation between the Jacobi and the Jacobi-type polynomials [4].

$$S_n = (1-x)\frac{d}{dx}P_n^{(\alpha,0)}(x) + (n^2 + [\alpha+1]n + m)P_n^{(\alpha,0)}(x).$$

This differential equation may be solved for $P_n^{(\alpha,0)}(x)$ to yield

$$P_n^{(\alpha,0)}(x) = \int_{-\infty}^{x} \frac{(1-x)^{(n^2+[\alpha+1]n+M)}S_n(\xi)}{(1-\xi)^{(n^2+[\alpha+1]n+M+1)}}\,d\xi.$$

The polynomials $\{S_n\}_{n=0}^{\infty}$ satisfy a three term recurrence relation

$$S_n(x) = (A_{n-1}x + B_{n-1})S_{n-1}(x) - C_{n-1}S_{n-2}(x),$$

where

$$A_{n-1} = \frac{(2n+\alpha)(2n+\alpha-1)(n^2+\alpha n+M)}{(n+\alpha)([n-1]^2+\alpha[n-1]+M)},$$

$$B_{n-1} = \frac{(2n+\alpha-1)P(n)}{(n+\alpha)(n)(2n+\alpha-2)([n-1]^2+\alpha[n-1]+M)^2},$$

where

$$\begin{aligned}
P(n) =&(-2)n^6 + (-6\alpha+6)n^5 + (-6\alpha^2+15\alpha-8-4M)n^4 \\
&+ (-2\alpha^3+12\alpha^2-16\alpha-8\alpha M+8M+6)n^3 \\
&+ (3\alpha^3-9\alpha^2+9\alpha-4\alpha^2 M+12\alpha M-4M-2M^2-2)n^2 \\
&+ (-\alpha^3+3\alpha^2-2\alpha+4\alpha^2 M-4\alpha M^2+2M^2)n \\
&+ (\alpha M^2),
\end{aligned}$$

and

$$C_{n-1} = \frac{(n+\alpha-1)(n-1)(2n+\alpha)(n^2+\alpha n+M)^2}{(n+\alpha)(n)(2n+\alpha-2)([n-1]^2+\alpha[n-1]+M)^2}.$$

It is possible to show that

$$\int_0^1 S_n(x)S_m(x)w\,dx = \frac{(n^2 + \alpha n + M)([n+1]^2 + [n+1]\alpha + M)}{(2n + \alpha + 1)}\,\delta_{mn}\,.$$

Again we shall highlight the appropriate differential operator for which these polynomials form a complete set of eigenfunctions. Unlike the other two cases preceding, however, Messrs. Everitt, Littlejohn and Krall ran out of "gas," and have not filled in the technical details. The problem remains open.

Lagrange's formula has the appearance of

$$\bar{z}L_4 y - y\overline{L_4 z} = [(q_2 y'')'\bar{z} - (q_2 y'')\bar{z}' - (q_2 \bar{z}'')'y + (q_2 \bar{z}'')y'$$
$$+ q_1(y'z - yz')]'\,,$$

where

$$q_2 = [(1-x)^{\alpha+4} - 2(1-x)^{\alpha+3} + (1-x)^{\alpha+2}]\,,$$
$$q_1 = [(2\alpha + 2 + 2M)(1-x)^{\alpha+2} - (2\alpha + 4 + 2M)(1-x)^{\alpha+1}]\,.$$

If the method of Frobenius is applied at $x = 1$, the indicial roots are $0, 1, -\alpha$ and $-\alpha + 1$. Three situations can arise. First, if $-1 < \alpha < 1$, there are four square integrable solutions (the limit 4 case), and so two boundary conditions are needed. If $1 \le \alpha < 3$, there are three square integrable solutions (the limit 3 case), and so only one boundary condition is needed. If $3 \le \alpha < \infty$, there are only two square integrable solutions (the limit 2 case), and no boundary conditions are needed.

In the limit 4 case the two boundary conditions are given by $\lim_{x\to 1}[y, S_0] = 0$ and $\lim_{x\to 1}[y, S_1] = 0$, where $[y, z]$ is the differentiated term on the right side of Lagrange's formula. In the limit 3 case the sole boundary condition is given by $\lim_{x\to 1}[y, S_0] = 0$. The other, here, is automatic.

If the method of Frobenius is applied at $x = 0$, the indicial roots are $-1, 0, 1, 2$. There are, therefore, three square integrable solutions and the limit 3 case holds. With one boundary condition required, the other two are used to generate a λ-dependent boundary condition at $x = 0$, equivalent to the differential equation $L_4 y = \lambda w y$ at $x = 0$. It is $L_4 y(0) = -2My'(0) = \lambda y(0)$.

The boundary value problem is then

$$L_4 y = \lambda y\,, \qquad 0 < x < 1\,,$$
$$-2My(0) = \lambda y(0)\,, \qquad \lambda = 0\,,$$

and, in the limit 4 case, $\lim_{x\to 1}[y, S_0] = 0$, $\lim_{x\to 1}[y, S_1] = 0$. In the limit 3 case, $\lim_{x\to 1}[y, S_0] = 0$. In the limit 2 case, no boundary condition is needed.

As an alternative we again consider a product space. Let $H = L^2(0, 1; (1 - x)^\alpha) \otimes R$, where Y and Z in H are given by

$$Y = (y(x), y_0)^T\,, \qquad Z = (z(x), z_0)^T\,,$$

and the inner product of Y and Z is

$$\langle Y, Z \rangle = \int_0^1 y(x)\overline{z}(x)(1-x)^\alpha dx + \frac{1}{M} y_0 \overline{z_0} .$$

The operator A for which the Jacobi-type polynomials form a complete set of eigenfunctions is defined as follows.

XV.7.1. Definition. We denote by D_y those elements $Y = (y(x), y_0)$ satisfying the following:

(1) $y(x)$ is in $L^2(0, 1; (1-x)^\alpha)$.

(2) y', y'', y''' exist and y''' is absolutely on $(0, 1)$.

(3) $L_4 y$ exists a.e. and is in $L^2(0, 1, ; 1-x)^\alpha)$.

(4) $y_0 = y(0)$.

(5) In the limit 4 case, $\lim\limits_{x \to 1}[y, S_0] = 0$, $\lim\limits_{x \to 1}[y, S_1] = 0$, while in the limit 3 case $\lim\limits_{x \to 1}[y, S_0] = 0$.

We define the Jacobi-type operator A by setting $AY = \begin{pmatrix} L_4 y \\ -2My'(0) \end{pmatrix}$ for all Y in A.

XV.7.2. Theorem. *The operator A is self adjoint.*

The proof is the same as that given for Theorem XV.6.2.

As in the previous section, we denote by H_0 spanned by the collection $\{(S_n(x), M)^T\}_0^\infty$.

XV.7.3. Theorem. *Let $F = (f, f_0)^T$ be orthogonal to H_0. Then $F = 0$.*

The proof is the same as that for the Legendre-type polynomials.

XV.7.4. Corollary. *If $F = (f(x), f_0)^T$ is in H, then*

$$F = \sum_{n=0}^\infty c_n (S_n(x), M)^T ,$$

where

$$c_n = \frac{[\int_0^1 f(x)S_n(x)(1-x)^\alpha dx + \frac{1}{M} f_0 S_n(0)]}{\int_0^1 S_n^2 \, w \, dx} .$$

This concludes the discussion of those operators generated by orthogonal polynomials satisfying fourth order differential equations. Their spectra are discrete eigenvalues. Their spectral resolutions are eigenfunction expansions. The classical problems have been known for a century, more or less, while the three new problems, now deemed semiclassical, are relatively new, different because of the Stieltjes jumps in the measures, and, as a result very interesting. We have H. L. Krall [7], [8] to thank for them.

References

[1] W. N. Everitt and L. L. Littlejohn, *Differential operators and the Legendre Type Polynomials*, Diff. and Int. Eq. **1** (1988), 97–116.

[2] W. N. Everitt, A. M. Krall and L. L. Littlejohn, *On some properties of the Legendre-type differential expression*, Quaes. Math. **13** (1990), 83–106.

[3] W. N. Everitt, A. M. Krall, L. L. Littlejohn and V. P. Onyango-Otieno, *Differential operators and the Laguerre-type polynomials*, SIAM J. Math. Anal. **23** (1992), 722–736.

[4] A. M. Krall, *Orthogonal polynomials satisfying fourth order differential equations*, Proc. Roy. Soc. Edin **87A** (1981), 271–288.

[5] ______, $M(\lambda)$ *theory for singular Hamiltonian systems with one singular point*, SIAM J. Math **20** (1989), 664–700.

[6] ______, , $M(\lambda)$ *theory for singular Hamiltonian systems with two singular points*, SIAM J. Math. **20** (1989), 701–715.

[7] H. L. Krall, *Certain differential equations for Tchebycheff polynomials*, Duke Math. J. **4** (1938), 705–718.

[8] ______, , *On orthogonal polynomials satisfying a certain fourth order differential equation*, The Pennsylvania State College Studies, No. 6, The Pennsylvania State College, State College PA, 1940.

[9] L. L. Littlejohn and A. M. Krall, *Orthogonal polynomials and higher order singular Sturm-Liouville systems* Acta Applic. **17** (1989), 97–170.

[10] D. Race, *Same strong limit -2 and Dirichlet criteria for fourth order differential expressions*, Math. Proc. Cambridge Phil. Soc. **108** (1990), 409–416.

Chapter XVI

Orthogonal Polynomials Satisfying
Sixth Order Differential Equations

We remind the reader that every even ordered formally symmetric differential operator can be rewritten as a real symmetric linear Hamiltonian system.

The embedding for a sixth order problem

$$-(p_0 y''')''' + (p_1 y'')'' - (p_2 y')' + (p_3 y) = \lambda\, wy + wf$$

as a linear Hamiltonian system is given at the beginning of Chapter IV, and is so cumbersome we will not repeat it here.

Nonetheless we can use the results to determine the boundary forms needed to impose and define boundary conditions in order to describe the two known new problems which have orthogonal polynomials as solutions.

There are, indeed, five new known sets of orthogonal polynomials satisfying differential equations of sixth order, but not lower order equations, in addition to the seven which satisfy fourth order equations as well. We shall discuss each separately. There may be more, but as of this writing, the author is unaware of them.

XVI.1   The H. L. Krall Polynomials

These polynomials were discovered by H. L. Krall about 1940, but, because of the disruption caused by World War II, were never fully explored nor were any results published. In the late 1970's he indirectly suggested to L. L. Littlejohn [1] that he investigate them for his PhD dissertation. By then all had been lost except the weight function

$$w = \frac{1}{A}\delta(x+1) + \frac{1}{B}\delta(x-1) + C[H(x+1) - H(x-1)],$$

$A, B, C > 0$. Note that if $A = B$, the weight and polynomials are the Legendre-type polynomials.

Littlejohn [1] found the sixth order differential operator is

$$L_6 y = ((x^2 - 1)^3 y''')''' - ((x^2 - 1)((-3AC - 3\,BC - 6)x^2$$
$$+ (3AC + 3BC + 18))y'')'' + [((6AC + 6BC + 12ABC^2)x^2$$
$$+ (12BC - 12AC)x - (12ABC^2 + 18AC + 18BC + 24))y']',$$

or

$$L_6 y = (x^2 - 1)^3 y^{(vi)} + 18x(x^2 - 1)^2 y(v)$$
$$+ [(3AC + 3BC + 96)x^4 + (-6AC - 6BC - 132)x^2$$
$$+ (3AC + 3BC + 36)]y^{(iv)}$$
$$+ [(24AC + 24BC + 168)x^3 + (-24AC - 24BC - 168)x]y'''$$
$$+ [(12ABC^2 + 42AC + 42BC + 72)x^2 + (12BC - 12AC)x$$
$$+ (-12ABC^2 - 30AC - 30BC - 72)]y''$$
$$+ [(24ABC^2 + 12AC + 12BC)x + (12BC - 12AC)]y'.$$

The eigenvalues are

$$\lambda_n = (24ABC^2 + 12AC + 12BC)n$$
$$+ (12ABC^2 + 42AC + 42BC + 72)n(n - 1)$$
$$+ (24AC + 24BC + 168)n(n - 1)(n - 2)$$
$$+ (3AC + 3BC + 96)n(n - 1)(n - 2)(n - 3).$$

These polynomials satisfy a three term recurrence relation

$$K_n(x) = (A_n x + B_n)K_{n-1}(x) - C_n K_{n-2}(x) = 0,$$

where

$$A_n = \frac{(2n - 1)(A(n)B(n - 1))x}{n(B(n)A(n - 1))},$$

$$B_n = \frac{(2n - 1)(2BC - 2AC)(C(n)B(n - 1))}{n(B(n)A(n - 1)^2)},$$

and

$$C_n = \frac{(n - 1)(B(n - 2)A(n)^2)}{n(B(n)A(n - 1)^2)},$$

in which

$$A(n) = n^4 + (2AC + 2BC - 1)n^2 + 4ABC^2,$$
$$B(n) = n^2 + n + AC + BC,$$
$$C(n) = -3n^4 + 6n^3 + (-2AC - 2BC - 3)n^2$$
$$+ (2AC + 2BC)n + 4ABC^2.$$

Initial values for K_n are

$$K_{-1} = 0 \quad , \quad K_0 = 2ABC^2/(AC + BC).$$

The moment equations are

$$(m - 5)[\mu_m - 3\mu_{m-2} + 3\mu_{m-4} - \mu_{m-6}]$$
$$+ [18\mu_m - 36\mu_{m-2} + 18\mu_{m-4}] = 0 \quad , \quad m \geq 5,$$
$$5(m - 3)(m - 4)(m - 5)[\mu_m - 3\mu_{m-2} + 3\mu_{m-4} - \mu_{m-6}]$$
$$- 2(m - 3)[(3AC + 3BC)\mu_m - (6AC + 6BC + 132)\mu_{m-2}$$
$$+ (3AC + 3BC + 36)\mu_{m-4}]$$
$$- [(24AC + 24BC + 168)\mu_m - (24AC + 24BC + 168)\mu_{m-2}] = 0,$$
$$m \geq 3,$$

and

$$3(m - 1)(m - 2)(m - 3)(m - 4)(m - 5)[\mu_m - 3\mu_{m-2} + 3\mu_{m-4} - \mu_{m-6}]$$
$$-(m - 1)(m - 2)(m - 3)[(3AC + 3BC + 96)\mu_m - (6AC + 6BC + 132)\mu_{m-2}$$
$$+(3AC + 3BC + 36)\mu_{m-4}]$$
$$+(m - 1)[(12ABC^2 + 42AC + 42BC + 72)\mu_m + (12BC - 12AC)\mu_{m-1}$$
$$-12ABC^2 + 30AC + 30BC + 72)\mu_{m-2}]$$
$$+[(24ABC^2 + 12AC + 12BC)\mu_m + (12BC - 12AC)\mu_{m-1}] = 0, \quad m \geq 1.$$

The moments are

$$\mu_{2m} = \frac{1}{A} + \frac{1}{B} + \frac{2C}{2m + 1} \quad , \quad \mu_{2m+1} = -\frac{1}{A} + \frac{1}{B} \quad , \quad m = 0, 1, \ldots .$$

The weight equations are

$$(x^2 - 1)w' = 0,$$
$$5(x^2 - 1)w''' + 90x(x^2 - 1)w''$$
$$+ [(-6AC - 6BC + 258)x^2 + (6AC + 6BC - 18)](x^2 - 1)w' = 0,$$

and

$$(x^2 - 1)^3 w^{(v)} + 30(x^2 - 1)^2 w^{(iv)}$$
$$+ [(-AC - BC + 268)x^2 + (AC + BC)(AC + BC - 48)](x^2 - 1)w'''$$
$$+ [(-12AC - 12BC + 816)x^3 + (12AC + 12BC - 456)x]w''$$
$$+ [(4ABC^2 - 22AC - 22BC + 672)x^2 + (4BC - 4AC)x$$
$$+ (-4ABC^2 + 2AC + 2BC - 120)]w' = 0.$$

The simultaneous solution of these equations, which vanishes at $\pm\infty$, is

$$w = \frac{1}{A}\delta(x+1) + \frac{1}{B}\delta(x-1) + C[H(x+1) - H(x-1)],$$

where δ is the Dirac delta function and H is the Heaviside function.

If the differential equations, or the recurrence relations, are solved for the polynomials K_n, $n = 0, 1, \ldots$, they are found to be

$$K_n = \sum_{j=0}^{n} \frac{[-1]^{j/2}(2n-j)!\, Q(n,j)x^{n-j}}{2^{n+1}(n - [\frac{j+1}{2}])!\, [\frac{j}{2}]!(n-j)!(n^2 + n + AC + BC)},$$

$n = 0, 1, \ldots$, where

$$Q(n,j) = \frac{1 + (-1)^j}{2}\{(n^4 + (2AC + 2BC - 1)n^2 + 4ABC^2) \\ + 2j(n^2 + n + AC + BC)\},$$

and [ ] denotes the greatest integer contained within.

We note that when $A = B$, $K_n = P_n^{(\alpha)}$, the Legendre-type polynomial of degree n, and if $C = 1$

$$\lim_{A \to \infty} K_n/A = P_n,$$

the Legendre polynomial of degree n.

If the recurrence relation is used, it is easy to see that

$$\int_{-1}^{1} K_n^2 w\, dx = C\{(n^4 + (2AC + 2BC) - 1)n^2 + 4ABC^2\} \\ \times \{(n+1)^4 + (2AC + 2BC - 1)(n+1)^2 + 4ABC^2\} \\ \times \left(2(2n+1)(n^2 + n + AC + BC)^2\right)^{-1}.$$

Finally we note that there is a generating function as well. If we let

$$M_1 f(x,t) = (AC + BC)f(x,t) + t\frac{\partial^2}{\partial t^2}(t\, f(x,t)),$$

and

$$M_2 f(x,t) = \left(\frac{AC + BC}{2}\right) - x\frac{\partial f(x,t)}{\partial x} + \frac{1}{2}t\frac{\partial^2}{\partial t^2}(t\, f(x,t)),$$

we find tediously that

$$\sum_{n=0}^{\infty}(n^2 + n + AC + BC)K_n(x)t^n \\ = (BC - AC)\,t\,(1 - 2xt + t^2)^{-3/2} \\ -(AC - BC)^2(1 - 2xt + t^2)^{-\frac{1}{2}} + M_1 \circ M_2(1 - 2xt + t^2)^{-\frac{1}{2}}.$$

If the method of Frobenius is applied at $x = \pm 1$, the indical equation is found to be

$$\rho(\rho - 1)^2(\rho - 2)(\rho - 3)(\rho + 1) = 0.$$

This indicates that there are five solutions to $L_6 y = \lambda w y$ in $L^2(-1, 1)$. Therefore two boundary conditions are needed each at ± 1. These are described by the right side term in Lagrange's formula. With

$$p_0 = -(x^2 - 1)^3,$$
$$p_1 = -((x^2 - 1)((-3AC - 3BC - 6)x^2 + (3AC + 3BC + 18))),$$
$$p_2 = ((12BC - 12AC)x - (12ABC^2 + 18AC + 18BC + 24)).$$

The Lagrange bilinear term is given by

$$\begin{aligned}
= \{&[-(p_0 y''')'' + (p_1 y'')' - (p_2 y')]\bar{z} \\
&- [-(p_0 y''')' + (p_1 y'')]\bar{z}' - (p_0 y''')\bar{z}'' \\
&+ (p_0 \bar{z}''')y'' + [-(p_0 \bar{z}''')' + (p_1 \bar{z}'')]y' \\
&+ [(p_0 z'')' + (p_2 \bar{z}')]y\}.
\end{aligned}$$

Boundary conditions are then given by (at $x = 1$)

$$\begin{aligned}
B_{11}(y) &= [y, 1](1) & &= -24[(AC + 1)y'(1) + y''(1)], \\
B_{12}(y) &= [y, (1 - x)](1) & &= -24(AC + 1)y(1), \\
B_{13}(y) &= [y, (1 - x)\log(1 - x)](1), \\
B_{14}(y) &= [y, (1 - x)^2](1) & &= 48y(1), \\
B_{15}(y) &= [y, (1 - x)^3](1),
\end{aligned}$$

and (at $x = -1$)

$$\begin{aligned}
B_{-11}(y) &= [y, 1](-1) & &= -24[(BC + 1)y')(-1) - y''(-1)], \\
B_{-12}(y) &= [y, (1 + x)](-1) & &= 24(BC + 1)y(-1), \\
B_{-13}(y) &= [y_3, (1 + x)\log(1 + x)](-1), \\
B_{-14}(y) &= [y, (1 + x)^2](-1) & &= -48y(-1), \\
B_{-15}(y) &= [y, (1 + x)^3](-1),
\end{aligned}$$

where the evaluations hold for "reasonable" functions. Further

$$L_6 y(1) = 24BC[(AC + 1)y'(1) + y''(1)]$$

and

$$L_6 y(-1) = -24AC[(BC + 1)y'(-1) - y''(-1)].$$

From this we see that

$$L_6 y(1) = \lambda y(1) \text{ and } L_6 y(-1) = \lambda y(-1)$$

are equivalent to

$$-BC[y,1](1) = -(\lambda/48)[y,(1-x^2)](1),$$
$$AC[y,1](-1) = -(\lambda/48)[y,(1+x)^2](-1).$$

Setting the two expressions for $y(1)$ and $y(-1)$ equal to each other results in

$$(AC+1)[y,(1-x)^2](1) = 2[y,(1-x)](1),$$
$$(BC+1)[y,(1+x)^2](-1) = -2[y,(1+x)](-1).$$

These four boundary conditions, together with $L_6 y = \lambda y$, $-1 < x < 1$, results in a λ- dependent boundary value problem for which the Krall polynomials are the eigenfunctions.

Just as with the Legendre-type polynomials, the problem can be set in $H = L^2(-1,1;1) \otimes R \otimes R$, where Y and Z in H are given by $Y = (y(x), y_1, y_{-1})^T$, $Z = (z(x), z_1, z_{-1})^T$, and where the inner product of Y and Z is

$$\langle Y, Z \rangle = C \int_{-1}^{1} \bar{z}(x) y(x) dx + (1/B)\bar{z}_1 y_1 + (1/A)\bar{z}_{-1} y_{-1}.$$

XVI.1.1. Definition. We denote by D_A those elements $Y = (y(x), y_1, y_{-1})^T$ satisfying the following:

(1) $y(x)$ is in $L^2(-1,1)$.

(2) $y', y'', y''', y^{(iv)}, y^{(v)}$ exist and $y^{(v)}$ is absolutely continuous.

(3) $L_6 y$ exists a.e. and is in $L^2(-1,1)$.

(4) $y_1 = y(1)$.

(5) $y_{-1} = y(-1)$.

We define the Krall operator A by setting

$$AY = \begin{pmatrix} L_6 y \\ L_6 y(1) \\ L_6 y(-1) \end{pmatrix} = \begin{pmatrix} L_6 y \\ -BC[y,1](1) \\ AC[y,1](-1) \end{pmatrix}.$$

XVI.1.2. Theorem. *The operator A is self-adjoint.*

Proof. See Theorem XV.5.2. $\qquad\qquad\qquad\qquad\qquad\qquad\qquad\qquad\qquad\qquad\qquad$ □

XVI.1.3. Theorem. *If $F = (f(x), f_1, f_{-1})^T$, then*

$$F = \sum_{n=0}^{\infty} C_n (K_n(x), K_n(1), K_n(-1))^T,$$

where

$$C_n = \frac{C \int_{-1}^{1} f(x)K_n(x)dx + (1/A)f_1 K_n(1) + (1/B)f_{-1}K_n(-1)}{\int_{-1}^{1} K_n^2 w\, dx}.$$

Unlike developers of the Legendre-type polynomials, no one has filled in the many technical details required to show that the limits employed here do in fact exist.

XVI.2   The Littlejohn Polynomials

These polynomials, discovered by L. L. Littlejohn [2] in 1986, are a generalization of the Laguerre polynomials. They are orthogonal with respect to the weight function

$$w(x) = x\, e^{-x} H(x) + (1/R)\delta(x),$$

and satisfy sixth order differential equations

$$\begin{aligned}
L_6 y = &- x^3\, y^{(vi)} + (3x^2 - 12x^2)y^{(v)} + (-3x^3 + 30x^2 - 30x)y^{(iv)} \\
&+ (x^3 - 24x^2 + 60x)y''' + (6x^2 - (36 + 6R)x)y'' \\
&+ ((6 + 6R)x - 12R)y' = \lambda_n y,
\end{aligned}$$

where

$$\lambda_n = n^3 + 3n^2 + (6R + 2)n, \qquad n = 0, 1, \dots.$$

If $L_6 y$ is multiplied by $x\, e^{-x}$, the symmetric form is found.

$$\begin{aligned}
x\, e^{-x} L_6 y = &-(x^4 e^{-x} y''')''' + 6((x^3 + x^2)e^{-x}y'')'' \\
&- (([6 + 6R]x^2 + 12x + 12)e^{-x}y')' \\
&= x\, e^{-x} \lambda_n y.
\end{aligned}$$

The polynomials satisfy a three term recurrence relation

$$L_n(x) = (A_n x + B)L_{n-1}(x) - C_n L_{n-2}(x),$$

where

$$A_n = \frac{-(n^2 + n + 2R)}{n(n^2 - n + 2R)},$$

$$B_n = \frac{\{2n^4 + (8R - 2)n^2 + (8R^2 - 8R)}{(n^2 - n + 2R)^2},$$

$$C_n = \frac{(n^2 + n + 2R)^2}{(n^2 - n + 2R)^2}.$$

Initial values for L_n are

$$L_{-1} = 0 \quad, \quad L_0 = 2R.$$

There are three moment equations, which we leave to the reader. The moments themselves are early found via the weight function: $\mu_0 = (R+1)/R$, $\mu_n = (n+1)!$, $n > 0$.

Likewise there are three weight equations for which w, previously given, is the common solution.

No one has actually derived the polynomials themselves. There is a generating function for them, however,

$$L_n(x) = \frac{e^x}{n!} D^n \{(n^2 + n + 2R)e^{-x}x^n + (-n^3 + n^2 - 2Rn)e^{-x}x^{n-1}$$
$$+ (-2n^3 + 2n)e^{-x}x^{n-2}\}\,.$$

If we set

$$p_0 = x^4 e^{-x}\,,$$
$$p_1 = 6(x^3 + x^2)e^{-x}\,,$$
$$p_2 = ([6 + 6R]x^2 + 12x + 12)\,,$$

and

$$= \{[-(p_0 y''')'' + (p_1 y'')' - (p_2 y')]\bar{z}$$
$$- [-(p_0 y''')' + (p_1 y'')]\bar{z}' - (p_0 y''')\bar{z}''$$
$$+ (p_0 \bar{z}''')y'' + E(p_0 \bar{z}''')' + (p_1 \bar{z})'']y'$$
$$+ [(p_0 \bar{z}'')' + (p_2 z')]y\}\,,$$

as in the previous section, then boundary conditions are:

At $x = 0$, the limit 4 case holds. Hence $[y, x^2] = 0$, and $[y, x^3] = 0$ automatically.

$$= 12y'(0)\,.$$
$$[y, x] = -12y(0)\,.$$

Since $L_6 y(0) = 12Ry'(0)$, we impose the λ dependent boundary condition

$$R[y, 1](0) = -\lambda[y, x](0)\,,$$

which is equivalent to $L_6 y(0) = \lambda y(0)$. At $x = \infty$, the limit 3 case holds, and no boundary conditions are required.

The boundary value problem is, therefore,

$$L_6 y = \lambda y \quad , \quad 0 \leq x < \infty\,.$$

There is an embedding of the problem in $H = L^2(0, \infty; xe^{-x}) \times R$, where, if $Y = (y(x), y_0)^T$ and $Z = (z(x), z_0)$ are in H, then

$$\langle Y, Z \rangle = \int_0^\infty y(x)\bar{z}(x)xe^{-x}dx + \frac{1}{R}y_0 \bar{z}_0\,.$$

The Littlejohn operator A, for which the Littlejohn polynomials form a complete set of eigenfunctions, is defined as follows.

XVI.2.1. Definition. We denote by D_A those elements Y in H satisfying the following:

(1) y is in $L^2(0, \infty; x\, e^{-x})$.

(2) $y, y', y'', y''', y^{(iv)}, y^{(v)}$ exist and $y^{(v)}$ is absolutely continuous.

(3) $L_6 y$ exists a.e. and is in $L^2(0, \infty; x\, e^{-x})$.

(4) $y_0 = -(1/12)\,[y, x](0)$.

We define the Littlejohn operator A by setting $AY = \begin{pmatrix} L_6 y \\ R[y, 1](0) \end{pmatrix}$.

XVI.2.2. Theorem. *The operator A is self-adjoint.*

XVI.2.3. Theorem. *Let $\{p_n\}_{n=0}^{\infty}$ denote the Littlejohn polynomials, and let their embedding in H be denoted by $\{(p_n(x), p_n(0))^T\}_{n=0}^{\infty}$. If $F = (f(x), f_0)^T$ is in H, then*

$$F = \sum_{n=0}^{\infty} c_n (p_n(x), p_n(0))^T$$

where

$$c_n = \frac{\int_0^{\infty} f(x) p_n(x) x\, e^{-x} dx + \frac{1}{R} f_0 p_n(0)}{\int_0^{\infty} p_n(x)^2 w\, dx}.$$

As is true with the Krall polynomials problem, no one has filled in the many technical details required to show that the limits employed do in fact exist.

XVI.3 The Second Littlejohn Polynomials

In 1994 [4] R. Koekoek exhibited yet a third sixth order differential equation with orthogonal polynomial solutions. He attributes these to L. L. Littlejohn. The weight function is

$$w(x) = \frac{3}{4}(1 - x^2)[H(x+1) - H(x-1)] + M\delta(x+1) + M\delta(x-1).$$

The differential equation is

$$-\frac{1}{36}M(1 - x^2)^3 y^{(vi)} + \frac{2}{3}Mx(1 - x^2)^2 y^{(v)}(x)$$

$$+\frac{5}{3}M(1 - x^2)(1 - 3x^2)y^{(iv)}(x) - \frac{40}{3}Mx(1 - x^2)y'''(x)$$

$$-(10M + 1)(1 - x^2)y''(x) + 4xy'(x) = \lambda_n y(x),$$

where

$$\lambda_n = \frac{1}{36}n(n+3)[(n-1)(n+1)(n+2)(n+4)M + 36], \qquad n = 0, 1, \ldots.$$

These are a special case of a more general situation. No further details are given. Obviously there is an embedding in $H = L^2(-1, 1; \frac{3}{4}(1 - x^2)) \otimes R \otimes R$, but until boundary conditions are developed, little else can be said.

We leave further detail to the reader.

XVI.4   Koekoek's Generalized Jacobi Type Polynomials

In the same article [4] which gives Littlejohn's sixth order differential equations, Koekoek gives two examples in a more general setting which, with the parameters set appropriately, give two additional differential equations associated with weights

$$w(x) = \frac{\Gamma(5/2)}{2^{3/2}\Gamma(2)\Gamma(\frac{1}{2})}(1 - x)(1 + x)^{-\frac{1}{2}}[H(x + 1) - H(x - 1)] + 2M\delta(x - 1),$$

and

$$w(x) = \frac{\Gamma(7/2)}{2^{5/2}\Gamma(2)\Gamma(\frac{3}{2})}(1 - x)(1 + x)^{-\frac{1}{2}}[H(x + 1) - H(x - 1)] + N\delta(x - 1).$$

We refer the reader to the next chapter where we discuss the more general setting.

References

[1]   L. L. Littlejohn, *The Krall polynomials: A new class of orthogonal polynomials*, Quaes. Math. **5** (1982), 255–265.

[2]   ______, *An application of a new theorem on orthogonal polynomials and differential equations*, Quaes. Math. **10** (1986), 49–61.

[3]   L. L. Littlejohn and A. M. Krall, *Orthogonal polynomials and higher order singular Sturm-Liouville systems*, Acta Applic. **17** (1989), 97–170.

[4]   R. Koekoek, *Differential equations for symmetric generalized ultraspherical polynomials*, Trans. Amer. Math. Soc. **345** (1994), 47–72.

Chapter XVII

Orthogonal Polynomials Satisfying
Higher Order Differential Equations

There are various possible iterative schemes available to produce differential equations of any even order. Multiples of the second order operators, fourth order operators and sixth order operators will generate a host of examples [10], but these "do not count" in a sense. Indeed their spectral resolutions all look like

$$A - \int \lambda^n dE(\lambda)$$

where $n > 1$, and $E(\lambda)$ is the spectral measure for the lowest order operator.

There appear to be two major generalizations of the previous chapters which lead to orthogonal polynomials satisfying any even order operator. (There may be additional sets, but as this is written they remain to be discovered.)

One is a generalization of the Jacobi, Jacobi-type, Legendre-type and Krall polynomials. The other is a generalization of the Laguerre, Laguerre-type and Littlejohn polynomials. We label these two sets the generalized Jacobi-type polynomials and the generalized Laguerre-type polynomials. Their beginnings lie in the paper of Koornwinder [8]. The differential equations were found by Koekoek and Koekoek [5], symmetrized by Everitt, Littlejohn and Wellman [4], and in the paper by Koekoek [7].

XVII.1   The Generalized Jacobi-Type Polynomials

Koornwinder [8] considered polynomials that are orthogonal with respect to the generalized weight function

$$w(x) = \frac{\Gamma(\alpha + \beta + 2)}{2^{\alpha+\beta+1}\Gamma(\alpha + 1)\Gamma(\beta + 1)}(1 - x)^\alpha(1 + x)^\beta[H((x + 1)$$
$$- H(x - 1)] + M\delta(x + 1) + N\delta(x - 1).$$

He showed that the polynomials are given by

$$P_n^{\alpha,\beta,M,N}(x) = ((\alpha+\beta+1)_n/n!)^2[(\alpha+\beta+1)^{-1}(B_nM(1-x)$$
$$- A_nN(1+x))\frac{d}{dx} + A_nB_n]P_n^{\alpha,\beta}(x) \quad , \quad n = 0,1,\ldots$$

where $\{P_n^{\alpha,\beta}(x)\}_{n=0}^{\infty}$ are the Jacobi polynomials,

$$A_n = \frac{(1+\alpha)_n n!}{(\beta+1)_n(\alpha+\beta+1)_n} + \frac{n(n+\alpha+\beta+1)M}{(\beta+1)(\alpha+\beta+1)} \, ,$$

and

$$B_n = \frac{(1+\beta)_n n!}{(\alpha+1)_n(\alpha+\beta+1)_n} + \frac{n(n+\alpha+\beta+1)N}{(\alpha+1)(\alpha+\beta+1)} \, .$$

R. Koekoek states the following in [5].

XVII.1.1. Theorem. *For all $\alpha > -1$, $\beta > -1$, $M \geq 0$, $N \geq 0$ the generalized Jacobi-type polynomials $\{P^{\alpha,\beta,M,N}(x)\}_{n=0}^{\infty}$ satisfy differential equations of infinite order of the form*

$$M\sum_{i=0}^{\infty} a_i(x)y^{(i)}(x) + N\sum_{i=0}^{\infty} b_i(x)y^{(i)}(x) + MN\sum_{i=0}^{\infty} C_i(x)y^{(i)}(x)$$
$$+(1-x^2)y''(x) + [(\beta-\alpha) - (\alpha+(\beta+2)x]y'(x) + n(n+\alpha+\beta+1)y(x) = 0 \,,$$

where

$$y(x) = P_n^{\alpha,\beta,M,N}(x) = A_0 P_n^{(\alpha,\beta)}(x) + [A_1(1-x) - A_2(1+x)]\frac{d}{dx}P_n^{(\alpha,\beta)}(x),$$

and

$$A_0 = 1 + M\frac{\binom{n+\beta}{n-1}\binom{n+\alpha+\beta+1}{n}}{\binom{n+\alpha}{n}} + N\frac{\binom{n+\alpha}{n-1}\binom{n+\alpha+\beta+1}{n}}{\binom{n+\beta}{n}}$$
$$+ MN\frac{(\alpha+\beta+2)^2}{(\alpha+1)(\beta+1)}\binom{n+\alpha+\beta+1}{n-1}^2 \,,$$

$$A_1 = \frac{M}{(\alpha+\beta+1)}\frac{\binom{n+\beta}{n}\binom{n+\alpha+\beta}{n}}{\binom{n+\alpha}{n}} + \frac{MN}{(\alpha+1)}\binom{n+\alpha+\beta}{n-1}\binom{n+\alpha+\beta+1}{n} \,,$$

$$A_2 = \frac{N}{\alpha+\beta+1}\frac{\binom{n+\alpha}{n}\binom{n+\alpha+\beta}{n}}{\binom{n+\beta}{n}} + \frac{MN}{(\beta+1)}\binom{n+\alpha+\beta}{n-1}\binom{n+\alpha+\beta+1}{n} \,.$$

There is no proof given. (The author is sure one will be forthcoming in the near future.) Implied is that when α and β are integers, the order of the differential equation is finite.

In the special case $\beta = \alpha$, $N = M$, there is considerable simplification.

XVII.1.2. Theorem. *Let*

$$y(x) = P_n^{\alpha,\alpha,M,M}(x) = C_0 P_n^{(\alpha,\alpha)}(x) + C_1 x \frac{d}{dx} P_n^{(\alpha,\alpha)}(x)$$

where

$$C_0 = 1 + M\frac{2n}{(\alpha+1)}\binom{n+2\alpha+1}{n} + 4M^2\binom{n+2\alpha+1}{n-1}^2,$$

$$C_1 = \frac{2M}{(2\alpha+1)}\binom{n+2\alpha}{n} + \frac{2M^2}{(\alpha+1)}\binom{n+2\alpha}{n}\binom{n+2\alpha+1}{n}.$$

The polynomials $\{P_n^{\alpha,\alpha,M,M}(x)\}_{n=0}^{\infty}$ satisfy the infinite order differential equations

$$M\sum_{i=0}^{\infty} C_i(x)y^{(i)}(x) + (1-x^2)y'' - 2(\alpha+1)xy' + n(n+2\alpha+1)y(x) = 0,$$

$n = 0, 1, \ldots$, where

$$C_0(x) = 4(2\alpha+3)\binom{n+2\alpha+2}{n-2} \quad, \quad n = 0, 1, \ldots,$$

and $C_i(x) - (2\alpha+3)(1-x^2)C_i^*(x) \quad, \quad i = 1, 2, \ldots$ where

$$C_1^*(x) = 0,$$

$$C_i^*(x) = \frac{2^i}{i!}\sum_{k=0}^{i-2}\binom{\alpha+1}{i-k-2}\binom{i-2\alpha-5}{k}\left(\frac{1-x}{2}\right)^k, \quad i = 2, 3, 4, \ldots.$$

When α is a nonnegative integer, the order of the equation is finite, $2\alpha+4$.

The proof, found in [5], is exceedingly long and tedious. We leave it as an "exercise."

Koekoek [7] also has details when $\beta = \pm\frac{1}{2}$, $M = 0$, (α and N are arbitrary), which follow from the previous case.

XVII.1.3. Theorem. *The polynomials $\{P_n^{\alpha,\frac{1}{2},O,N}(t)\}_{n=0}^{\infty}$ satisfy the infinite order differential equations*

$$N\sum_{j=0}^{\infty} d_i^*(t)y^{(j)}(t) + (1-t^2)y''(t)$$

$$-\frac{1}{2}[(2\alpha+1)+(2\alpha+3)t]y'(t) + \frac{1}{2}n(n+2\alpha+1)y(t) = 0,$$

$$n = 0, 1, \ldots,$$

where

$$d_0^* = \frac{1}{2}(2n+3)\binom{2n+2\alpha+2}{2n-2},$$

$$d_j^*(t) = \frac{1}{8}d_j\left(\sqrt{\frac{1+t}{2}}\right),$$

$$d_j(x) = \sum_{i=j}^{2j} \frac{i!2^{3j-i}}{(2j-i)!(i-j)!}c^{2j-i}C_i(x),$$

$$2x^2 - 1 = t,$$

and $C_i(x)$ are the coefficients in Theorem XVII.1.2.

When α is a nonnegative integer, the order of the differential equations is $2\alpha + 4$.

The proof, found in [7] is again a long, tedious computation. It is based on the fact

$$\frac{P_{2n}^{\alpha,\alpha,M,M}(x)}{P_{2n}^{\alpha,\alpha,M,M}(1)} = \frac{P_n^{\alpha,\frac{1}{2},0\,2M}(2x^2-1)}{P_n^{\alpha,\frac{1}{2},0,2M}(1)}.$$

XVII.1.4. Theorem. *The polynomials $\{P_n^{\alpha\frac{1}{2}ON}(t)\}_{n=0}^{\infty}$ satisfy the infinite order differential equations*

$$N\sum_{j=0}^{\infty} e_j^*(t)y^{(j)}(t) + (1-t^2)y''(t)$$

$$-\frac{1}{2}[(2\alpha-1)+(2\alpha+5)t]y'(t) + \frac{1}{2}n(2n+2\alpha+3)y(t) = 0,$$

$$n = 0, 1, \ldots,$$

where

$$e_0^* = \frac{1}{2}\binom{2n+2\alpha+3}{2n-1},$$

$$e_j^*(t) = \frac{1}{8(2\alpha+3)}e_j\left(\sqrt{\frac{1+t}{2}}\right),$$

$$e_j(x) = \sum_{i=j}^{2j+1} \frac{(i+1)!2^{3j-i}}{(2j-i+10!(i-j)!}x^{2j-i}C_i(x),$$

$$2x^2 - 1 = t,$$

and $C_i(x)$ are the coefficients in theorem XVII.1.2.

When α is a nonnegative integer, the order of the differential equations is $2\alpha + 4$.

Remark. Bavinck [1] has generalized these results in part. He has found the polynomials and the differential equations that are orthogonal with respect to

$$\langle f,g\rangle = \frac{\Gamma(2\alpha+2)}{2^{2\alpha+1}\Gamma(\alpha+1)^2}\int_{-1}^{1} f(x)g(x)(1-x^2)^\alpha dx$$
$$+M[f(-1)g(-1)+f(1)g(1)]+N[f'(-1)g'(-1)+f'(1)g'(1)].$$

The term involving N is new. The order of the differential equation is

$$\begin{array}{rcl}
2 & \text{if} & M=N=0\,, \\
2\alpha+4 & \text{if} & M>0, N=0\,, \\
2\alpha+8 & \text{if} & M=0, N>0\,, \\
4\alpha+10 & \text{if} & M>0, N>0\,.
\end{array}$$

The weight function generates a Sobolev space, which is different from those discussed in chapter.

Further results may also be found in [3].

XVII.2   The Generalized Laguerre-Type Polynomials $\{L_n^{\alpha,M}(x)\}_{n=0}^\infty$

Koornwinder [8] also considered briefly the polynomials, which are orthogonal with respect to the generalized weight function

$$w(x) = \frac{1}{\Gamma(\alpha+1)}x^\alpha e^{-x}H(x) + M\,\delta(x).$$

In 1991 there followed the remarkable paper by Koekoek and Koekoek [5] which showed that these polynomials, named $\{L_n^{\alpha,M}(x)\}_{n=0}^\infty$ satisfied differential equations of infinite order, but when α is a nonnegative integer the order is $2\alpha+4$.

They showed that

$$L_n^{\alpha,M}(x) = \left[1+M\binom{n+\alpha}{n-1}\right]L_n^{(\alpha)}(x)+M\binom{n+\alpha}{n}\frac{d}{dx}L_n^{(\alpha)}(x)\,,\qquad n=0,1,\ldots,$$

where $\{L_n^{(\alpha)}(x)\}_{n=0}^\infty$ are the classical Laguerre polynomials.

XVII.2.1. Theorem. *For all $\alpha>-1$, $M>0$ the generalized Laguerre-type polynomials $\{L_n^{\alpha,M}(x)\}_{n=0}^\infty$ satisfy differential equations of infinite order of the form*

$$N\sum_{i=0}^\infty a_i(x)y^{(i)}(x)+xy'(x)+(\alpha+1-x)y'(x)+ny(x)=0\,,$$

$n = 0, 1, \ldots,$ *where the coefficients are*

$$\alpha_0 = \binom{n + \alpha + 1}{n - 1},$$

$$\alpha_1(x) = -x,$$

$$\alpha_2(x) = -\frac{1}{2}(\alpha + 1)x^2 + \frac{1}{2}(\alpha + 2)(\alpha + 3)x,$$

$$\alpha_3(x) = -\frac{1}{12}\alpha(\alpha + 1)x^3 + \frac{1}{6}(\alpha + 1)(\alpha + 2)(\alpha + 3)x^2$$
$$- \frac{1}{12}(\alpha + 1)(\alpha + 2)(\alpha + 3)(\alpha + 4)x,$$

and in general

$$\alpha_i(x) = \frac{1}{i!}\sum_{j=1}^{i}(-1)^{i+j+1}\binom{\alpha + 1}{j - 1}\binom{a + 2}{i - j}(\alpha + 3)_{i-j}x^i,$$

$i = 1, 2, 3, \ldots.$

The proof is quite tedious and sheds little light, so we again leave it as an "exercise."

The symmetric forms of the differential equations are available when α is a nonnegative integer [4]. Multiplied by $x^\alpha e^{-x}$, they become

$$\sum_{k=0}^{\alpha+2}(-1)^k (b_{k\alpha N}(x)\, y^{(k)}(x))^{(k)} = 0,$$

where $b_{0\alpha N} = \left(-N\binom{n+\alpha-1}{n-1} - n\right)x^\alpha e^{-x},$

$$b_{k\alpha n} = \sum_{r=2k-2}^{\alpha+k}\left(\frac{(\alpha + 1)!\, N}{(k - 1)!\, k!(r - 2k + 2)!} + \delta_{k,1}\delta_{\alpha+1,r}\right)x^r e^{-x},$$

$k = 1, 2, \ldots, \alpha + 2,$ where $\delta_{y,z}$ is the Kronecker delta function defined by

$$\delta_{y,z} = \begin{cases} 1 & \text{if } y = z, \\ 0 & \text{if } y \neq z. \end{cases}$$

Unfortunately the authors of [4] were unable to give a proof of their statement.

XVII.3　The Generalized Laguerre-Type Polynomials $\{L_n^{2(1/R)}(x)\}_{n=0}^{\infty}$

Before Koekoeks' general result had been found, Littlejohn [9] had very nicely found the eighth order differential equations associated with the weight

$$w(x) = x^2 e^{-x} H(x) + \frac{1}{R}\,\delta(x).$$

(The author's name appears on the article, but the work is all Littlejohn's.) As the ultimate example of the trend of the past fifteen years, it is worth looking at.

The differential equations are

$$L_8 y = x^4 y^{(8)} + (-4x^4 + 24x^3)y^{(7)} + (6x^4 - 84x^3 + 168x^2)y^{(6)}$$
$$+ (-4x^4 + 108x^3 - 504x^2 + 336x)y^{(5)} + (x^4 - 60x^3 + 540x^2 - 840x)y^{(4)}$$
$$+ (12x^3 - 240x^2 + 720x)y^{(3)} + (36x^2 - (48R + 240)x)y''$$
$$+ ((24 + 48R)x - 144R)y' = (n^4 + 11n^2 + (6 + 48R)n)y, \quad n = 0, 1, \ldots.$$

The polynomials are

$$\phi_n(x) = \frac{(n+2)!}{2} \sum_{k=0}^{n} \frac{(-1)^k [12R(k+3) + (n+1)(n+2)(n+3)]x^k}{k!(k+3)(n-k)!},$$

and have norms and inner product given by

$$\int_0^{\infty} \phi_n(x)\phi_m(x)x^2 e^{-x}dx + \frac{1}{k}\phi_n(0)\phi_m(0)$$
$$= \frac{(N+1)(n+2)[12R + (n+1)(n+2)(n+3)][12R + n(n+1)(n+2)]}{4} \delta_{nm}.$$

There is a three term recurrence relation,

$$\phi_{-1} = 0, \quad \phi_1 = 6R,$$
$$\phi_n(x) = \phi_{n-1}(x)\left[\frac{-[12R + n(n+1)(n+2)]}{n[12R + (n-1)n(n+1)]}x\right.$$
$$+ \left.\frac{[24R(n-1)n(n+1)(n+2) + 144R^2(2n+1) + (n-1)n^2(n+2)(2n+1)]}{n[12R + (n-1)n(n+1)]^2}\right]$$
$$- \phi_{n-2}(x)\frac{(n+1)[12R + n(n+1)(n+2)]^2}{n[(12R + n-1)n(n+1)]^2}.$$

The equation

$$L_8 y = \lambda y$$

is in the limit 6 case at $x = 0$ and in the limit 4 (?) case at ∞. Consequently two boundary conditions are needed at 0. Letting $[\cdot, \cdot]'$ denote the right side of Lagrange's identity, we find as boundary terms

$$(0) = -144y'(0),$$
$$[y, x](0) = 144y(0),$$
$$L_8 y(0) = 144Ry'(0) = \lambda y(0).$$

Hence one boundary condition at $x = 0$ is

$$144R[y, 1] = \lambda[y, x](0).$$

The other is, we think, but are not sure,

$$\lim_{x\to 0}[y, x^2](x) = 0.$$

We define $H = L^2(0, \infty; x^2 e^{-x}) \otimes R$, (Here R denotes the real line.), where elements of H are given by $Y = (y(x), y_0)^T$ and $Z = (z(x), z_0)^T$, and the inner product on H is

$$\langle Y, Z \rangle = \int_0^\infty y(x)\overline{z(x)}x^2 e^{-x} + \frac{1}{R}y_0\overline{z_0}.$$

XVII.3.1. Definition. We denote by D_A those elements $Y = (y(x), y_0)^T$ in H satisfying:

(1) y is in $L^2((0, \infty' x^2 e^{-x})$.

(2) $y^{(i)}(x)$, $i = 0, \ldots, 7$ exist and $y^{(7)}$ is absolutely continuous on compact subsets of $[0, \infty)$.

(3) $L_8 y$ exists a.e. and is in $L^2(0, \infty; x^2 e^{-x})$.

(4) $y_0 = [y, x](0)$.

(5) $\lim_{x\to 0}[y, x^2](x) = 0$.

We define the operator A by setting $AY = \binom{L_8 y(x)}{144R[y, 1](0)}$.

XVII.3.2. Theorem. *The operator A is self-adjoint.*

XVII.3.3. Theorem. *If $F = (f(x), f_0)^T$, then*

$$F = \sum_{n=0}^\infty \frac{c_n(\phi_n(x), \phi_n(0))^T}{\int_{-\infty}^\infty \phi_n(x)^2 w\, dx},$$

where c_n is the usual coefficient

Remark. J. Koekoek, R. Koekoek and H. Barinck [6] have generalized these results. They have found the polynomials and differential equations that are orthogonal with respect to

$$\langle f, g \rangle = \frac{1}{\Gamma(\alpha + 1)} \int_{-\infty}^\infty x^\alpha e^{-x} f(x)g(x)dx$$
$$+ M f(0)g(0) + N f'(0)g'(0).$$

The term involving N is new. The order of the differential equation is

$$
\begin{array}{lll}
2\alpha + 4 & \text{if} & M > 0,\ N = 0, \\
2\alpha + 8 & \text{if} & M = 0,\ N > 0, \\
4\alpha + 10 & \text{if} & M > 0,\ N > 0.
\end{array}
$$

The weight function generates a Sobolev space, which is different from those discussed in the chapters to come.

Further results may also be found in [2].

References

[1] H. Bavinck, *Differential operators having Sobolev-type Gegenbauer polynomials as eigenfunctions*, J. Comp. Appl. Math., to appear.

[2] ______, *Differential operators having Sobolev-type Laguerre polynomials as eigenfunctions* Proc. Amer. Math. Soc. **125** (1997), 3561–3567.

[3] H. Bavinck and J. Koekoek, *Differential operators having symmetric orthogonal polynomials as eigenfunctions.*

[4] W. N. Everitt, L. L. Littlejohn and R. Wellman, *The symmetric form of Koekoeks' Laguerre-type differential equations*, J. Comp. Appl. Math., **57** (1995), 115–121.

[5] J. Koekoek and r. Koekoek, *On a differential equation for Koornwinder's generalized Laguerre polynomials*, Proc. Amer. Math. Soc. **4** (1991), 1045–1054.

[6] J. Koekoek, R. Koekoek and H. Bavinck, *On differential equations for Sobolev-type Laguerre polynomials*, Trans. Amer. Math. Soc. **350** (1998), 347–393.

[7] R. Koekoek, *Differential equations for symmetric generalized ultra spherical polynomials*, Trans. Amer. Math. Soc. **345** (1994), 47–72.

[8] T. H. Koornwinder, *Orthogonal polynomials with a weight function $(1 - x)^{\alpha}(1 + x)^{\beta} + M\delta(x) = 1 + N\delta(x - 1)$*, Canad. Math. Bull. **27** (1984), 205–214.

[9] L. L. Littlejohn and A. M. Krall, *Sturm-Liouville operators and orthogonal polynomials*, CMS Conf. Proc. **8** (1986), 247–260.

[10] L. L. Littlejohn, *The Krall polynomials as solutions to a second order differential equation*, Canad. Math. Bull. **26** (1983), 410–417.

Chapter XVIII

Differential Operators in Sobolev Spaces

Perhaps the subject of this chapter was motivated by the solution of the unitless equation of vibrating motion

$$y'' + y = 0.$$

Of course everyone who has ever taken a course in elementary differential equations can immediately write the solution as

$$y = A\sin(x + B),$$

where A and B are arbitrary constants.

Or the other hand, many physicists prefer a different approach. If the differential equation is multiplied by y', the result

$$y'y'' + yy' = 0$$

is immediately integrable, giving

$$E \equiv \frac{1}{2}(y')^2 + \frac{1}{2}(y)^2 = \frac{1}{2}A^2,$$

where E is the total, kinetic plus potential, energy of the system.

Solving for y', we have

$$\frac{y'}{\sqrt{A^2 - y^2}} = 1.$$

If $y = A\sin\theta$, this reduces to $\frac{d\theta}{dx} = 1$. Thus $\theta = x + B$ and $y = A\sin(x + B)$.

The point of all this is that the total energy of the system

$$E = \frac{1}{2}(y')^2 + \frac{1}{2}(y^2)$$

is exhibited. This term can arise in a very natural way by using the *left* side of the standard Sturm–Liouville operator, and its integral generates a norm new to boundary value problems. The Sobolev spaces they generate therefore also have differential operators which turn out to be self-adjoint, as well as a host of orthogonal polynomial examples.

The theory we are about to consider is sometimes called "left-definite" because the norm is derived from the left side of the differential equation. That designation, however, is slowly disappearing.

XVIII.1   Regular Second Order Sobolev Boundary Value Problems [2]

Consider the boundary value problem over the finite interval (a, b),

$$-(py')' + qy = \lambda\, w\, y + wf\,,$$

$$\begin{pmatrix} \alpha_{11} & \alpha_{12} \\ \alpha_{21} & \alpha_{22} \end{pmatrix} \begin{pmatrix} y(a) \\ py'(a) \end{pmatrix} + \begin{pmatrix} \beta_{11} & \beta_{12} \\ \beta_{21} & \beta_{22} \end{pmatrix} \begin{pmatrix} y(b) \\ py'(b) \end{pmatrix} = 0\,,$$

where $1/p, q, w > 0$ are continuous and rank $\begin{pmatrix} \alpha_{11} & \alpha_{12} & \beta_{11} & \beta_{12} \\ \alpha_{21} & \alpha_{22} & \beta_{21} & \beta_{22} \end{pmatrix} = 2$. Also assume the standard self-adjointness requirement for problems set in $L^2(a, b; w)$,

$$\begin{pmatrix} \alpha_{11} & \alpha_{12} \\ \alpha_{21} & \alpha_{22} \end{pmatrix} \begin{pmatrix} 0 & -1 \\ 1 & 0 \end{pmatrix} \begin{pmatrix} \alpha_{11} & \alpha_{21} \\ \alpha_{12} & \alpha_{22} \end{pmatrix} = \begin{pmatrix} \beta_{11} & \beta_{12} \\ \beta_{21} & \beta_{22} \end{pmatrix} \begin{pmatrix} 0 & -1 \\ 1 & 0 \end{pmatrix} \begin{pmatrix} \beta_{11} & \beta_{21} \\ \beta_{12} & \beta_{22} \end{pmatrix}\,.$$

(The α's and β's are real).

XVIII.1.1. Definition. We denote by D those elements y in $L^2(a, b; w)$ which satisfy:

(1) y is absolutely continuous.

(2) $ly = [(-by')' + qy]/w$ is in $L^2(a, b; w)$.

(3) $\begin{pmatrix} \alpha_{11} & \alpha_{12} \\ \alpha_{21} & \alpha_{22} \end{pmatrix} \begin{pmatrix} y(a) \\ py'(b) \end{pmatrix} + \begin{pmatrix} \beta_{11} & \beta_{12} \\ \beta_{21} & \beta_{22} \end{pmatrix} \begin{pmatrix} y(b) \\ py'(b) \end{pmatrix} = 0.$

We define the operator L by setting $Ly = ly$ for all y in D.

XVIII.1.2. Theorem. *L is self-adjoint. Its spectrum consists only of eigenvalues with ∞ as their only limit. The associated eigenfunctions form a complete orthogonal set.*

This, of course, is all well known.

Consider now the expression

$$ly = [(-py')' + qy]/w\,.$$

If the inner product

$$\int_a^b (ly)\bar{z}\, w\, dx = \int_a^b [(-py')' + qy]\bar{z}\, dx$$

is integrated by parts in the first term,

$$\langle ly, z\rangle = -py'\overline{z}\big|_a^b + \int_a^b [py'\overline{z}' + qy\overline{z}]dx\,.$$

The question arises can $py'\overline{z}\big|_a^b$ be reevaluated using the boundary conditions so that $\langle \cdot, \cdot\rangle$ is an inner product?

Rearranging the boundary conditions and introducing a minus sign with the $py'(b)$ term, we have

$$\begin{pmatrix} \alpha_{11} & \beta_{11} \\ a_{21} & \beta_{21} \end{pmatrix} \begin{pmatrix} y(a) \\ y(b) \end{pmatrix} = \begin{pmatrix} -\alpha_{12} & \beta_{12} \\ -\alpha_{22} & \beta_{22} \end{pmatrix} \begin{pmatrix} py'(a) \\ -py'(b) \end{pmatrix} = \begin{pmatrix} 0 \\ 0 \end{pmatrix}\,.$$

In trying to solve for $\begin{pmatrix} py'(a) \\ py'(b) \end{pmatrix}$, we must consider the coefficient matrix

$$\begin{pmatrix} -\alpha_{12} & \beta_{12} \\ -\alpha_{22} & \beta_{22} \end{pmatrix}\,.$$

Three situations can arise. The matrix may be nonsingular. The matrix may be singular but nonzero. The matrix may be zero.

The Nonsingular Case

Without loss of generality we may assume that $\alpha_{22}\beta_{12} - \alpha_{12}\beta_{22} = 1$. This simplifies the formulas. We solve for $\begin{pmatrix} py'(a) \\ py'(b) \end{pmatrix}$.

$$\begin{pmatrix} py'(a) \\ -py'(b) \end{pmatrix} = \begin{pmatrix} \alpha_{11}\beta_{22} - \alpha_{21}\beta_{12} & \beta_{11}\beta_{22} - \beta_{12}\beta_{22} \\ \alpha_{11}\alpha_{22} - \alpha_{12}\alpha_{21} & \beta_{11}\alpha_{22} - \beta_{21}\alpha_{12} \end{pmatrix} \begin{pmatrix} y(a) \\ y(b) \end{pmatrix}\,.$$

Inserting this in

$$(-py'\overline{z})\big|_a^b = (\overline{z}(a)\overline{z}(y)) \begin{pmatrix} py'(a) \\ -py'(b) \end{pmatrix}\,,$$

we have

$$\langle ly, z\rangle = \int_a^b [py'\overline{z}' + qy\overline{z}]dx$$

$$+(\overline{z(a)}, \overline{z(b)}) \begin{pmatrix} \alpha_{11}\beta_{22} - \alpha_{21}\beta_{12} & \beta_{11}\beta_{22} - \beta_{12}\beta_{21} \\ \alpha_{11}\alpha_{22} - \alpha_{12}\alpha_{21} & \beta_{11}\alpha_{22} - \beta_{21}\alpha_{12} \end{pmatrix} \begin{pmatrix} y(a) \\ y(b) \end{pmatrix}\,.$$

The matrix is symmetric by the self-adjointness criterion. If it is positive, then the right side above serves to define a Sobolev inner product, which we denote by $\langle \cdot, \cdot\rangle_{H^1}$.

The Singular Nonzero Case

When $\begin{pmatrix} -\alpha_{12} & \beta_{12} \\ -\alpha_{22} & \beta_{22} \end{pmatrix}$ is singular but not zero, the matrix rows are linearly dependent. There exists a number k such that $k\alpha_{12} = \alpha_{22}$, $k\beta_{12} = \beta_{22}$. Row manipulation in the rearranged boundary conditions gives

$$\begin{pmatrix} \alpha_{11} & \beta_{11} \\ \alpha_{21} - k\alpha_{11} & \beta_{21} - k\beta_{11} \end{pmatrix} \begin{pmatrix} y(a) \\ y(b) \end{pmatrix} - \begin{pmatrix} -\alpha_{12} & \beta_{12} \\ 0 & 0 \end{pmatrix} \begin{pmatrix} py'(a) \\ -py'(b) \end{pmatrix} = \begin{pmatrix} 0 \\ 0 \end{pmatrix}.$$

Assume without loss of generality that $\alpha_{12}^2 + \beta_{12}^2 = 1$. Define y_a, y_b, y_a', y_b' by

$$\begin{pmatrix} -\alpha_{12} & \beta_{12} \\ \beta_{12} & \alpha_{12} \end{pmatrix} \begin{pmatrix} y(a) \\ y(b) \end{pmatrix} = \begin{pmatrix} y_a \\ y_b \end{pmatrix} ,$$

$$\begin{pmatrix} -\alpha_{12} & \beta_{12} \\ \beta_{12} & \alpha_{12} \end{pmatrix} \begin{pmatrix} py'(a) \\ -py'(b) \end{pmatrix} = \begin{pmatrix} y_a' \\ y_b' \end{pmatrix} .$$

Then, since the coefficient matrix is unitary,

$$\begin{pmatrix} -\alpha_{12} & \beta_{12} \\ \beta_{12} & \alpha_{12} \end{pmatrix} \begin{pmatrix} y_a \\ y_b \end{pmatrix} = \begin{pmatrix} y(a) \\ y(b) \end{pmatrix} ,$$

$$\begin{pmatrix} -\alpha_{12} & \beta_{12} \\ \beta_{12} & \alpha_{12} \end{pmatrix} \begin{pmatrix} y_a' \\ y_b' \end{pmatrix} = \begin{pmatrix} py'(a) \\ -py'(b) \end{pmatrix} .$$

The coefficient matrix becomes

$$\begin{pmatrix} -\alpha_{11}\alpha_{12} + \beta_{11}\beta_{12} & \alpha_{11}\beta_{12} + \beta_{11}\alpha_{12} \\ -\alpha_{12}[\alpha_{21} - k\alpha_{11}] + \beta_{12}[\beta_{21} - k\beta_{11}] & \beta_{12}[\alpha_{21} - k\alpha_{11}] + \alpha_{12}[\beta_{21} - k\beta_{11}] \end{pmatrix} \begin{pmatrix} y_a \\ y_b \end{pmatrix}$$

$$- \begin{pmatrix} 1 & 0 \\ 0 & 1 \end{pmatrix} \begin{pmatrix} y_a' \\ y_b' \end{pmatrix} = \begin{pmatrix} 0 \\ 0 \end{pmatrix} .$$

This gives two constraints. If we denote the first matrix by $\begin{pmatrix} P & Q \\ R & S \end{pmatrix}$, then

$$P y_a + Q y_b = y_a' \quad , \quad R y_a + S y_b = 0 .$$

The boundary terms

$$(\bar{z}(a)\,\bar{z}(b)) \begin{pmatrix} py'(a) \\ -py'(b) \end{pmatrix} = (\bar{z}(a)\,\bar{z}(b)) \begin{pmatrix} -\alpha_{12} & \beta_{12} \\ \beta_{12} & \alpha_{12} \end{pmatrix} \begin{pmatrix} -\alpha_{12} & \beta_{12} \\ \beta_{12} & \alpha_{12} \end{pmatrix} \begin{pmatrix} y_a' \\ y_b' \end{pmatrix} ,$$

$$= (\bar{z}_a \bar{z}_b) \begin{pmatrix} y_a' \\ y_b' \end{pmatrix} ,$$

$$= \bar{z}_a y_a' + \bar{z}_b y_k' .$$

Since substitution can only be made for y_a', we must require that the coefficient of $y_b' = 0$. Thus we restrict ourselves to a subspace where $z_b = 0$ (and $y_b = 0$). The boundary terms become $\bar{z}_a(P y_a + Q y_b)$, but since $y_b = 0$, we are left with $\bar{z}_a P y_a$.

The self-adjointness criterion

$$\alpha_{11}\alpha_{22} - \alpha_{12}\alpha_{21} = \beta_{11}\beta_{22} - \beta_{12}\beta_{21}\,,$$

together with the equations $ka_{12} = a_{22}$ and $kb_{12} = b_{22}$, yield

$$\alpha_{12}(\alpha_{11}k - \alpha_{21}) = \beta_{12}(\beta_{11}k - \beta_{21})\,.$$

Hence there is a parameter j such that

$$j\alpha_{12} = \beta_{11}k - \beta_{21}\quad,\quad j\beta_{12} = \alpha_{11}k - \alpha_{21}\,.$$

If these are used to simplify R, it is found that $R = 0$. The $R - S$ constraint vanishes. Thus we have a Dirichlet formula

$$\langle ly, z\rangle = \int_a^b [py'\overline{z}' + qy\overline{z}]dx$$
$$+(-\alpha_{12}\overline{z}(a) + \beta_{12}\overline{z}(b))(-\alpha_{11}\alpha_{12} + \beta_{11}\beta_{12})(-\alpha_{12}y(a) + \beta_{12}y(b)),$$

where y and z satisfy

$$\beta_{12}y(a) + \alpha_{12}y(b) = 0\quad,\quad \beta_{12}z(a) + \alpha_{12}z(b) = 0\,.$$

Again the right side serves to define a Sobolev inner product, again denoted by $\langle\cdot,\cdot\rangle_{H^1}$. It acts only on elements y and z which lie in the subspace generated by the constraint generated by the coefficients α_{12} and β_{12}.

The Singular Zero Case

In this case $\begin{pmatrix} -\alpha_{12} & \beta_{12} \\ -\alpha_{22} & \beta_{22} \end{pmatrix}$ is the zero matrix. The boundary conditions are reduced to

$$\begin{pmatrix} \alpha_{11} & \beta_{11} \\ \alpha_{21} & \beta_{21} \end{pmatrix}\begin{pmatrix} y(a) \\ y(b) \end{pmatrix} = \begin{pmatrix} 0 \\ 0 \end{pmatrix}\,.$$

Since the coefficient matrix has rank 2, $y(a) = 0$ and $y(b) = 0$. If z also satisfies these constraints, then

$$\langle ly, z\rangle = \int_a^b [py'\overline{z}' + qyz]dx\,.$$

The boundary terms disappear. Here again the right side generates a Sobolev norm denoted by $\langle\cdot,\cdot\rangle_{H^1}$. It acts only on functions which vanish at a and b.

In every case, therefore, we have

$$\langle ly, z\rangle = \langle y, z\rangle_{H^1}\,.$$

XVIII.1.2. Definition. We denote by H^1 the Sobolev space generated by the inner product $\langle\cdot,\cdot\rangle_{H^1}$ in each of the three cases.

XVIII.1.3. Definition. We denote by $\mathcal{D}$ those elements y in H^1 which satisfy:

  (1) y is absolutely continuous on closed subintervals of $[a, b]$.

  (2) py' is absolutely continuous on closed subintervals of $[a, b]$.

  (3) $ly = [-(py')' + qy]/w$ exists $\mathcal{L}y = ly$ for all y in $\mathcal{D}$.

We remind the reader that the matrices $A = (\alpha_{ij})$ and $B = (\beta_{ij})$ satisfy $AJA^* = BJB^*$, where $J = \begin{pmatrix} 0 & -1 \\ 1 & 0 \end{pmatrix}$.

We assume that $q > \epsilon w$ for some $\epsilon > 0$.

XVIII.1.4. Theorem. *L, acting in $L^2(a, b; w)$, is bounded below by ϵ.*

Proof. $\langle Ly, y \rangle_{L^2} = \langle y, y \rangle_{H^1} > \epsilon \langle y, y \rangle_{L^2}$.

This implies that $\langle (L - \epsilon I)y, y \rangle_{L^2} > 0$. $\square$

XVIII.1.5. Corollary. *L^{-1} exists and is given by a Green's function*

$$L^{-1}f(x) = \int_a^b G(x, \xi) f(\xi) w(\xi) d\xi \,.$$

L^{-1} is bounded by $1/\epsilon$.

Proof. Let $Ly = f$, $L^{-1}f = y$, and

$$\langle f, L^{-1}f \rangle_{L^2} \geq \epsilon \langle L^{-1}f, L^{-1}f \rangle_{L^2} \,.$$

Apply Schwarz's inequality on the left to get

$$\|L^{-1}f\|_{L^2} \leq (1/\epsilon)\|f\|_{L^2} \,.$$

We can now say more about $\mathcal{L}$. $\square$

XVIII.1.6. Theorem. *$\mathcal{L}$ is symmetric.*

Proof. The Dirichlet formula

$$\langle Ly, z \rangle_{L^2} = \langle y, z \rangle_{H^1}$$

holds for y in $\mathcal{D}$, z in H^1. Assume that z is also in $\mathcal{D}$ and replace z by $\mathcal{L}z$. Then

$$\langle Ly, Lz \rangle_{L^2} = \langle y, \mathcal{L}z \rangle_{H^1}.$$

But this also implies that

$$\langle Ly, Lz \rangle_{L^2} = \langle Ly, z \rangle_{H^1} \,.$$ $\square$

XVIII.1.7. Theorem. *$\mathcal{L}^{-1}$ exists and is bounded.*

Proof. We can solve $\mathcal{L}y = f$ by the same methods as $Ly = f$. The same Green's function also generates $\mathcal{L}^{-1}$. Dirichlet's formula yields

$$\langle f, L^{-1}f \rangle_{L^2} = \langle \mathcal{L}^{-1}f, \mathcal{L}^{-1}f \rangle_{H^1}\,.$$

Apply Schwarz's inequality on the left,

$$\|\mathcal{L}^{-1}f\|_{H^1}^2 < f\|_{L^2}\|f\|_{L^2}(1/\epsilon) \leq (1/\epsilon)^2\|f\|_{H^1}^2\,.$$

So $\|\mathcal{L}^{-1}\|_{H^1} \leq (1/\epsilon)$. $\qquad\qquad\qquad\qquad\qquad\qquad\qquad\qquad\square$

XVIII.1.8. Theorem. *$\mathcal{L}$ is self-adjoint in H^1.*

Proof. The range of $\mathcal{L}$ is all of H^1. So $\mathcal{L}$ is maximally extended. This says $\mathcal{L}$ is self-adjoint [1, p. 85]. $\qquad\qquad\qquad\qquad\qquad\qquad\qquad\qquad\qquad\square$

Finally, we can say a great deal about the (point) spectrum of $\mathcal{L}$ and its spectral resolution.

XVIII.1.9. Theorem. *The spectrum of $\mathcal{L}$ consists of the same eigenvalues as L, $\{\lambda_i\}_1^\infty$, with the same eigenfunctions $\{y_i\}_{i=1}^\infty$. Since $\|y_i\|_{H^1}^2 = \lambda_i\|y_i\|_{L^2}^2$, $i = 1, 2, \ldots$. They must be renormalized. These eigenfunctions form a complete orthogonal set in H^1.*

We leave the proof to the reader.

XVIII.2   Regular Sobolev Boundary Value Problems for Linear Hamiltonian Systems [3]

As one might expect, the results of the previous section can be extended to regular Hamiltonian systems. The same situations arise, so in order not to bore the reader we will give only the highlights.

If the reader recalls the problem is concerned with the $2n$-dimensional system

$$JY' = [\lambda\mathcal{A} + \mathcal{B}]Y + \mathcal{A}F\,,$$

with boundary conditions

$$AY(a) + BY(b) = 0\,.$$

Here $\mathcal{A} \geq 0$ and $\mathcal{B}$ are symmetric $n \times n$ matrices,

$$J = \begin{pmatrix} 0 & -I \\ I & 0 \end{pmatrix}$$

is an $n \times n$ matrix, as are A and B, which satisfy

$$AJA^* = BJB^*\,.$$

Y, F, Z, G are n-dimensional vectors.

XVIII.2.1. Definition. We denote by D those elements Y in $L^2_{\mathcal{A}}(a,b)$ satisfying:

(1) Y is absolutely continuous on $[a,b]$.

(2) $ly = JY' - \mathcal{B}Y = \mathcal{A}F$ exists a.e. and is in $L^2_{\mathcal{A}}(a,b)$.

(3) $AY(a) + BY(b) = 0$.

We define the operator L by setting $LY = F$ for all F in D when $JY' - \mathcal{B}Y = \mathcal{A}F$.

XVIII.2.2. Theorem. *L is self-adjoint. The spectrum of L consists only of eigenvalues, which are in a complete orthonormal set in $L^2_{\mathcal{A}}(a,b)$. If $F(x)$ is in $L^2_{\mathcal{A}}(a,b)$, then*

$$F(x) = \sum_{i=1}^{\infty} Y_i(x) \int_a^b Y_i^*(\xi)\mathcal{A}(\xi)F(\xi)d\xi \,,$$

where $\{Y_i\}_{i=1}^{\infty}$ are the eigenfunctions associated with the eigenvalues $\{\lambda_i\}_{i=1}^{\infty}$.

For elements $Y(x)$ in D,

$$LY(x) = \sum_{i=1}^{\infty} \lambda_i Y_i(x) \int_a^b Y_i^*(\xi)\mathcal{A}(\xi)F(\xi)d\xi \,.$$

We now assume that $\mathcal{A} = \begin{pmatrix} E & 0 \\ 0 & 0 \end{pmatrix}$, where the components are all $n \times n$ matrices. We likewise decompose $\mathcal{B} = \begin{pmatrix} -B_{11} & B_{12} \\ B_{12}^* & B_{22} \end{pmatrix}$ into $n \times n$ matrices and assume $-B_{11} \leq 0 \leq B_{22}$. Further we assume $\rho E \leq B_{11}$ for some $\rho > 0$. Lastly we decompose Y into $\begin{pmatrix} Y_1 \\ Y_2 \end{pmatrix}$, Z into $\begin{pmatrix} Z_1 \\ Z_2 \end{pmatrix}$, where the components are $n \times 1$ matrices.

We use the notation of Schneider and Niessen [6], [7]. Note that

$$\langle Y, Z \rangle_{\mathcal{A}} = \int_a^b Z_1^* E Y_1 d\xi$$

$$= \int_a^b Z^* \mathcal{A} Y \, d\xi$$

generates $L^2_{\mathcal{A}}(a,b)$. Secondly, we can introduce a Sobolev space through the inner product

$$\langle Y, Z \rangle_{\mathcal{B}} = \int_a^b [Z_1^* B_{11} Y_1 + Z_2^* B_{22} Y_2] d\xi$$

$$= \int_a^b Z^* \begin{pmatrix} B_{11} & 0 \\ 0 & B_{22} \end{pmatrix} Y \, d\xi \,.$$

In $L^2_{\mathcal{A}}(a,b)$ the inner product of LY and Z is given by

$$\langle LY, Z \rangle_{\mathcal{A}} = \int_a^b Z^* \mathcal{A}(LY) d\xi$$

$$= \int_a^b Z^* \mathcal{A} F d\xi$$

$$= \int_a^b (Z_1^* Z_2^*) \begin{pmatrix} E & 0 \\ 0 & 0 \end{pmatrix} \begin{pmatrix} F_1 \\ F_2 \end{pmatrix} d\xi$$

$$= \int_a^b Z_1^* E\, F_1 d\xi \,.$$

If $JY' - \mathcal{B}Y = \mathcal{A}F$ is decomposed into components, we find

$$-Y_2' + B_{11}Y_1 - B_{12}Y_2 = EF_1 \,,$$
$$Y_1' - B_{12}^* Y_1 - B_{22}Y_2 = 0 \,.$$

We continue with the inner product.

$$\langle LY, Z \rangle_{\mathcal{A}} = \int_a^b Z_1^* \{ -Y_2' + B_{11}Y_1 - B_{12}Y_2 \} d\xi$$

$$= -Z_1^* Y_2 \big|_a^b + \int_a^b \{ Z_1^{*'} Y_2 + Z_1^* B_{11}Y_1 - Z_1^* B_{12}Y_2 \} d\xi$$

$$= -Z_1 Y_2 \big|_a^b + \int_a^b [Z_1^* B_{11}Y_1 + Z_2^* B_{22}Y_2 \} d\xi \,,$$

provided we define the second component of Z through the equation

$$Z_1' - B_{12}^* Z_1 = B_{22}Z_2 \,.$$

This component does not contribute to the $L^2_{\mathcal{A}}(a,b)$ norm, and so we lose nothing by doing so.

We have

$$\langle LY, Z \rangle_{\mathcal{A}} = -Z_1^* Y_2 \big|_a^b + \langle Y, Z \rangle_{\mathcal{B}} \,.$$

We must expand on the first term by using the boundary condition $AY(a) + BY(b) = 0$. It can be written as

$$(A_1 A_2 B_1 - B_2) \begin{pmatrix} Y_1(a) \\ Y_2(a) \\ Y_1(b) \\ -Y_2(b) \end{pmatrix} = 0 \,,$$

or as

$$(A_1 B_1 A_2 - B_2) \begin{pmatrix} Y_1(a) \\ Y_1(b) \\ Y_2(a) \\ -Y_2(b) \end{pmatrix} = 0 \,.$$

If we set $(A_1 B_1) = M$, $(-A_2 B_2) = N$,

$\begin{pmatrix} Y_1(a) \\ Y_1(b) \end{pmatrix} = y_1$, $\begin{pmatrix} Y_2(a) \\ -Y_2(b) \end{pmatrix} = y_2$, we have

$$M y_1 - N y_2 = 0.$$

The rank of $M : N$ is $2n$. Let the rank of N be j, $0 \le j \le 2n$. The sizes of A_1, A_2, B_1, B_2 are all $2n \times n$. The sizes of M and N are $2n \times 2n$.

We solve $M y_1 - N y_2 = 0$ for as many of the y_2 components as possible. There are unitary matrices P and Q such that

$$M y_1 - M y_2 = M P P^* y_1 - N P P^* y_2$$

$$= \begin{pmatrix} M_1 \\ M_2 \end{pmatrix} P^* y_1 - \begin{pmatrix} N_1 \\ 0 \end{pmatrix} P^* y_2 = 0,$$

and

$$Q \begin{pmatrix} M_1 \\ M_2 \end{pmatrix} P^* y_1 - Q \begin{pmatrix} N_1 \\ 0 \end{pmatrix} P^* y_2 = 0,$$

$$\begin{pmatrix} M_{11} & M_{12} \\ M_{21} & M_{22} \end{pmatrix} \begin{pmatrix} y_{11} \\ y_{12} \end{pmatrix} - \begin{pmatrix} N_{11} & 0 \\ 0 & 0 \end{pmatrix} \begin{pmatrix} y_{21} \\ y_{22} \end{pmatrix} = 0,$$

where N_{11} is nonsingular and, of course,

$$MP = \begin{pmatrix} M_{11} \\ M_{12} \end{pmatrix} \quad , \quad NP = \begin{pmatrix} N_1 \\ 0 \end{pmatrix},$$

$$Q \begin{pmatrix} M_{11} \\ M_{12} \end{pmatrix} = \begin{pmatrix} M_{11} & M_{12} \\ M_{21} & M_{22} \end{pmatrix} \quad , \quad Q \begin{pmatrix} N_1 \\ 0 \end{pmatrix} = \begin{pmatrix} N_{11} & 0 \\ 0 & 0 \end{pmatrix},$$

$$P^* y_1 = \begin{pmatrix} y_{11} \\ y_{12} \end{pmatrix} \quad , \quad P^* y_2 = \begin{pmatrix} y_{21} \\ y_{22} \end{pmatrix}.$$

We see, therefore that the boundary condition has been decomposed into two parts

$$y_{21} = N_{11}^{-1}(M_{11} y_1 + M_{12} y_2) = N_{11}^{-1} M_1 P^* y_1,$$

and

$$M_{21} y_{11} + M_{22} y_{12} = M_2 P^* y_1 = 0.$$

We shall use the first, then we shall show that the second vanishes automatically. In the boundary term of the Dirichlet formula

$$-Z_1 Y_2 \Big|_a^b = (Z_1^*(a) Z_1^*(b)) \begin{pmatrix} Y_2(a) \\ -Y_2(b) \end{pmatrix}$$

$$= z_1^* y_2 = (P^* z_1)^* (P^* y_2)$$

$$= (z_{11}^* z_{12}^*) \begin{pmatrix} y_{21} \\ y_{22} \end{pmatrix}$$

$$= z_{11}^* (N_{11}^{-1}(M_{11} M_{12})) \begin{pmatrix} y_{11} \\ y_{12} \end{pmatrix} + z_{12}^* y_{22}$$

$$= z_{11}^* (N_1^{-1} M_{11}) y_{11} + z_{11}^* N_{11}^{-1} M_{12} y_{12} + z_{12}^* y_{22}.$$

We have no control over y_{22}. We therefore require the constraint $y_{12} = 0$, $z_{12} = 0$ be put into effect. Then

$$-Z_1^* Y_2 \Big|_a^b = z_{11}^* N_{11}^{-1} M_{11} y_{11}$$

$$= (z_{11}^* z_{12}^*) \begin{pmatrix} N_{11}^{-1} M_{11} & 0 \\ 0 & 0 \end{pmatrix} \begin{pmatrix} y_{11} \\ y_{12} \end{pmatrix}$$

$$= z_1^* P \begin{pmatrix} N_{11}^{-1} M_{11} & 0 \\ 0 & 0 \end{pmatrix} P^* y_1 .$$

We must assume $N_{11}^{-1} M_{11} \geq 0$. We shall show it is symmetric.

Next consider the constraint $M_2 P^* y_1 = 0$. Since

$$A = (M - N) \begin{pmatrix} I & 0 \\ 0 & 0 \\ 0 & I \\ 0 & 0 \end{pmatrix} \quad , \quad B = (M - N) \begin{pmatrix} 0 & 0 \\ I & 0 \\ 0 & 0 \\ 0 & -I \end{pmatrix} ,$$

we see that $A J A^* = B J B^*$ is equivalent to

$$M N^* = N M^* .$$

Inserting the unitary matrix P,

$$(MP)(NP)^* = (NP)(MP)^* .$$

This is the same as

$$\begin{pmatrix} M_1 \\ M_2 \end{pmatrix} \begin{pmatrix} N_1 \\ 0 \end{pmatrix}^* = \begin{pmatrix} N_1 \\ 0 \end{pmatrix} \begin{pmatrix} M_1 \\ M_2 \end{pmatrix}^* .$$

Multiply on the left by Q on the right by Q^*, and

$$\begin{pmatrix} M_{11} & M_{12} \\ M_{21} & M_{22} \end{pmatrix} \begin{pmatrix} N_{11}^* & 0 \\ 0 & 0 \end{pmatrix} = \begin{pmatrix} N_{11} & 0 \\ 0 & 0 \end{pmatrix} \begin{pmatrix} M_{11}^* & M_{21}^* \\ M_{12}^* & M_{22}^* \end{pmatrix} .$$

This yields

$$M_{11} N_{11}^* = N_{11} M_{11}^* \quad \text{and} \quad N_{11}^{-1} M_{11} = (N_{11}^{-1} M_{11})^* ,$$

the symmetry promised, and

$$M_{21} N_{11}^* = 0 .$$

Since N_{11} is nonsingular, $M_{21} = 0$.

The second boundary constraint is thus $M_{22} y_{12} = 0$. Hence the boundary conditions are

$$\begin{pmatrix} M_{11} & M_{12} \\ 0 & M_{22} \end{pmatrix} \begin{pmatrix} y_{11} \\ y_{12} \end{pmatrix} - \begin{pmatrix} N_{11} & 0 \\ 0 & 0 \end{pmatrix} \begin{pmatrix} y_{21} \\ y_{22} \end{pmatrix} = 0 .$$

Since the rank of $\begin{pmatrix} M_{11} & M_{12} & N_{11} & 0 \\ 0 & M_{22} & 0 & 0 \end{pmatrix}$ is $2n$, M_{22} is nonsingular. This shows that $y_{12} = 0$.

In summary, the Dirichlet formula is

$$\langle LY, Z \rangle_A = \langle Y, Z \rangle_B + z_1^* P \begin{pmatrix} N_{11} & 0 \\ 0 & 0 \end{pmatrix} P^* y \,,$$

where $P^* y_1 = \begin{pmatrix} y_{11} \\ y_{12} \end{pmatrix}$, $P^* z_1 = \begin{pmatrix} z_{12} \\ z_{22} \end{pmatrix}$. The right side acts as a Sobolev inner product on a subspace determined by the constraint $y_{12} = z_{12} = 0$. This is the setting we seek.

We now extend the operator theory to the Sobolev space.

We assume that if $\|Y\|_A = 0$, then $\|Y\|_B = 0$ as well. This is true of second order scalar embeddings.

XVIII.2.3. Definition. We denote by $\mathcal{D}$ those elements Y in H^1, the Sobolev space generated by $\langle \cdot, \cdot \rangle_B$ satisfying:

(1) $lY = JY' - \mathcal{B}Y = \mathcal{A}F$, where F is in H^1.

(2) $AY(a) + BY(b) = 0$.

We define the operator $\mathcal{L}$ by setting $\mathcal{L}Y = F$ for all Y in $\mathcal{D}$.

In order for $\mathcal{L}$ to be well defined, we must assume that whenever $\|Y\|_B = 0$, then $\|F\|_B = 0$ as well. This is more subtle then might first seem. We have little control over the matrices B_{11} and B_{22}. In the scalar embedding of fourth order operators

$$(py'')'' = (qy')' + ry = \lambda wy + wf \,,$$
$$B_{11} = \begin{pmatrix} r & 0 \\ 0 & q \end{pmatrix} \quad , \quad B_{22} = \begin{pmatrix} 0 & 0 \\ 0 & 1/p \end{pmatrix} \,.$$

It is quite possible for either or both B_{11} and B_{12} to be singular. Here the assumption holds because of the details concerning the embeddings.

XVIII.2.4. Theorem. *$\mathcal{L}$ is self-adjoint. The spectrum of $\mathcal{L}$ consists of the same eigenvalues as L with the same eigenfunctions renormed. The spectral expansions of L also hold for $\mathcal{L}$.*

See [3] for details.

XVIII.3   Singular Second Order Sobolev Boundary Value Problems [4], [5]

As we saw in Section 1 of this chapter, it was necessary to somehow incorporate the term $-py'\bar{z}\big|_a^b$. This was done by using boundary conditions. Since these are really not available for singular problems, something else must be done [4], [5].

The setting to begin with is $L^2(a, b; w)$, the weight w appearing in the differential equation

$$(-py')' + qy = \lambda wy,$$

where p, q, w are positive, measurable functions over (a, b). The inner product is given by

$$\langle f, g \rangle_{L^2} = \int_a^b f\, \overline{g}\, w\, dx\,.$$

Secondly we are interested in the Sobolev space H^1 with inner product

$$\langle f, g \rangle_{H^1} = \int_a^b [pf'\overline{g}' + qf\overline{g}]dx.$$

We assume that $p > 0$ and $q > 0$ are continuous and that further there exist ϵ_1 and $\epsilon_2 > 0$ so that $\epsilon_1 w \le q \le \epsilon_2 w$. We further assume that a and b are singular points for

$$(-py')' + qy = \lambda wy.$$

In view of the ϵ-assumptions, $H^1 \subset L^2$.

To make a long story short, we will show that in virtually all the interesting cases $\lim\limits_{x \to a} py'\overline{z} = 0$, and $\lim\limits_{x \to b} py'\overline{z} = 0$, and we can ignore the boundary term in the Dirichlet formula.

In order to proceed we need to modify the Weyl procedure, which shows there is an $L^2(c, b; w)$ solution and an $L^2(a, c; w)$ solution of $-(py')' + qy = \lambda wy$, λ complex, to show instead there is a solution in $H^1\langle c, b, p, q \rangle$ and $H^1\langle a, c, p, q \rangle$, $a < c < b$.

XVIII.3.1. Theorem. *Under the assumptions above with λ having a nonzero imaginary past, $-(py')' + qy = \lambda wy$ has a solution $\psi_b(x, \lambda)$ in $H^1\langle c, b, p, q \rangle$, and $\psi_a(x, \lambda)$ in and $H^1\langle a, c, p, q \rangle$ with inner products*

$$\langle y, z \rangle_{H^1\langle c, b, p, q \rangle} = \int_c^b [p\overline{y}'z' + qy\overline{z}]dx\,,$$

and

$$\langle y, z \rangle_{H^1\langle a, c, p, q \rangle} = \int_a^c [p\overline{y}'z' + qy\overline{z}]dx\,,$$

respectively.

Proof. If the Dirichlet formula,

$$\int_c^{b'} [-py' + qy]\overline{y}dx = \lambda \int_c^{b'} |y|^2 w\, dx$$

is integrated by parts, we find

$$\int_c^{b'} [p|y'|^2 + q|y|^2]dx - p\overline{y}y\Big|_c^{b'} = \lambda \int_c^{b'} |y|^2 w\, dx\,.$$

The term $py'\overline{y}$ is troublesome. It is convenient to try to make it vanish at b'. We do so by ultimately requiring $p(b')y(b') = 0$. Before pursing that direction, however, let us examine the situation a bit more thoroughly.

If we impose a general boundary condition at b',

$$\cos\delta y(b') + \sin\delta y'(b') = 0$$

for some real δ, then the solution of $-(py')' + qy = \lambda wy$ satisfying it must be of the form

$$\psi(x,\lambda) = \theta(x,\lambda) + m(\lambda)\phi(x,\lambda),$$

where θ and ϕ are likewise solutions satisfying for some real γ,

$$\theta(c,\lambda) = \cos\gamma, \quad p(c)\theta'(c,\lambda) = -\sin\gamma,$$
$$\phi(c,\lambda) = \sin\gamma, \quad p(c)\phi'(c,\lambda) = \cos\gamma.$$

Note that $W[\theta,\phi] = p[\theta\phi' - \theta'\phi] \equiv 1$, and that

$$\sin\gamma\theta(c,\lambda) + \cos\gamma p(c)\theta'(c,\lambda) = 0,$$
$$\cos\gamma\phi(c,\lambda) - \sin\gamma p(c)\phi'(c,\lambda) = 0.$$

We can easily see that if $\mathrm{Im}\,\lambda \neq 0$,

$$m(\lambda) = \frac{\cos\delta\theta(b',\lambda) + \sin\delta p(b)\theta'(b',\lambda)}{\cos\delta\phi(b',\lambda) + \sin\delta p(b')\phi'(b',\lambda)}\,.$$

The denominator cannot be zero, for if it were, λ would be a complex eigenvalue for the regular Sturm–Liouville problem with boundary conditions

$$\cos\gamma y(c) - \sin\gamma p(c)y'(c) = \theta,$$
$$\cos\delta y(b') + \sin\delta p(b')y'(b') = 0.$$

Further note that $\tan\delta$ is given by

$$\tan\delta = -\frac{\theta(b',\lambda) + m(\lambda)\phi(b',\lambda)}{p(b)(\theta'(b',\lambda) + m(\lambda)\phi'(b',\lambda))}\,.$$

We will need this shortly.　　　　　　　　　　　　　　　　　　　　　$\square$

We know that as δ varies over real values from 0 to π, $m(\lambda)$ describes a circle in the complex plane. As b' approaches b the circles contract, either approaching a limit circle or a limit point. If m is on the limit circle or is the limit point, then $\psi(x,\lambda)$ is in $L^2(c,b;w)$.

Let us now take a different approach. Since

$$\psi(c,\lambda) = \cos\gamma + m(\lambda)\sin\gamma,$$
$$p(c)\psi'(c,\lambda) = -\sin\gamma + m(\lambda)\cos\gamma,$$

we also have

$$\overline{\psi}(c,\lambda) = \cos\gamma + \overline{m(\lambda)}\sin\gamma,$$
$$p(c)\overline{\psi'(c,\lambda)} = -\sin\gamma + \overline{m(\lambda)}\cos\gamma.$$

Therefore

$$p(c)\psi'(c,\lambda)\overline{\psi}(c,\lambda) = [|m|^2 - 1]\sin\gamma\cos\gamma + m\cos^2\gamma - \overline{m}\sin^2\gamma.$$

Further $\psi(b',\lambda) = K\sin\delta$,

$$p(b')\psi'(b',\lambda) = -K\cos\delta,$$

and

$$\overline{\psi(b',\lambda)} = \overline{K}\sin\delta,$$
$$p(b')\overline{\psi'(b',\lambda)} = -\overline{K}\cos\delta,$$

so that

$$p(b')\psi'(b',\lambda)\overline{\psi}(b',\lambda) = -|K|^2\sin\delta\,\cos\delta,$$

where

$$|K|^2 = |\psi(b',\lambda)|^2 + |p(b')\psi'(b',\lambda)|^2.$$

Replace y in the Dirichlet formula by ψ, and insert these boundary values, above, in the result.

We have

$$\int_c^{b'} [p|\psi'|^2 + q|\psi|^2]dx + |K|^2\sin\delta\cos\delta$$
$$+[|m|^2 - 1]\sin\gamma\cos\gamma + m\cos^2\gamma - \overline{m}\sin^2\gamma$$
$$= (\mu + i\nu)\int_c^{b'} |\psi|^2 w\,dx,$$

where $\lambda = \mu + i\nu$.

The imaginary part of this is what leads in the Weyl–Titchmarsh limit point–limit circle theory:

$$\text{Im}\,(m) = \nu\int_c^{b'} |\psi|^2 w\,dx.$$

The real part is

$$\int_c^{b'} [p|\psi'|^2 + q|\psi|^2]dx + K\sin\delta\cos\delta,$$
$$= -[|m|^2 - 1]\sin\gamma\cos\gamma - \text{Re}\,(m)\cos^2\gamma + \text{Re}\,(m)\sin^2\gamma$$
$$+\mu\text{Im}\,(m)/\nu.$$

If we now let b' approach b, m approaches the limit point or a point on the limit circle. However, only if δ is in $[0,\pi/2]$ will we be able to deduce that ψ is in

$H^1[c, b, p, q]$, since, otherwise, the two terms on the left could become infinite. In anticipation, we fix $\delta = \pi/2$. In this case we have $p(b')\psi'(b', \lambda) = 0$ and $m(\lambda) = -\theta'(b', \lambda)/\phi'(b', \lambda)$. We then find

$$\int_c^{b'} p[|\psi'|^2 + q|\psi|^2]dx \leq [1 - |m|^2]\sin\gamma\cos\gamma$$
$$-\mathrm{Re}\,(m)\cos^2\gamma + \mathrm{Re}\,(m)\sin^2\gamma + \mu\mathrm{Im}\,(m)/\nu,$$

where, now, all m's are on the limit point or limit circle. We can now let the limit of the integral approach b, yielding $\psi_b(x, \lambda)$.

We see that in general, if the limit circle case holds with two solutions in $L^2(c, b; w)$, there need not be more than one solution $\psi_b(c, \lambda)$ in $H^1[c, b, p, q]$. Indeed, the example of the Legendre polynomials is in this situation. What happens is interesting; the points m on the b' circle migrate as b' approaches b so that ultimately δ is not in $[0, \pi/2]$. Only if δ is frozen at $\pi/2$, so that $p(b')\psi'_b(b', \lambda) \equiv 0$ and $m(b) = -\theta(b', \lambda)/\phi(b', \lambda)$, does $m(\lambda)$ approach $m_b(\lambda)$, through subsequences if necessary, with $\psi_b(x, \lambda)$ in $H^1[c, b, p, q]$.

We can establish that $\lim\limits_{x \to b} p(b)\psi'_b(x, \lambda) = 0$.

XVIII.3.2. Theorem. *Let $\psi_b(x, \lambda)$ be the solution of $-(py')'+qy = \lambda wy$, $\mathrm{Im}\,\lambda \neq 0$, in $H^1[c, b, p, q]$ generated by solutions $\psi_{b'}(x, \lambda)$ satisfying $p(b')\psi_{b'}(b', \lambda) = 0$. Then $\lim\limits_{x \to b} p(x)\psi'_b(x) = 0$.*

Proof. First note that

$$p(b')\psi'_b(b', \lambda) = p(b')\psi'_b(b', \lambda) - p(b')\psi'_{b'}(b', \lambda)$$
$$= (m_b - m_{b'})p(b')\phi'(b', \lambda).$$

Now

$$p(b')\phi'(b', \lambda) = p(c)\phi'(c, \lambda) + \int_0^{b'} (p\phi')'dx$$
$$= p(c)\phi'(c, \lambda) + \int_0^{b} [q(x) - \lambda w(x)]\phi(x, \lambda)dx.$$

Thus

$$|p(b')\phi'(b', \lambda)| \leq |p(c)\phi'(c, \lambda)| + K\int_0^{b'} w|\phi|dx$$
$$\leq p(c)\phi'(c, \lambda)| + K\left[\int_c^{b'} |^2 w\,dx\right]^{\frac{1}{2}}\left[\int_c^{b'} |\phi|^2 w\,dx\right]^{\frac{1}{2}}.$$

In the limit point case

$$|m_b - m_{b'}| < \frac{2}{|\nu|}\int_c^{b'} |\phi|^2 w\,dx.$$

Thus

$$|m_b - m_{b'}||p(b')\phi(b', \lambda)| < \frac{A + B\left[\int_c^{b'} |\phi|^2 w\, dx\right]^{\frac{1}{2}}}{\int_c^{b'} |\phi|^2 w\, dx},$$

which approaches 0 as $b' \to b$.

In the limit circle cases

$$|p(b')\phi'(b', \lambda)| < K.$$

Since $m_{b'} \to m_b$,

$$\lim_{b' \to b} [m_{b'} - m_b] p(b')\phi(b', \lambda) = 0. \qquad \square$$

In order to proceed further we need a technical theorem which will allow us to remove the boundary terms in the Dirichlet formula. Note as we proceed that the "z" terms are arbitrary in $H^1(c, b, p, q)$.

XVIII.3.3. Theorem. *Let y and z be in $H^1(c, b, p, q)$ and $\lim\limits_{x \to b} p(x)y'(x) = 0$. If $\lim p(x)y'(x)\bar{z}(x)$ exists, then $\lim_{x \to b} p(x)y'(x)\bar{z}(x) = 0$.*

Proof. Note that $p^{\frac{1}{2}}y'$ and $p^{\frac{1}{2}}z'$ are in $L^2(c, b)$. Let $\alpha = \lim p(x)y'(x)\bar{z}(x)$, and suppose $\alpha \neq 0$. Then near b, $p^{\frac{1}{2}}y' \sim \alpha/p^{\frac{1}{2}}\bar{z}$. However $p^{\frac{1}{2}}y'$ is in $L^2(c, b)$, so $(p^{\frac{1}{2}}\bar{z})^{-1}$ is in $L^2(c_0, b)$ for some c_0 in (c, b).

Now $p^{\frac{1}{2}}z'$ is in $L^2(c, b)$, so

$$z/z' = (p^{\frac{1}{2}}z')(p^{\frac{1}{2}}z)^{-1}$$

is in $L^1(c_0, b)$. But then

$$(\ln |z|)' = \mathrm{Re}\ (z'\bar{z})/(z\,\bar{z})$$

is in $L^1(c_0, b)$. Thus $\lim\limits_{x \to b} \ln |z|$ exists and is finite. Therefore z is bounded as $z \to b$.

This yields a contradiction, since z is bounded, and $\lim\limits_{x \to b} p(x)y'(x) = 0$. $\qquad \square$

As a corollary we state

XVIII.3.4. Theorem. *Let y and z be in $H^1(a, c, p, q)$, let py' be locally absolutely continuous and $\lim\limits_{x \to a} p(x)y'(x) = 0$. If $\lim\limits_{x \to a} p(x)y'(x)\bar{z}(x)$ exists, then*

$$\lim_{x \to a} p(x)y'(x)\bar{z}(x) = 0.$$

XVIII.3.5. Definition. We denote by D those elements y in $L^2(a, b; w)$ satisfying:

(1) y' is absolutely continuous on every compact subinterval of (a, b).

(2) py' is absolutely continuous on every compact subinterval of (a, b), and

$$ly = ((-py')' + qy)/w$$

is in $L^2(a, b; w)$.

(3) For every λ, Im $(b) \neq 0$,

$$\lim_{x \to a} p(x)[y'(x)\psi_a(x,\lambda) - y'(x)\psi_a(x,\lambda)] = 0,$$

$$\lim_{x \to b} p(x)[y'(x)\psi_b(x,\lambda) - y'(x)\psi_b(x,\lambda)] = 0,$$

where ψ_a and ψ_b are solutions of $ly = \lambda y$ in $H^1(a,c,p,q)$ and $H^1(c,b,p,q)$, respectively.

We define the operator L by setting $Ly = ly$ for all y in D_0.

XVIII.3.6. Theorem.

(1) *L is self-adjoint in $L^2(a,b;w)$.*

(2) *L is bounded below by $\epsilon_1 > 0$.*

(3) *For all complex λ, $(L-\lambda)^{-1}$ exists and is bounded.*

$$(L-\lambda)^{-1} f(x) = -\psi_b(x,\lambda) \int_a^x \frac{\psi_a(\xi,\lambda)f(\xi)w(\xi)}{m_b(\lambda) - m_a(\lambda)}\, d\xi$$

$$- \psi_a(x,\lambda) \int_x^b \frac{\psi_b(\xi,\lambda)f(\xi)w(\xi)}{m_b(\lambda) - m_a(\lambda)}\, d\xi.$$

(4) *There exists a spectral measure $P(\lambda)$ such that*

$$f(x) = \int_{-\infty}^{\infty} Y(x,\lambda)d\, P(\lambda)G(\lambda),$$

where $Y(x,\lambda) = (\theta(x,\lambda), \phi(x,\lambda),$ and

$$G(\lambda) = \int_a^b \begin{pmatrix} \theta(x,\lambda)f(x)w(x) \\ \phi(x,\lambda)f(x)w(x) \end{pmatrix} dx$$

for all $f(x)$ in $L^2(a,b;w)$. Further for all F,

$$L^{-1}F(x) = \int_{-\infty}^{\infty} (1/\lambda)\, Y(x,\lambda)dP(\lambda)G(\lambda),$$

and for f in $L^2(a,b,w)$

$$LF(x) = \int_{-\infty}^{\infty} \lambda Y(x,\lambda)d\, P(\lambda G(\lambda).$$

Proof. This, of course, is well known. It is a restatement of the results in Chapter XI. We list them because we wish to reproduce them in $H^1(a,b;p,q)$.

Instead of using $\lim_{x \to a} W[y,\psi_a] = 0$ and $\lim_{x \to b} W[y,\psi_b] = 0$, we use

$$\lim_{x \to c} p(x)y'(x) = 0 \quad \text{and} \quad \lim_{x \to b} p(x)y'(x) = 0$$

to describe boundary conditions. $\qquad\qquad\qquad\qquad\qquad\qquad\qquad\qquad\square$

XVIII.3.7. Theorem. *Let y be in $H^1(a,b,p,q)$ and also lie in the domain of the maximal operator generated by l in $H^1(a,b,p,q)$, that is, ly is in $H^1(a,b,p,q)$ with no mention of boundary conditions. Then $\lim\limits_{x\to a} p(x)y'(x) = 0$ and $\lim\limits_{x\to b} p(x)y'(x) = 0$ if and only if for Im $(\lambda) \neq 0$,*

$$\lim_{x\to a} p(x)[y(x)\psi_a'(x,\lambda) - y'(x)\psi_a(x,\lambda)] = 0\,,$$

$$\lim_{x\to b} p(x)[y(x)\psi_b'(x,\lambda) - y'(x)\psi_b(x,\lambda)] = 0.$$

Proof. We observe that q is in the domain of the maximal operator associated with l in $L^2(a,b;w)$ and $l(1)$ is also in $L^2(a,b;w)$. Therefore if $(l-\lambda)1 = f_1$, then in the limit-circle case at b,

$$1 = -\psi_b \int_c^x \phi f_1 w d\xi - \phi \int_x^b \psi_b f_1 w d\xi + \alpha\psi_b + \beta\phi.$$

For all y in the domain of the maximal operator in $L^2(a,b;w)$,

$$W[y,1] = -W[y,\psi_b] \int_0^x \phi f_1 dd\xi - W[y,\phi] \int_x^b \psi_b f w d\xi + \alpha W[y,\psi_b] + \beta W[y,\phi]\,.$$

If we let $y = \psi_b$ and let x approach b, we find $\beta = 0$. So in general, as x approaches b,

$$\lim_{x\to b} W[y,1](x) = \left[-\int_a^b \phi f_1 w d\xi\right] \lim_{x\to b} [y,\psi_b].$$

This implies that if $\lim\limits_{x\to b} W[y,\psi_b] = 0$, then $\lim\limits_{x\to b} W[y,1] = 0$. Since this holds for y in the domain of the maximal operator, it holds for y in the domain of the maximal operator in H^1 as well.

Conversely, we assume that y is in $\mathcal{D}$, as defined below. Since as x approaches b, $\lim\limits_{x\to b} p(x)y'(x) = 0$ implies $\lim\limits_{x\to b} p(x)y(x)\psi_b(x,\lambda) = 0$, and $\lim\limits_{x\to b} p(x)\psi_b'(x)$ implies $\lim\limits_{x\to b} p(x)\psi_b'(x,\lambda)y(x) = 0$, we find by taking their difference that

$$\lim_{x\to b} W[y,\psi_b](x) = 0.$$

If the limit point case holds at b, then $\lim\limits_{x\to b} W[y,z] = 0$ for all y,z in the domain of the maximal operator in L^2. The limiting Wronskians with $z = 1$ or $z = \psi_b$ are always both 0.

The situation at $x = a$ is the same. $\square$

We are now in a position to define a differential operator $\mathcal{L}$ on $\mathcal{D}$ in $H^1(a,b,p,q)$ which is, in effect a restriction of L.

XVIII.3.8. Definition. We denote by $\mathcal{D}$ those elements y in $H^1(a,b,p,q)$ satisfying:

(1) py' is absolutely continuous on every compact subinterval of (a,b), and

$$ly = (-(py')' + qy)/w$$

is in $H^1(a,b,p,q)$.

(2)

$$\lim_{x \to a} p(x)y'(x) = 0,$$
$$\lim_{x \to b} p(x)y'(x) = 0.$$

We define the operator $\mathcal{L}$ by setting $\mathcal{L}y = ly$ for all y in $\mathcal{D}$.

We see that $\mathcal{D} \subset D \subset L^1(a,b;w)$ and that $\mathcal{D} \subset H^1(a,b;p,q) \subset L^2(a,b;w)$.

The proof that $\mathcal{L}$ is self-adjoint in $H^1(a,b,p,q)$ follows a different path from that required of L in $L^2(a,b;w)$. We first need a Dirichlet formula. Eventually we shall need an extended version of this Dirichlet formula.

XVIII.3.9. Theorem. *The Dirichlet formula. Let y be in $\mathcal{D}, z$ in H^1. Then*

$$\int_a^b [-(py')' + qy]\bar{z}\,dx = \int_a^b [py'\bar{z}' + qy\bar{z}]dx.$$

Proof. The left equals the right with $-(py')\bar{z}\big|_a^b$ added. These terms are, however, 0.

Another way of stating the Dirichlet formula is

$$\langle Ly, z \rangle_{L^2} = \langle y, z \rangle_{H^1}. \qquad\qquad \square$$

XVIII.3.10. Theorem. $\mathcal{D}$ *is dense in* $H^1(a,b,p,q)$.

Proof. If $\mathcal{D}$ is not dense, there exists an f perpendicular to $\mathcal{D}$. That is for all y in $\mathcal{L}$,

$$\langle y, f \rangle_{H^1} = 0.$$

Therefore

$$0 = \langle y, f \rangle_{H^1} = \langle Ly, f \rangle_{L^2}.$$

We choose y so that $Ly = f$ in L^2 and $\mathcal{L}y = f$ in H^1. Then

$$0 = \langle f, f \rangle_{L^2},$$

so $f = 0$, a.e., and $f = 0$ in H^1. $\qquad\qquad \square$

XVIII.3.11. Theorem. $\mathcal{L}$ *is symmetric.*

Proof. Let y, z be in $\mathcal{D}$ and replace z by $\mathcal{L}z$. Then

$$\langle y, \mathcal{L}z \rangle_{H^1} = \langle Ly, Lz \rangle_{L^2}.$$

Reverse y and z and conjugate

$$\langle \mathcal{L}y, z \rangle_{H^1} = \langle Ly, Lz \rangle_{L^2}.$$

Subtract to find

$$\langle y, \mathcal{L}z \rangle_{H^1} = \langle \mathcal{L}y, z \rangle_{H^1} \qquad\qquad \square$$

XVIII.3.12. Theorem. *The inverse operator $\mathcal{L}^{-1}$ exists and is bounded on $H^1(a, b, p, q)$. The range of $\mathcal{L}$ is all of $H^1(a, b, p, q)$.*

Proof. We consider a subinterval (a', b') (a, b). On (a', b'), define D' just as D was defined earlier, but, since the ends are regular use instead the boundary conditions

$$p(a')y'(a') = 0 \quad , \quad p(b')y'(a') = 0.$$

We define L' by setting $L'y = ly$ for all y in D'. Then

$$
\begin{aligned}
\langle L'y, y \rangle_{L^2} &= \int_{a'}^{b'} (L'y)\overline{y}\, w\, dx \\
&= \int_{a'}^{b'} [p|y'|^2 + q|y|^2]dx \\
&\geq \epsilon_1 \int_{a'}^{b'} |y|^2 w\, dx \\
&= \epsilon_1 \|y\|_{L^2}^2.
\end{aligned}
$$

This implies $L' \geq \epsilon_1$, which, in turn, implies $\|(L')^{-1}\| \leq 1/\epsilon_1$ $(L')^{-1}$ is given by the standard formula with the Green's function kernel replaced by substituting a' for a, b' for b in the limits of integration, $\psi_{a'}$ and $\psi_{b'}$ in place of ψ_a and ψ_b, and $M_{a'}$ and $M_{b'}$ in place of m_a and m_b.

The statement $\|(L')^{-1}\| \leq 1/\epsilon_1$ is equivalent to

$$\int_{a'}^{b'} \left| \int_{a'}^{b'} G'(x, \xi)f(\xi)w(\xi)d\xi \right|^2 w(x)dx$$

$$\leq (1/\epsilon_1)^2 \int_{a'}^{b'} |f(\xi)|^2 w(\xi)d\xi\,,$$

where G' is the Green's function for (a', b'). We can now let (a', b') approach (a, b), first on the right, then within the expression for G', then on the inner integral on the left, and finally on the outer integral on the left. The result is

$$\int_c^b \left| \int_a^b G'(x,\xi)f(\xi)w(\xi)d\xi \right|^2 w(x)dx$$

$$\leq (1/\epsilon_1)^2 \int_a^b |f(\xi)|^2 w(\xi)d\xi,$$

or $\|L^{-1}\|_{L^2} \leq 1/\epsilon_1$.

Again return to (a',b'). Define $\mathcal{D}'$ over (a',b') just as $\mathcal{D}$, but again replace the boundary conditions by $p(a')y'(a') = 0$, $p(b')y'(b') = 0$. Define $\mathcal{L}'$ by setting $\mathcal{L}y = ly$ for all y in $\mathcal{D}'$. The Dirichlet formula for f in $H^1(a',b',p,q)$ shows

$$\langle (\mathcal{L}')^{-1}f, (\mathcal{L}')^{-1}f \rangle_{H^1} = \langle f, (L')^{-1}f \rangle_{L^2}.$$

Apply Schwarz's inequality on the right.

$$\|(\mathcal{L}')^{-1}f\|^2 \leq \|f\|_{L^2}\|(L')^{-1}\|_{L^2}$$
$$\leq (q/\epsilon_1)\|f\|_{L^2}^2$$

Note that since $q \geq \epsilon_1 w$,

$$\|f\|_{L^2}^2 \leq (1/\epsilon_1)\|f\|_{H^1}^2.$$

Thus

$$\|(\mathcal{L}')^{-1}f\|_{H^1}^2 \leq (1/\epsilon_1)^2\|f\|_{H^1}^2 ,$$

or

$$\|(\mathcal{L}')^{-1}f\|_{H^1} \leq (1/\epsilon_1)\|f\|_{H^1}$$

over $(a'b')$. Rewritten in integral form, this is

$$\left[\int_{a'}^{b'} \left[p\left|\{(\mathcal{L}')^{-1}f\}'\right|^2 + q\left|\{(\mathcal{L}')^{-1}f\}\right|^2 \right] dx \right]^{\frac{1}{2}}$$

$$\leq (1/\epsilon_1) \left[\int_{a'}^{b'} [p|f'|^2 + q|f|^2]dx \right]^{1/2}.$$

Again let (a',b') approach (a,b) on the right, then on the left within the integrals represented by $(\mathcal{L}')^{-1}$, then on the outer integral limits. The result is

$$\|\mathcal{L}^{-1}f\|_{H^1} \leq (1/\epsilon_1)\|f\|_{H^1} .$$

We can now put all this together.   □

XVIII.3.13. Theorem. *$\mathcal{L}$ is self-adjoint. So is $\mathcal{L}^{-1}$.*

Proof. $\mathcal{L}$ is symmetric with equal deficiency indices. So it has a self-adjoint extension. Since the range of $\mathcal{L}$ is all of H^1, it is already maximally extended, and is self-adjoint. [1, p. 85].

We can also present another, rather simple criterion which makes the bound term $-py'\bar{z}$ vanish at a singular end point. It bypasses a great deal of the complications just presented. $\qquad\qquad\square$

XVIII.3.14. Theorem. *Assume that $(q/p)^{\frac{1}{2}}$ is not in $L^1(a,c)$ and that $(q/p)^{\frac{1}{2}}$ is not in $L^1(c,b)$, where $a < c < b$. Then $\lim\limits_{x\to a} p\overline{yz} = 0$ and $\lim\limits_{x\to b} py'\bar{z} = 0$ for all elements y in the domain of the maximal operator generated by $ly = (-(py')' + qy)/w$ in $H^1(a,b,p,q)$ and for z also in $H^1(a,b,p,q)$.*

Proof. The general Dirichlet formula

$$\int_0^b [(-py') + qy]\bar{z}\, dx = -py'\bar{z}\Big|_0^b + \int_0^b py'\bar{z}' + qy\bar{z}]dx$$

shows that as x approaches b, $\lim\limits_{x\to b} py'\bar{z} = A$ exists. Suppose A is not 0. Then for $x > \alpha$, α near b, $|py'\bar{z}| > |A|/2$.

Now $p^{\frac{1}{2}}y'$ and $q^{\frac{1}{2}}z$ are in $L^2(\alpha,p)$.

But

$$\int_\alpha^b |p^{\frac{1}{2}}y'|\, |q^{\frac{1}{2}}z|dx = \int_\alpha^b |p\bar{y}'\bar{z}| \left(\frac{q}{p}\right)^{\frac{1}{2}} dx$$

$$> (|A|/2) \int_\alpha^b (q/p)^{\frac{1}{2}}dx$$

$$= \infty.$$

Therefore $A = 0$.

In any of the situations where the Dirichlet formula

$$\int_a^b [(-py') + qy]\bar{z}\, dx = \int_a^b [py'\bar{z}' + qy\bar{z}]dx,$$

so that $\lim\limits_{x\to a} py'\bar{z} = 0$ and $\lim\limits_{x\to b} py'\bar{z} = w$, we can show that the resulting operators L and $\mathcal{L}$, in $L^2(a,b;w)$ and $H^1(a,b;p,q)$ respectively, have the same spectral resolution [4].

It is easier to work with L^{-1} and $\mathcal{L}^{-1}$. We focus our attention on them. Let $z = L^{-1}f$ in L^2; let $z = \mathcal{L}^{-1}f$ in H^1; and denote L^{-1} by R, $\mathcal{L}^{-1}$ by $\mathcal{R}$.

The Dirichlet formula with any y in $\mathcal{D}$

$$\langle Ly, z\rangle_{L^2} = \langle y, z\rangle_{H^1}$$

becomes

$$\langle Ly, Rf\rangle_{L^2} = \langle y, \mathcal{R}f\rangle_{H^1}.$$

Since R and $\mathcal{R}$ are bounded, we find

$$\langle Ly, p(R)f \rangle_{L^2} = \langle y, p(\mathcal{R})f \rangle_{H^1},$$

where p is a polynomial. This implies through limits (they definitely exist on the left, so they do on the right as well) that

$$\langle Ly, u(R)f \rangle_{L^2} = \langle y, u(\mathcal{R})f \rangle_{H^1},$$

where u is piecewise continuous. In particular, if

$$e_\mu(\lambda) = \begin{cases} 1 &, & \lambda \le \mu, \\ 0 &, & \lambda > \mu, \end{cases}$$

the corresponding projection operators $E(\lambda)$ and $\mathcal{E}(\lambda)$, which are the spectral measures of R and $\mathcal{R}$, respectively, are also equal,

$$\langle Ly, E(\lambda)f \rangle_{L^2} = \langle y, \xi(\lambda)f \rangle_{H^1}.$$

Integrating with respect to $E(\lambda)$ we $\mathcal{E}(\lambda)$, we have

$$\left\langle Ly, \int_0^{1/\epsilon_1} \lambda^n dE(\lambda)f \right\rangle_{L^2} = \left\langle y, \int_0^{1/\epsilon_1} \lambda^n d\xi(\lambda)f \right\rangle_{H^1},$$

$n \ge -1$. Further

$$\left\langle Ly, \int_0^{1/\epsilon_1} dE(\lambda)f \right\rangle_{L^2} = \left\langle y, \int_0^{1/\epsilon_1} d\mathcal{E}(\lambda)f \right\rangle_{H^1}.$$

We can now make the substitution $\lambda = 1/\mu$, $E(\lambda) = F(\mu)$, $\mathcal{E}(\lambda) = \mathcal{F}(\mu)$. F and $\mathcal{F}$ are spectral measures for L and $\mathcal{L}$. Thus

$$\left\langle Ly, \int_{\epsilon_1}^\infty \mu \, dF(\mu)z \right\rangle_{L^2} = \left\langle y, \int_{\epsilon_1}^\infty \mu \, d\mathcal{F}(\mu)z \right\rangle_{H^1},$$

when z is in $\mathcal{D}$ and

$$\left\langle Ly, \int_{\epsilon_1}^\infty dF(\mu)z \right\rangle_{L^2} = \left\langle y, \int_{\epsilon_1}^\infty d\mathcal{F}(\mu)z \right\rangle_{H^1},$$

when z is in H^1. $\qquad\qquad\qquad\qquad\qquad\qquad\qquad\qquad\qquad\qquad\qquad\qquad$ $\square$

Now let the right sides of any of the inner products be denoted by X and Y, respectively, so that we have

$$\langle Ly, X \rangle_{L^2} = \langle y, Y \rangle_{H^1}.$$

The second, on the right, equals $\langle Ly, Y \rangle_{L^2}$. So

$$\langle Ly, X - Y \rangle_{L^2} = 0.$$

Since the range of L is all of L^2, $X = Y$ in L^2 and $X = Y$ in H^1 as well.

References

[1] N. I. Akhiezer and I. M. Glazman, **Theory of Linear Operators in a Hilbert Space**, vol. I, Fredrick Ungar, New York, 1963.

[2] A. M. Krall, *Left Definite Theory for Second Order Differential Operators with Mixed Boundary conditions*, J. Diff. Eq. **118** (1995), 153–165.

[3] ______, *Left Definite Regular Hamiltonian Systems*, Math. Nachr. **174** (1995), 203–217.

[4] ______, *Singular Left-Definite Boundary Value Problems*, Indian J. Pure Appl. Math. **29** (1998), 29–36.

[5] A. M. Krall and D. Race *Self-Adjointness for the Weyl Problem Under an Energy Norm*, Quaes. Math. **18** (1995), 407–426.

[6] A. Schneider and H. D. Niessen, *Linksdefinite singuläre kanonische Eigenwertprobleme, I*, J. f. d. reine ang. Math. **281** (1976), 13–52.

[7] ______, *Linksdefinite singuläre kanonische Eigenwertprobleme, II*, J. f. d. reine ang. Math. **289** (1977), 62–84.

Chapter XIX

Examples of Sobolev Differential Operators

From each section of the previous chapter we list at least one example. For the singular problems there are several.

XIX.1 Regular Second Order Problems [3]

We keep the general $p > 0$, $q > 0$ and $r > 0$ when considering the operator

$$ly = [-(py')' \mid qy]/w$$

under the usual constraints. We introduce boundary conditions at the ends a and b:

$$\cot \gamma y(a) - 1p(a)y'(a) = 0,$$
$$\cot \delta y(b) + 1p(b)y'(b) = 0.$$

The appropriate Sobolev space is determined by

$$\langle y, z \rangle_{H^1} = \int_a^b [py'\bar{z}' + qy\bar{z}]dx$$
$$+ (\bar{z}(a)\bar{z}(b)) \begin{pmatrix} \cot \gamma & 0 \\ 0 & \cot \delta \end{pmatrix} \begin{pmatrix} y(a) \\ y(b) \end{pmatrix},$$

whenever $\cos \gamma, \cos \delta > 0$.

In the case s when $\cot \gamma = \infty$ or $\cot \delta = \infty$, the terms $\bar{z}(a) \cot \gamma y(a)$ or $\bar{z}(b) \cot \delta y(b)$ are missing from the inner product formula, and the Sobolev space must be constrained by requiring $y(a)$ and $z(a)$ be zero, or by requiring $y(b)$ and $z(b)$ be zero.

A second, more complicated, example is found by setting

$$\alpha_{11} = 0 \quad , \quad \alpha_{12} = -1 \quad , \quad \beta_{11} = 0 \quad , \quad \beta_{12} = -1,$$
$$\alpha_{21} = 0 \quad , \quad \alpha_{22} = -1 \quad , \quad \beta_{21} = 0 \quad , \quad \beta_{22} = -1.$$

Here

$$\begin{pmatrix} py'(a) \\ -py'(b) \end{pmatrix} = \begin{pmatrix} 1 & 1 \\ 1 & 1 \end{pmatrix} \begin{pmatrix} y(a) \\ y(b) \end{pmatrix},$$

and the Sobolev inner product is

$$\langle yz, \rangle_{H^1} = \int_a^b (py'\bar{z}' + qy\bar{z})dx + (\bar{z}(a), \bar{z}(b)) \begin{pmatrix} 1 & 1 \\ 1 & 1 \end{pmatrix} \begin{pmatrix} y(a) \\ y(b) \end{pmatrix}.$$

Yet another example is found by setting

$$\begin{aligned}
\alpha_{11} &= .6 \quad , \quad \alpha_{12} = .6, \\
\alpha_{21} &= .4 \quad , \quad \alpha_{22} = 1.0, \\
\beta_{11} &= .8 \quad , \quad \beta_{12} = .8, \\
\beta_{21} &= 1.0 \quad , \quad \beta_{22} = 1.6.
\end{aligned}$$

The Dirichlet formula, determining the inner product, is

$$\int_a^b (ly)\bar{z}dx = \int_a^b [py'\bar{z}' + qy\bar{z}]dx$$
$$+ (-.6z(a) + .8z(b))^*(-.6y(a) + .8y(b)).$$

y and z must satisfy the constraints

$$.8y(a) + .6y(b) = 0 \quad , \quad .8z(a) + .6z(b) = 0.$$

Here the matrix $\begin{pmatrix} -\alpha_{12} & \beta_{12} \\ -\alpha_{22} & \beta_{22} \end{pmatrix}$ is singular but not 0.

XIX.2   Regular Hamiltonian Systems [4]

Perhaps the simplest new problem which can be expressed as a linear Hamiltonian system is the system representation of a self-adjoint problem of fourth order. If we let

$$\begin{aligned}
y_1 &= y & , \quad y_2 &= y', \\
y_3 &= -(py'')' + qy' & , \quad y_4 &= py'',
\end{aligned}$$

the fourth order problem

$$(py'')'' - qy')' + ry = \lambda wy + wf$$

is represented by the four-dimensional Hamiltonian system

$$\begin{pmatrix} 0 & 0 & -1 & 0 \\ 0 & 0 & 0 & -1 \\ 1 & 0 & 0 & 0 \\ 0 & 1 & 0 & 0 \end{pmatrix} \begin{pmatrix} y_1 \\ y_2 \\ y_3 \\ y_4 \end{pmatrix}' =$$

$$\lambda \begin{pmatrix} w & 0 & 0 & 0 \\ 0 & 0 & 0 & 0 \\ 0 & 0 & 0 & 0 \\ 0 & 0 & 0 & 0 \end{pmatrix} \begin{pmatrix} y_1 \\ y_2 \\ y_3 \\ y_4 \end{pmatrix} + \begin{pmatrix} -r & 0 & 0 & 0 \\ 0 & q & 1 & 0 \\ 0 & 1 & 0 & 0 \\ 0 & 0 & 0 & 1/p \end{pmatrix} \begin{pmatrix} y_1 \\ y_2 \\ y_3 \\ y_4 \end{pmatrix} + \begin{pmatrix} w & 0 & 0 & 0 \\ 0 & 0 & 0 & 0 \\ 0 & 0 & 0 & 0 \\ 0 & 0 & 0 & 0 \end{pmatrix} \begin{pmatrix} f \\ 0 \\ 0 \\ 0 \end{pmatrix}.$$

The Dirichlet formula is

$$\int_a^b [(py'')'' + (qy')' + (ry)]\bar{z}\,dx$$

$$= \int_a^b [py''\bar{z}'' + qy'\bar{z}' + ry\bar{z}]\,dx$$

$$+ \cot\alpha y_1(a)\bar{z},\ (a) + \cot\beta y_2(a)\bar{z}_2(a)$$

$$+ \cot\gamma y_3(b)\bar{z}_3(b) + \cot\delta y_4(b)\bar{z}_4(b),$$

where we have imposed the boundary conditions

$$y_3(a) = \cot\alpha y_1(a) \quad , \quad y_4(a) = \cot\beta y_2(a)$$
$$y_3(b) = \cot\gamma y_1(b) \quad , \quad y_4(b) = \cot\delta y_2(b).$$

z_1, z_2, z_3, z_4 likewise satisfy the same boundary conditions at a and b.

In the case $\alpha = 0$, we must restrict ourselves to a subspace determined by $y_1(a)$, $z_1(a) = 0$. The α term above is excluded from the formula. The other cases $\beta = 0$, $\gamma = 0$, $\delta = 0$ are handled similarly.

If $\alpha, \beta, \gamma, \delta = \pi/2$, then the various terms above disappear, but there is no subspace restriction.

Another, more general, system example again is given by a scalar fourth order operator, but with the boundary condition

$$\begin{pmatrix} .6I & .6I \\ .4I & 1.2I \end{pmatrix} \begin{pmatrix} Y_1(a) \\ Y_2(a) \end{pmatrix} + \begin{pmatrix} .8I & .8I \\ 1.0I & 1.6I \end{pmatrix} \begin{pmatrix} Y_1(a) \\ Y_2(b) \end{pmatrix} = \begin{pmatrix} 0 \\ 0 \end{pmatrix},$$

where $I = \begin{pmatrix} 1 & 0 \\ 0 & 1 \end{pmatrix}$, $Y_1 = \begin{pmatrix} y_1 \\ y_2 \end{pmatrix}$, $Y_2 = \begin{pmatrix} y_3 \\ y_4 \end{pmatrix}$. The subspace constraint is $y_{12} = 0$ and $y_{21} = .28y_{11} - .96y_{12}$, where

$$\begin{pmatrix} y_{11} \\ y_{12} \end{pmatrix} = \begin{pmatrix} -.6I & -.8I \\ -.8I & -.6I \end{pmatrix} \begin{pmatrix} Y_1(a) \\ Y_1(b) \end{pmatrix},$$

and

$$\begin{pmatrix} y_{21} \\ y_{22} \end{pmatrix} = \begin{pmatrix} -.6I & -.8I \\ -.8I & -.6I \end{pmatrix} \begin{pmatrix} Y_2(a) \\ Y_2(b) \end{pmatrix} .$$

The Dirichlet formula is

$$\int_a^b [(py'')'' - q(y')' + ry]\bar{z}dx$$

$$= \int_a^b [py''\bar{z}'' + qy'\bar{z}' + ry\bar{z}]dx$$

$$+ (\bar{z}(a)\bar{z}'(a)z(b)\bar{z}'(b)) \begin{bmatrix} .1008 & .1008 & .1344 & .1344 \\ .1008 & .1008 & .1344 & .1344 \\ .1344 & .1344 & .1792 & .1792 \\ .1344 & .1344 & .1792 & .1792 \end{bmatrix} \begin{pmatrix} y(a) \\ y'(a) \\ y(b) \\ y'(b) \end{pmatrix} .$$

XIX.3　Singular Second Order Sobolev Boundary Value Problems

There are five main examples of singular Sobolev boundary problems: those associated with Fourier integrals (associated with the ordinary Laplacian), Bessel functions, the Jacobi and Legendre polynomials, the Laguerre polynomials, and Hermite polynomials. We give a brief description of each.

XIX.3.1　The Laplacian Operators [5]

First let us examine the differential operator

$$ly = -y'' + ky \quad , \quad k > 0$$

on $(-\infty, \infty)$. At $\pm\infty$, the boundary term in the Dirichlet formula

$$\int_{-\infty}^{\infty} (-y'' + ky)\bar{z}dx = -y'z\big|_{-\infty}^{\infty} + \int_{-\infty}^{\infty} [y'\bar{z}' + ky\bar{z}]dx$$

vanishes according to Theorem XVIII.3.13. The Dirichlet formula is, accordingly

$$\int_{-\infty}^{\infty} (-y'' + ky)\bar{z}dx = \int_{-\infty}^{\infty} [y'\bar{z}' + ky\bar{z}]dx .$$

It is a very straightforward procedure to see that l generates a self-adjoint differential operator on $L^2(-\infty, \infty)$ and $H^1(-\infty, \infty; 1, k)$. The spectral resolution of the operators is

$$f(x) = \frac{1}{\sqrt{2\pi}} \int_{-\infty}^{\infty} g(\lambda)e^{i\lambda x}d\lambda ,$$

where

$$g(\lambda) = \frac{1}{\sqrt{2\pi}} \int_{-\infty}^{\infty} f(\xi)e^{-i\lambda\xi}d\xi .$$

These are the complex inverse Fourier transform and the transform itself.

If the interval $[0, \infty)$ is considered instead, and a boundary condition $y(0) = 0$ is imposed, then the Sobolev operator must be constrained by the subspace in which $y(0) = 0$ — in $H^1(0, \infty; 1, k)/\{y(0) = 0\}$. The Dirichlet formula is, again

$$\int_{-\infty}^{\infty} (-y'' + ky)\bar{z}dx = \int_0^{\infty} [y'\bar{z}' + ky\bar{z}]dx\,.$$

The spectral resolution of the self-adjoint operator in $L^2(0, \infty)$ or $H^1(0, \infty; 1, k)$ given by $-y'' + ky$ is

$$f(x) = \sqrt{\frac{2}{\pi}} \int_0^{\infty} g(\lambda) \sin \lambda x d\lambda\,,$$

where

$$g(\lambda = \sqrt{\frac{2}{\pi}} \int_0^{\infty} f(\xi) \sin \lambda \xi d\xi\,.$$

These are the inverse Fourier sine transform and the sine transform itself.

If, again, the interval $[0, \infty)$ is considered, and a boundary condition $y'(0) = 0$ is imposed, then there is no subspace constraint. The Dirichlet formula is

$$\int_0^{\infty} (-y'' + ky)\bar{z}dx = \int_0^{\infty} [y'\bar{z}' + ky\bar{z}]dx\,.$$

The spectral resolution of the self adjoint operator in $L^2(0, \infty)$ or $H^1(0, \infty; 1, k)$ is given by

$$f(x) = \sqrt{\frac{2}{\pi}} \int_0^{\infty} g(\lambda) \cos \lambda x d\lambda\,,$$

where

$$g(\lambda) = \sqrt{\frac{2}{\pi}} \int_0^{\infty} f(\xi) \cos \lambda \xi d\xi\,.$$

These are the inverse Fourier cosine transform and the cosine transform itself.

XIX.3.2 The Bessel Operators [6]

The differential expression associated with the classic Bessel functions is

$$ly = \{-(xy')' + [n^2/x + 1x]y\}/x\,,$$

set in $L^2(0, 1; x)$ or $L^2(0, \infty; x)$, where n is a fixed parameter, which we fix ≥ 0.

Within the setting $(0, 1]$ a singular boundary condition at $x = 0$, while a regular condition is required at $x = 1$. Determining these, we arrive at an appropriate differential operator.

XIX.3.2.1. Definition. We denote by $D_{(0,1]}$ those elements y in $L^2(0,1;x)$ which satisfy:

(1) y is absolutely continuous on all closed subsets of $(0,1]$.

(2) (xy') is absolutely continuous on all closed subsets of $(0,1]$.

(3) ly exists a.e. and y in $L^2(0,1;x)$.

(4) $\lim\limits_{x\to 0} -\,xy'(x) = 0$.

(5) For $0 \le \delta \le \pi/2$,
$$\cos\delta y(1) + \sin\delta y'(1) = 0\,.$$

We define the Bessel operator L by setting $Ly = ly$ for all y in $D_{(0,1]}$.

The general theorem determines that the boundary condition (4) is required only when $0 \le n < 1$ while it is automatic for $n \ge 1$.

It is then easy to see that L is self-adjoint in $L^2(0,1;x)$. Eigenvalues are determined by $\{\lambda_k\}_{k=1}^\infty$, the zeros of
$$\left[\cos\delta J_n\left(\sqrt{\lambda-1}x\right) + \sin\delta J_n'\left(\sqrt{\lambda-1}x\right)\right]_{x=1}.$$

The spectrum is discrete and the eigenfunctions $\{J_n(\sqrt{\lambda_k-1}x)\}_{n=1}^\infty$ are complete. The spectral resolution of L, therefore, is the eigenfunction expansion, for arbitrary f in $L^2(0,1;x)$
$$f(x) = \sum_{k=1}^\infty C_k J_n(\sqrt{\lambda_k-1}x)/\|J_n(\sqrt{\lambda_k-1}x)\|\,,$$

where
$$C_k = \int_0^1 f(x)J_n(\sqrt{\lambda_k-1}x)dx/\|J_n(\sqrt{\lambda_k-1}x)\|\,.$$

On $(0,\infty)$, $ly = \lambda y$ is in the limit point case at ∞, and the boundary condition (5) disappears. The resulting differential operator is also self-adjoint on $(0,\infty)$.

The spectrum of M, however, is continuous on $(0,\infty)$. The resulting spectral resolution is given via the Hankel transform. If f is an arbitrary element in $L^2(0,\infty;x)$, then
$$f(x) = \int_0^\infty F_n(x)J_n(sx)ds\,,$$

where
$$F_n(s) = \int_0^\infty f(\xi)J_n(s\xi)\xi d\xi.$$

The H^1 theory follows a similar route. The Dirichlet formula on $(0,1]$ is
$$\int_0^1 \{-(xy')' + [(n^2/x) + 1x]y\}\bar{z}\,dx$$
$$= -xy'\bar{z}\big|_0^1 + \int_0^1 (xy'\bar{z}' + (n^2/x)y\bar{z} + xy\bar{z})dx.$$

The limit of the integrated term at $x = 0$ is 0, since $\lim\limits_{x \to 0} - xy'(x) = 0$, by Theorem XVIII.3.3 if $n = 0$, and by Theorem XVIII.3.13 if $n \neq 0$.

The term $xy'(x)\bar{z}(x)$ at $x = 1$ yields $\cot(\delta)y(1)z(1)$ since y satisfies the boundary condition (5). If $\delta = 0$, the boundary condition at $x = 1$ is $y(1) = 0$. Therefore we are faced with operating within the subspace where all elements y vanish at $x = 1$. The term $-xy'(x)\bar{z}(x)$ at $x = 1$ does not appear in the Dirichlet formula (since $z(1) = 0$ as well). If $\delta = \frac{\pi}{2}$, then the boundary condition is $y'(1) = 0$. The term $y'(1) = 0$ causes the term $-xy'(x)\bar{z}(\alpha)$ at $x = 1$ to vanish.

Thus the Dirichlet formula yielding the inner product is

$$\int_0^1 \{-(xy')' + [(n^2/x) + 1x]\bar{z}dx$$

$$= \int_0^1 [xy'\bar{z}' + (n^2/x)y\bar{z} + xy\bar{z}]dx + \cot \delta y(a)\bar{z}(1),$$

with the proviso that if $\delta = 0$ the last term is not present.

The L^2 operator L, restricted to the Sobolev space $H^1(0, x; x, \frac{n^2}{x} + x)$ or its subspace $y(0) = 0$ when $\delta = 0$, call it $\mathcal{L}$, remains self-adjoint. The spectrum of M is the same as that of L, and the spectral resolution is likewise unchanged.

On $(0, \infty)$ the Dirichlet formula is

$$\int_0^1 \{-(xy')' + [(n^2/x) + 1x]y\}\bar{z}\, dx$$

$$= \int_0^1 [xy'\bar{z}' + (n^2/x)y\bar{z} + xy\,\bar{z}]dx .$$

The boundary term at $x = 0$ vanishes as before. At $x = \infty$ it also vanishes according to Theorem XVII.3.13.

The L^2 operator M, restricted to the Sobolev space $H^1(0, \infty; x, \frac{n^2}{x} + x)$, call it M, remains self-adjoint. Its spectrum is continuous on $(0, \infty)$. The Hankel transform also gives its spectral resolution.

XIX.3.3 The Jacobi Operator [1], [8]

The Legendre differential expression

$$ly = -((1 - x^2)y')' + \frac{1}{4}y,$$

set in $L^2(-1, 1)$, is self-adjoint with the boundary constraints

$$\lim_{x \to 1} (1 - x^2)y'(x) = 0 \text{ and } \lim_{x \to -1} (1 - x^2)y'(x) = 0 .$$

(Note the component multiplied by $\frac{1}{4}$). Therefore, when set in $H^1(-1,1;(1-x^2),1)$, according to theorem XVIII.3.3. The boundary term $-(1-x^2)y'(x)\bar{z}(x)$ vanishes at ± 1, and subsequently the Dirichlet formula is

$$\int_{-1}^{1}(-((1-x^2)y')'+\frac{1}{4}y)\bar{z}dx = \int_{-1}^{1}[(1-x^2)y'\bar{z}'+\frac{1}{4}y\bar{z}]dx.$$

The differential expression

$$ly = -((1-x^2)y')'+\frac{1}{4}y$$

with boundary conditions

$$\lim_{x\to 1}(1-x^2)y'(x) = 0 \quad , \quad \lim_{x\to -1}(1-x^2)y'(x) = 0$$

remains self-adjoint in $H^1(-1,1;(1-x^2),1)$.

More generally, the general Jacobi operator

$$ly = [-((1-x)^{A+1}(1+x)^{B+1}y')'+\frac{1}{4}(1-x)^A(1+x)^B y]/(1-x)^A(1+x)^B,$$

set in $L^2(-1,1;(1-x)^A(1+x)^B)$, is self-adjoint when constrained by boundary conditions

$$\lim_{x\to 1}(1-x)^{A+1}(1+x)^{B+1}y'(x) = 0,$$
$$\lim_{x\to -1}(1-x)^{A+1}(1+x)^{B+1}y'(x) = 0,$$

when $A > -1$, $B > -1$. (The cases $A \le -1$, $B \le -1$ were discussed earlier in Chapter XIV. See [1].)

According to Theorem XVIII.3.3, when set in $H^1(-1,1;(1-x)^{A+1}(1+x)^{B+1}, \frac{1}{4}(1-x)^A(1+x)^B)$, the boundary term in the Dirichlet formula vanishes, and the formula is

$$\int_{-1}^{1}(-(1-x)^{A+1}(1+x)^{B+1}y')'+\frac{1}{4}(1-x)^A(1+x)^B y)\bar{z}\,dx$$
$$= \int_{-1}^{1}[(1-x)^{A+1}(1+x)^{B+1}\bar{y}'\bar{z}'+\frac{1}{4}(1-x)^A(1+\alpha)^B y\bar{z}]dx.$$

Set in H^1, above, the operator l remains self adjoint.

The spectral resolutions of the Legendre and Jacobi operators remain the standard eigenfunction expansion.

XIX.3.4   The Generalized Laguerre Operator [2]

The differential operator

$$ly = [-(x^{\alpha+1}e^{-x}y')'+x^\alpha e^{-x}y]/x^\alpha e^{-x}$$

constrained by

$$\lim_{x \to 0} x^{\alpha+1} e^{-x} y'(x) = 0$$

is self-adjoint in $L^2(0, \infty; x^\alpha e^{-x})$, $\alpha \geq -1$. (Note the $x^\alpha e^{-x} y$ component.) Since, at $x = 0$, according to Theorem XVIII.3.3 the Dirichlet boundary term

$$\lim_{x \to 0} x^{\alpha+1} e^{-x} y'(x) \bar{z}(x)$$

also vanishes, and at $x = \infty$, according to Theorem XVIII.3.13, it also vanishes, the Dirichlet formulais

$$\int_0^\infty (-(x^{\alpha+1} e^{-x} y')' + x^\alpha e^{-x} y) \bar{z} \, dx$$

$$= \int_0^\infty [(x^{\alpha+1} e^{-x}) y' \bar{z}' + x^\alpha e^{-x} y \bar{z}] dx \quad , \quad \alpha \geq 1.$$

The operator l with the boundary constraint $\lim\limits_{x \to 0} x^{\alpha+1} e^{-x} y'(x) = 0$ remains self-adjoint in $H^1(0, \infty; x^{\alpha+1} e^{-x}, x^\alpha e^{-x})$. (Again, the case $\alpha < -1$ was discussed in Chapter XIV. See [2].)

The spectral resolution remains the standard eigenfunction expansion.

Remark. L. L. Littlejohn and K. H. Kwon [9] have discussed the Laguerre polynomials $\{L_n^{-k}(x)\}_{n=0}^\infty$. They are found to be orthogonal with respect to the weight function

$$\langle p, q \rangle = p(0)q(0) + \int_0^\infty p'(x)q'(x)e^{-x} dx,$$

when $k = -1$,

$$\langle p, q \rangle = p(0)q(0) + [p'(0)q(0) + p(0)q'(0)]$$

$$+ 2p'(0)q'(0) + \int_0^\infty p''(x)q''(x)e^{-x} dx$$

when $k = -2$, and in general

$$\langle p, q \rangle = \sum_{m=0}^{k-1} \sum_{j=0}^{m} B_{mj}(k)[p^{(m)}(0)q^j(0) + p^{(j)}(0)q^{(m)}(0)]$$

$$+ \int_0^\infty p^{(k)}(x)q^{(k)}(x)e^{-x} dx,$$

where

$$B_{mj}(k) = \sum_{p=0}^{\lambda} (-1)^{m+j} \binom{k-1-p}{m-p} \binom{k-1-p}{j-p},$$

$$0 \leq j < m \leq k-1$$

$$= \frac{1}{2} \sum_{p=0}^{m} \binom{k-1-p}{m-p}^2 \quad , \quad 0 \leq j = m \leq k-1.$$

XIX.3.5 The Hermite Operator [7]

The differential operator

$$ly = (-(e^{-x^2}y')' + e^{-x^2}y)/e^{-x^2}$$

defined on $(-\infty, \infty)$ has a Dirichlet formula in which the boundary term $e^{-x^2}y'\bar{z}$ vanishes at $\pm\infty$ by Theorem XVIII.3.13. (Note the component $e^{-x^2}y$.) The Dirichlet formula is, therefore,

$$\int_{-\infty}^{\infty}(-(e^{-x^2}y')' + e^{-x^2}y)dx = \int_{-\infty}^{\infty}[e^{-x^2}y'\bar{z}' + e^{-x^2}y\bar{z}]dx.$$

The right side generates a Sobolev space $H^1(-\infty, \infty; e^{-x^2}, e^{-x^2})$ in which the expression l generates a self-adjoint differential operator whose spectral resolution is the standard Hermite polynomial eigenfunction expansion.

XIX.3.6 The Generalized Even Hermite Operator [7]

The differential expression

$$ly = \frac{1}{2}(-(x^{2\mu}e^{-x^2}y')' + 2kx^{2\mu}e^{-x^2}y)/x^{2\mu}e^{-x^2},$$

$\mu > -\frac{1}{2}$, with boundary condition

$$\lim_{x \to 0} x^{2\mu}e^{-x^2}y'(x) = 0$$

leads to the Dirichlet formula

$$\int_0^\infty (ly)\bar{z}x^{2\mu}e^{-x^2}\,dx = \frac{1}{2}\int_0^\infty [x^{2\mu}e^{-x^2}y'\bar{z}' + 2kx^{2\mu}e^{-x^2}y\bar{z}]dx.$$

(Note the k term.)

  The boundary terms vanish because of Theorem XVIII.3.4 when $x = 0$ and because of Theorem XVIII.3.13 when $x = \infty$. This means that the operator L in $L^2(0, \infty; x^{2\mu}e^{-x^2})$ when restricted to $H^1(0, \infty; \frac{1}{2}x^{2\mu}e^{-x^2}, k\,x^{2\mu}e^{-x^2})$ remains self-adjoint with the same spectral resolution: the even degree generalized Hermite polynomials.

XIX.3.7 The Generalized Odd Hermite Operator [7]

The differential expression

$$ly = \frac{1}{2}(-(x^{2\mu}e^{-x^2}y')' + (2\mu\,x^{2\mu-1} + 2k\,x^{2\mu})e^{-x^2}y)/x^{2\mu}e^{-x^2},$$

$\mu > -\frac{1}{2}$, with boundary condition

$$\lim_{x \to 0} x^{2\mu} e^{-x^2} \left(x \frac{dy}{dx} - y \right) = 0$$

leads to the Dirichlet formula

$$\int_0^\infty (ly) \overline{z}\, x^{2\mu} e^{-x^2}\, dx = \frac{1}{2} \int_0^\infty [x^{2\mu} e^{-x^2} y' \overline{z}' + (2\mu\, x^{2\mu-2} + 2k\, x^{2\mu}) e^{-x^2} y\overline{z}]dx.$$

(Note the k term.) The boundary terms vanish because of Theorem XVIII.3.4 when $x = 0$ and because of Theorem XVIII.3.13 when $x = \infty$. This means that the operator L in $L^2(0, \infty; x^{2\mu} e^{-x^2})$ when restricted to $H^1(0, \infty; \frac{1}{2}x^{2\mu} e^{-x^2}, (\mu\, x^{2\mu-2} + kx^{2\mu})e^{-x^2})$ remains self-adjoint with the same spectral resolution: the odd degree generalized Hermite polynomials.

The cases when $\mu < 0$ are transformable just as in Section XIV.6.

References

[1] M. Hajmirzaahmad, *Jacobi polynomial expansions*, J. Math. Appl. Anal. **181** (1994), 35–61.

[2] ______, *Laguerre polynomial expansions*, J. Comp. Appl. Math. **59** (1995), 25–37.

[3] A. M. Krall, *Left Definite Theory for Second Order Differential Operators with Mixed Boundary conditions*, J. Diff. Eq. **118** (1995), 153–165.

[4] ______, *Left Definite Regular Hamiltonian Systems*, Math. Nachr. **174** (1995), 203–217.

[5] ______, H^1 *convergence of Fourier integrals*, Indian J. Pure Appl. Math. **26** (1995), 41–50.

[6] ______, H^1 *Bessel expansions*, Indian J. Pure Appl. Math **26** (1995), 51–60.

[7] ______, *Spectral analysis for the generalized Hermite polynomials*, Trans. Amer. Math. Soc. **344** (1994), 155–172.

[8] A. M. Krall and L. L. Littlejohn, *The Legendre polynomials under a left definite energy norm*, Quaes. Math. **16** (1993), 393–403.

[9] L. L. Littlejohn and K. H. Kwon, *The orthogonality of the Laguerre polynomials* $\{L_n^{-k}(x)\}_{n=0}^\infty$ *for positive integers k*, Ann. Num. Math. **2** (1995), 289–303.

Chapter XX

The Legendre-Type Polynomials and the Laguerre-Type Polynomials in a Sobolev Space

We give only the highlights of these examples. That the polynomials are orthogonal is relatively easy, but showing that the differential operators, when restricted, remain self-adjoint is more difficult. We refer the interested reader to [3] and [4], for the details, which are quite complicated.

XX.1 The Legendre-Type Polynomials

We remind the reader that the appropriate setting for the Legendre-type operator is the Hilbert space $H = L^2(-1, 1; \frac{\alpha}{2}) \otimes R \otimes R$, where, with Y in H given by $Y = (y(x), y_1, y_{-1})^T$ and Z in H given by $Z = (z(x), z_1, z_{-1})^T$, have the inner product

$$\langle Y, Z \rangle_H = \int_{-1}^{1} y(x)\bar{z}(x)\frac{\alpha}{2}\,dx + \frac{1}{2}y_1\bar{z}_1 + \frac{1}{2}y_{-1}\bar{z}_{-1}.$$

Associated with the differential expression

$$L_4 y = [(1 - x^2)^2 y'']'' + 4[(\alpha(x^2 - 1) - 2)y']' + ky$$

we define the operator A by setting

$$AY = \begin{pmatrix} L_4 y \\ -8y'(-1) + ky(-1) \\ 8y'(1) + ky(1) \end{pmatrix} \quad , \quad k > 0,$$

where the first component is defined on $(-1, 1)$. Then A is self-adjoint in H.

The appropriate Sobolev space is given by [3],

$$\langle Y, Z\rangle_S = \frac{\alpha}{2}\int_{-1}^{1}[(1-x^2)y''\bar{z}'' + (8+4\alpha(1-x^2))y'\bar{z}']dx$$

$$+ k\left\{\frac{\alpha}{2}\int_{-1}^{1}y\bar{z}dx + \frac{1}{2}y(1)\bar{z}(1) + \frac{1}{2}y(-1)\bar{z}(-1)\right\}.$$

In essence it is generated by the Dirichlet formula for $\langle AY, Z\rangle_H$. We have

$$\int_{-1}^{1}[(1-x^2)y'')'' + 4[(\alpha(x^2-1)-2)y']' + ky]\bar{z}\frac{\alpha}{2}dx$$

$$= \int_{-1}^{1}[(1-x^2)^2y''\bar{z}'' + (8+4\alpha(1-x^2))y'\bar{z}' + ky\bar{z}]dx$$

$$+\frac{\alpha}{2}((1-x^2)^2y'')'\bar{z}\Big|_{-1}^{1} + \frac{\alpha}{2}((1-x^2)g'')'\bar{z}\Big|_{-1}^{1} - \frac{\alpha}{2}((1-x^2)^2y'')\bar{z}'\Big|_{-1}^{1}$$

$$-\frac{\alpha}{2}((8+4\alpha(1-x^2))y'\bar{z}'\Big|_{-1}^{1}.$$

Employing limits found in [1], [3], the integrated limiting terms are found to mostly vanish. Adding the appropriate jumps at ± 1 yields the H inner product and

$$\langle AY, Z\rangle_H = \langle Y, Z\rangle_S.$$

We then employ the usual practice of defining the restriction of A to the Sobolev space to be T. After a bit of work it is found that T is also self-adjoint, and that its spectrum is the same as A's. Likewise it has the same set of complete eigenfunctions.

XX.2   The Laguerre-Type Polynomials

The original setting for the Laguerre-type differential operator was $H = L^2(0, \infty; e^{-x}) \otimes R$ with elements $Y = \begin{pmatrix} y(x) \\ y_0 \end{pmatrix}$ and $Z = \begin{pmatrix} z(x) \\ z_0 \end{pmatrix}$ having inner product

$$\langle Y, Z\rangle_H = \int_{0}^{\infty}y(x)\bar{z}(x)e^{-x}dx + \frac{1}{R}y_0\bar{z}_0.$$

Associated with the Laguerre-type operator

$$L_4y = [(x^2e^{-x}y'')'' - (([2R+2]x+2)e^{-x}y')' + ke^{-x}y]/e^{-x},$$

we define the operator A by setting

$$AY = \begin{pmatrix} L_4y \\ -2Ry'(0) + ky(0) \end{pmatrix},$$

where the first component is defined on $(0, \infty)$.

The appropriate Sobolev space is given by [4]:

$$\langle Y, Z \rangle_S = \int_0^\infty [x^2 e^{-x} y'' \bar{z}'' + ((2R+2)x + 2)e^{-x} y' \bar{z}' \; dx$$
$$+ k \left[\int_0^\infty e^{-x} y \bar{z} dx + y(0)\bar{z}(0)/R \right].$$

Again it is the Dirichlet formula that generates the inner product. We have

$$\int_0^\infty [(x^2 e^{-x} y'')'' - (([2k+2]x + 2)e^{-x} y')' + k e^{-x}] \bar{z} dx$$
$$= \int_0^\infty [x^2 e^{-x} y'' \bar{z}'' + ((2k+2)x + 2)e^{-x} y' \bar{z}' + k e^{-x} y \bar{z}] dx$$
$$+ (x^2 e^{-x} y'')' \bar{z} \Big|_0^\infty - (x^2 e^{-x} y'') \bar{z}' \Big|_0^\infty$$
$$- ((2k+2)x + 2)e^{-x} y' \bar{z} \Big|_0^\infty.$$

Employing limits found in [2], [4], the integrated limiting terms are found to mostly vanish. Adding the appropriate jump at $x = 0$ yields the inner product

$$\langle AY, Z \rangle_H = \langle Y, Z \rangle_S.$$

We then employ the usual practice of defining the restriction of A to the Sobolev space to be T. After a bit of work it is found that T is also self-adjoint, and that its spectrum is the same as A's. Likewise it has the same set of complete eigenfunctions.

XX.3 Remarks

The Sobolev embedding of the Jacobi-type operator should be similar but it has not been worked out in detail. Indeed, it is anticipated that at $x = 1$, the various limit cases will cause difficulty. We invite the reader to work on it.

The Sobolev embeddings for the Legendre squared and cubed operatorsand all the powers of the Laguerre and Hermite operators have only just been worked out by Littlejohn et al. [5] [6]. The higher powers of the Legendre operator and all the Jacobi operators have proven to be harder nuts to crack but should fall soon.

The Legendre squared operator has a Dirichlet formula of

$$\int_{-1}^1 l^2 y \bar{z} dx = \int_{-1}^1 [k^2 y \bar{z} + (2k+1)(1-x^2) y' \bar{z}' + (1-x^2)^2 y'' \bar{z}'' dx,$$

while the Legendre cubed operator has a Dirichlet formula of

$$\int_{-1}^1 l^3 y \bar{z} dx = \int_{-1}^1 [k^3 y \bar{z} + (3k^2 + 6k + 4)(1-x^2) y' \bar{z}'$$
$$+ (3k+8)(1-x^2)^2 y'' \bar{z}'' + (1-x^2)^3 y''' \bar{z}'''] dx,$$

where $ly = -(1-x^2)y')' + ky$, $k > 0$.

We refer to [6] for further details.

In the Laguerre case with

$$ly = [-(x^{\alpha+1}e^{-x}y')' + ky]/x^{\alpha}e^{-x}$$

the Dirichlet formula for l^n is

$$\int_0^{\infty} l^n y \bar{z} x^{\alpha} e^{-x} dx = \sum_{j=0}^{n} b_j(n,k) \int_0^{\infty} y^{(j)} \bar{z}^{(j)} x^{a+j} e^{-x} dx \,,$$

where the numbers $b_j(n,k)$ are related to the Stirling numbers. We find

$$b_j(n,k) = \sum_{i=0}^{j} \frac{(-1)^{i+j}}{j!} \binom{j}{i} (k+i)^n \,.$$

If $(k+i)^n$ is expanded and the order of summation is reversed,

$$b_j(n,k) = \sum_{m=0}^{n} \left(\sum_{i=0}^{j} \frac{(-1)^{i+j}}{j!} \binom{j}{i} i^{n-m} \right) k^m,$$

$$= \sum_{m=0}^{n} \binom{n}{m} S^{(j)}_{n-m} k^m \,,$$

where $S^{(j)}_{n-m}$ is the Stirling number of the second kind. We refer to [5] for further details.

References

[1]   W. N. Everitt, A. M. Krall and L. L. Littlejohn, *On some properties of the Legendre-type differential expression*, Quaes. Math. **13** (1990), 83–108.

[2]   ———, *Differential operators and the Laguerre-type polynomials*, SIAM J. Math. Anal. **23** (1992), 722–736.

[3]   A. M. Krall, W. N. Everitt and L. L. Littlejohn, *The Legendre-type operator in a left definite Hilbert space*, Quaes. Math. **15** (1992), 467–475.

[4]   ———, *The Laguerre-type operator in a left definite Hilbert space*, J. Math. Anal. Appl. **192** (1995), 460–468.

[5]   L. L. Littlejohn and R. Wellman, *Left-definite spectral theory with applications to differential operators*, manuscript.

[6]   L. L. Littlejohn, J. Arvesu and F. Marcellan, *On the right-definite and left-definite spectral theory of the Legendre polynomials*, manuscript.

Closing Remarks

And so we have finished our excursion into Hilbert spaces, linear operators, differential operators and orthogonal polynomials, not because the theory is completed, but because this is where the road stops. The next section is still under construction.

There are some gaps and omissions. Linear Hamiltonian systemsneed improvement when it comes to boundary conditions, especially if they seem to be valid for all λ with Im $\lambda \neq 0$. Left definite singular Hamiltonian systems has no theory available. The Hinton–Shaw and the Niessen versions need to be made more compatible.

Someone must work out the details of the Jacobi-type problems of fourth order. Both right and left definite problems need to be discussed.

There is no discussion available with regard to the boundary conditions for the singular Jacobi-type and Laguerre-type problems of Koekoeks and Bavinck. In all probably these problems must be written as linear Hamiltonian systems in order to get a handle on them.

Finally, in retrospect, the author notes that the works of H. Langer, A. Dijksma and H. de Snoo, who worked on $\pi - \kappa$ spaces have not been mentioned. They closely parallel the Sobolev theory. The reader is encouraged to consult their works.

Likewise Alouf Jirari, who developed a complete theory of both right and left definite settings for orthogonal polynomials satisfying a second order *difference* equation, has not been previously mentioned. Her work can be easily extended, as have some Europeans, particularly H. Bavinck, been doing.

And so on. The reader is encouraged, if he (she) does not like a particular section, to improve on it himself (herself). That is the way mathematics is done. It has been fun. The author hopes that he, too, can join in the future.

Index

$A^{\frac{1}{2}}$, 30
A^*, adjoint of A, 19, 43
A^{**}, 43
$A^{**} = A$, 43
$(A^{-1})^*$, 43
A^{-1}, 43
$\| A \|$, operator norm of A, 18
$\mathcal{A}(x)$, 88
adjoint
    boundary condition, 83
adjoint operator A^*, 41
annihilator, 80, 83, 103
    boundary condition, 247
Atkinson, F.V., xiii
Atkinson's theory for singular
    Hamiltonian systems of even
    dimension, 73

Bavinck, 343
Bernoulli numbers, $\{B_{2r}\}_{r=1}^{\infty}$, 229
Bessel
    –'s inequality, 14
    functions, 148, 153
    operator, 331, 332
    polynomial equation, 239
boundary conditions,
    333, 334, 336, 337
    adjoint –, 83
    annihilator –, 247
    extension of the –, 122
    of linear Hamiltonian
        systems, 343
    other –, 102
    parametric –, 63, 83, 130
    singular –, 79, 117

boundary form, 80
boundary value problems, 97
    generalized Laguerre –, 152
    Hermite –, 152
    Jacobi –, 149
    Legendre –, 150
    Legendre polynomial –, 134
    ordinary Laguerre –, 152
    regular second order
        Sobolev –, 302
    regular Sobolev – for linear
        Hamiltonian systems, 307
    self-adjoint – with mixed
        boundary conditions, 131
    singular second order
        Sobolev –, 330
    singular Sturm–Liouville –, xiii
boundary values, 142
bounded
    linear operator, 17
    linear operator on a Hilbert
        space, 17
    monotonic sequence, 29
    operator, 17
Brown, Dick, xiii

Cauchy sequence, 6, 9, 10
    closed –, 17
Cayley transform of A, 47
compact set, 26
compact support, 207
compatibility condition, 262
complete orthogonal set, 248
complete orthonormal set, 25
complex linear space, 1

continuous spectrum $\sigma_c(A)$, 25
convergence
  operator –, 28
    strong –, 28
    uniform –, 28
    weak –, 28

D, 61, 142, 146
D^*, 62, 146
difference equation
  second order –, 343
differential equation
  distributional –, 215
  Legendre –, 133
  of the second Littlejohn
    polynomials, 289
  self-adjoint –
    of first order, 51
    of fourth order, 53
    of second order, 52
    of third order, 52
  self-adjoint, real – of sixth
    order, 54
differential expression, 336
differential operator, 334, 336, 343
  L, 118
  Jacobi –, 149
  self-adjoint –, 118
Dirac distribution, 209
Dirac systems, 51, 54
Dirichlet formula, 305, 306, 312,
  315, 323, 328–337, 340, 341
  of the Legendre cubed
    operator, 341
  of the Legendre squared
    operator, 341
distribution, 207
  Dirac –, 209
  Heaviside –, 209
  limits of –s, 211
  of compact support, E', 214
  of rapid decay, P', 213
  of rapid growth, 208
  of slow growth, S', 212

Schwarz –, xiii
  without constraint, D', 207
distributional
  differential equations, 215
  weight function, 261, 275
$D_{L_0^*}$, 60
D_M, 58, 79, 117
domain of A^*, D^*, 43
$D^\perp$, 18

$e(A)$, 33
eigenfunction, 66
  expansion, xiii
eigenvalues of $\mathcal{A}(x, \lambda)$, 91
$E_\lambda F(x)$, 183, 205
$E(\mu)$, 33
even Hermite operator, 254
Everitt, W.N., xiii
extended Green's formula with two
  singular points, 145
extension of the boundary
  conditions, 122

Fourier cosine transform, 331
Fourier sine transform, 185, 331
Fourier transform, 213, 330
Fulton, Charles, xiii
function
  Bessel –, 148
  distributional weight –, 261
  $M(\lambda)$ –, 138
  Riemann integrable –, 2
  support of a –, 207
  T, class of all piecewise
    continuous –s on $[m, M]$, 32
  test –
    of rapid decay, S, 212
    of slow growth, P, 213
    with compact support, D, 207
    without constraint, E, 214
  weight –, 281
$F(x)$, 182

Gegenbauer operator, 264
generalized
    even Hermite operator, 336
    Hermite polynomials, 252
        form a complete orthogonal
            set, 256
        of even degree, 253
        of odd degree, 255
    Jacobi-type polynomials, 291, 292
    Koekoek's – Jacobi-type
        polynomials, 290
    Laguerre
        -type polynomials, 295, 296
        boundary value problem, 152
        equation, 238
        operator, 334
        polynomials, 148, 245, 264
    odd degree – Hermite
        polynomials, 337
    odd Hermite operator, 256, 336
$G(\lambda)$, 174, 181, 182, 194, 197
Gram–Schmidt theorem, 13
graph of A, G_A, 42
graph of A^* in $\mathcal{H}$, G_{A^*}, 43
Green's formula, 129, 146, 272
    extended – with one singular
        point, 125
    extended – with two singular
        points, 145

$\mathcal{H} = H \otimes H$, set of ordered pairs
    $\{x, y\}$, 42
Hamiltonian systems
    Atkinson's theory for singular –
        of even dimension, 73
    boundary conditions of
        linear –, 343
    left definite singular –, 343
    Niessen approach to singular –, 87
    regular –, 328
Hankel transform, 332, 333
Hazmirzaahmad, Mojdeh, xiv
Heaviside distribution, 209

Hermite
    boundary value problem, 152
    equation, 239
    operator, 336
        even –, 254
        generalized even –, 336
        generalized odd –, 256, 336
    polynomial, 336
        odd degree generalized –, 337
        problems, 133
    polynomials, 148, 249, 265
        form a complete orthogonal
            set, 251
        generalized, 252
        generalized – form a complete
            orthogonal set, 256
        generalized – of even
            degree, 253
        generalized – of odd
            degree, 255
        series representation
            of the –, 250
Hermitian form, 3–5
    positive definite –, 6
Hilbert space, xiii, 6, 343
    isometric –, 15
Hinton, Don, xiii
Hinton and Shaw's
    extension of Weyl's $M(\lambda)$ theory
        to systems, 107
    extension with two singular
        points, 137
H_R, 208

identity
    operator, I, 18
    polarization –, 7
inequality
    Bessel's –, 14
    Minkowski's –, 6, 7
    Schwarz's –, 6, 7
inner product space, 9, 10
integral
    Lebesgue –, 3, 8

Lebesgue–Stieltjes –, 9
Riemann –, 8
isometric Hilbert space, 15

Jacobi
  boundary value problem, 149
  differential operator, 149
  equation, 238
  operator, 334
    spectral resolution of the –, 334
  polynomials, 148, 239, 263
    series representations
      of the –, 240
Jacobi-type
  operator, 274, 341
  polynomials, 274
    generalized –, 291, 292
    Koekoek's generalized –, 290
  problems of fourth order, 343
Jirari, Alouf, xiv

Kaper, Hans, xiii
Koekoek, 343
Koekoek's generalized Jacobi-type
  polynomials, 290
Krall, H.L., (1907–1994), xiv
H.L. Krall polynomials, 281
Krein space, 259

L, 64
  self-adjoint, 121
L^*, 64, 147
$(L - \lambda I)^{-1}$, 66
λ_0 fixed, 118
L_0^*, 58, 60
ℓ^2, 8
$L_A^2(a,b)$, 9
$L_A^2(a,b)$, 57
Laguerre
  equation
    generalized –, 238
  operator
    generalized –, 334

polynomials, 335
  generalized –, 148, 245, 264
  series representation
    of the –, 246
Laguerre-type
  operator, 270, 340
  polynomials, 270, 340
    generalized –, 295, 296
    in a Sobolev space, 339
  problem, 155
Laplacian operators, 330
Lebesgue integral, 3, 8
Lebesgue–Stieltjes
  integral, 9
  measure, 8
left-definite theory, 302
Legendre
  boundary value problem, 150
  differential equation, 133
  differential expression, 333
  operator, 264
    cubed –, 341
    higher powers of the –, 341
    spectral resolution of the –, 334
    squared –, 341
  polynomial boundary value
    problem, 134
  polynomials, 243
  squared problem, 154
Legendre-type
  operator, 266, 268
  polynomials, 265
    in a Sobolev space, 339
lemma
  Riemann–Lebesgue –, 14
limit
  circle case, 317
  point case, 316
linear
  Hamiltonian system, xiii
  manifold, 2, 10
  operators, 343
  space, 1
Littlejohn, Lance, xiii

Littlejohn
    operator, 288
    polynomials, 287
        second –, 289
$L_p^2(a,b)$, 9
LY, 142, 146
ℓY, 95

M-circle, 111
M_{11}, 204
M_{12}, 204
M_{21}, 204
M_{22}, 204
M_a, 203
manifold
    linear –, 2, 10
M_b, 203
m_b, 75, 77
measure
    Lebesgue–Stieltjes –, 8
Mingarelli, Angelo, xiii
Minkowski's inequality, 6, 7
mixing operators, U, V, 42
$M(\lambda)$ functions, 138
$M(\lambda)$ surface, 159
moment
    equations, 244, 263, 266, 271,
        275, 283
    relations, 263, 264
moments, 223, 226, 249, 264, 283

Niessen, Heinz-Dieter, xiii
Niessen approach to singular
    Hamiltonian systems, 87
normal, 19

operator
    $B = (I + A^*A)^{-1}$, 45
    $C = A(I + A^*A)^{-1}$, 45
    adjoint – of A^*, 41
    bounded, 17
    bounded linear –, 17
    bounded linear – on a Hilbert
        space, 17

convergence, 28
    strong –, 28
    uniform –, 28
    weak –, 28
Gegenbauer –, 264
I, identity –, 18
Laplacian –, 330
Legendre –, 264
Littlejohn –, 288
mixing –s, U, V, 42
norm of A, $\parallel A \parallel$, 18
P, projection –, 21
polynomial –, 23
projection –, 69
resolvent –, 18
self-adjoint –, xiii, 17
Tchebycheff –, 264
unbounded linear –, 41
ordinary Laguerre boundary value
    problem, 152
orthogonal, 7
orthogonal polynomials, xiii,
    223, 343
    basic properties of –, 223
    satisfying
        fourth order differential
            equations, 261
        higher order differential
            equations, 291
        second order differential
            equations, 237
        sixth order differential
            equations, 281
orthogonality, 6
orthonormal set, 7, 13
other boundary conditions, 102

P and Q, 61
$P^2 = P$, 21
$P = P^*$, 21
parallelogram law, 6, 7, 9
parametric boundary conditions,
    63, 83, 130
Parseval's equality, 15, 169, 192

$P_{b'}(\lambda)$, 169
$P_I(\lambda)$, 192
$\pi - \kappa$ spaces, 343
$P(\lambda)$, 170, 193
$P(\lambda)$, 203
P_M, projection of H onto M,
   12, 14, 21
point spectrum $\sigma_p(A)$, 25
polarization, 5
polarization identity, 7
polynomial operator, 23
polynomials
   generalized Laguerre –, 148
   Hermite –, 148
   Jacobi –, 148
   orthogonal –, xiii, 223, 343
   Tchenbycheff –, 223
Pontrjagin space, 240
positive definite Hermitian form, 6
projection, 21
   operator, P, 21, 69
   valued measure, $P(\lambda)$, 243, 252
Pythagorean theorem, 7, 22

$\rho(A)$, resolvent of A, 18
regular linear Hamiltonian
   systems, 51
regular second order Sobolev
   boundary value problems, 302
relativistic quantum mechanics, 55
residual spectrum $\sigma_r(A)$, 25, 27
resolution
   spectral –, 69
     of a bounded self-adjoint
      operator, 33
     of an unbounded self-adjoint
      operator, 46
     of bounded normal and unitary
      operators, 37
     of bounded normal
      operators, 37
     of unitary operators, 39
resolvent
   of A, $\rho(A)$, 18

of L, 121
   operator, 18
Riemann
   integrable function, 2
   integral, 8
Riemann–Lebesgue lemma, 14
Riesz–Fischer theorem, 14
Riesz–Frechet theorem, 12

S-Hermitian systems, 51, 56
$S(A)$, 32
$\sigma(A)$, spectrum of A, 18
Schwarz distribution, xiii
Schwarz's inequality, 6, 7
second order difference equation, 343
self-adjoint, 19, 20, 45, 144,
   333–335, 337, 339, 340
   real, differential equation
     of sixth order, 54
   boundary value problems
     with mixed boundary
     conditions, 131
   differential equation
     of first order, 51
     of fourth order, 53
     of second order, 52
     of third order, 52
   differential operator, 118, 336
   L is –, 121
   operator, xiii, 17
sequence
   bounded monotonic, 29
   Cauchy –, 6, 9, 10
   closed Cauchy –, 17
series representations
   of the Jacobi polynomials, 240
   of the Laguerre polynomials, 246
   of the Hermite polynomials, 250
set
   compact –, 26
   complete orthogonal –, 248
   complete orthonormal –, 25

of ordered pairs $\{x, y\}$,
   $\mathcal{H} = H \otimes H$, 42
orthonormal –, 7, 13
Shaw, Ken, xiii
Shaw
  Hinton and –'s
    extension of Weyl's $M(\lambda)$
      theory to systems, 107
    extension with two singular
      points, 137
singular
  boundary conditions, 79, 117
  point, 74
  Sturm–Liouville boundary value
    problems, xiii
sixth order differential
  equations, 287
  operator, 282
Sobolev
  boundary value problems
    regular – for linear
      Hamiltonian systems, 307
    singular second order –, 330
  inner product, 303, 305
Sobolev space, 295, 298, 302, 305,
  308, 327, 336, 340, 341
  Laguerre-type polynomials
    in a –, 339
  Legendre-type polynomials
    in a –, 339
solutions
  square inegrable –, 115
space
  complex linear –, 1
  Hilbert –, xiii, 6, 343
  inner product –, 9, 10
  isometric Hilbert –, 15
  Krein –, 259
  linear –, 1
  $\pi - \kappa$ –s, 343
  Pontrjagin –, 240
  Sobolev –, 295, 298, 302, 305,
    308, 327, 336, 340, 341
spectral expansion, 168

spectral resolution, 69, 335–337
  for linear Hamiltonian systems
    with one singular point, 167
  for linear Hamiltonian systems
    with two singular points, 189
  of a bounded self-adjoint
    operator, 33
  of an unbounded self-adjoint
    operator, 46
  of bounded normal and unitary
    operators, 37
  of the Jacobi operator, 334
  of the Legendre operator, 334
spectrum, 333
  continuous –, $\sigma_c(A)$, 25
  of A, $\sigma(A)$, 18
  point –, $\sigma_p(A)$, 25
  residual –, $\sigma_r(A)$, 25, 27
square integrable solutions, 115
Stieltjes measure, 70
Stirling
  number of the second kind, 342
  numbers, 342
subspace, 2
  expansions, 185
support
  compact, 207
  of a function, 207
symmetric, 45
symmetry factor, 226, 232
system
  Dirac –, 51, 54
  linear Hamiltonian –, xiii
  regular linear Hamiltonian –, 51
  S-Hermitian –, 51, 56

T, class of all piecewise continuous
  functions on $[m, M]$, 32
$T(A)$, 33
Tchebycheff
  operator, 264
  polynomials, 223, 261
  problem
    of the first kind, 150

of the second kind, 151
test function
  of rapid decay, S, 212
  of slow growth, P, 213
  with compact support, D, 207
  without constraint, E, 214
theorem
  Gram–Schmidt –, 13
  Pythagorean –, 7, 22
  Riesz–Fischer –, 14
  Riesz–Frechet –, 12
three term recurrence relation, 225,
  238, 244, 247, 250, 262, 267, 271,
  276, 282, 287, 297

unbounded linear operator, 41
unitary, 19

weight
  equation, 244, 246, 250, 263, 264
  function, 281
    distributional –, 261
  functional, 223
  problem, 257
Weyl circle, 165
Weyl–Titchmarsh limit point–limit
  circle theory, 315

Zettl, Tony, xiii